Horia-Nicolai Teodorescu, Junzo Watada, and Lakhmi C. Jain (Eds.)

Intelligent Systems and Technologies

Studies in Computational Intelligence, Volume 217

Editor-in-Chief
Prof. Janusz Kacprzyk
Systems Research Institute
Polish Academy of Sciences
ul. Newelska 6
01-447 Warsaw
Poland
E-mail: kacprzyk@ibspan.waw.pl

Further volumes of this series can be found on our homepage: springer.com

Vol. 196. Valentina Emilia Balas, János Fodor and Annamária R. Várkonyi-Kóczy (Eds.)
Soft Computing Based Modeling in Intelligent Systems, 2009
ISBN 978-3-642-00447-6

Vol. 197. Mauro Birattari
Tuning Metaheuristics, 2009
ISBN 978-3-642-00482-7

Vol. 198. Efrén Mezura-Montes (Ed.)
Constraint-Handling in Evolutionary Optimization, 2009
ISBN 978-3-642-00618-0

Vol. 199. Kazumi Nakamatsu, Gloria Phillips-Wren, Lakhmi C. Jain, and Robert J. Howlett (Eds.)
New Advances in Intelligent Decision Technologies, 2009
ISBN 978-3-642-00908-2

Vol. 200. Dimitri Plemenos and Georgios Miaoulis *Visual Complexity and Intelligent Computer Graphics Techniques Enhancements,* 2009
ISBN 978-3-642-01258-7

Vol. 201. Aboul-Ella Hassanien, Ajith Abraham, Athanasios V. Vasilakos, and Witold Pedrycz (Eds.)
Foundations of Computational Intelligence Volume 1, 2009
ISBN 978-3-642-01081-1

Vol. 202. Aboul-Ella Hassanien, Ajith Abraham, and Francisco Herrera (Eds.)
Foundations of Computational Intelligence Volume 2, 2009
ISBN 978-3-642-01532-8

Vol. 203. Ajith Abraham, Aboul-Ella Hassanien, Patrick Siarry, and Andries Engelbrecht (Eds.)
Foundations of Computational Intelligence Volume 3, 2009
ISBN 978-3-642-01084-2

Vol. 204. Ajith Abraham, Aboul-Ella Hassanien, and André Ponce de Leon F. de Carvalho (Eds.)
Foundations of Computational Intelligence Volume 4, 2009
ISBN 978-3-642-01087-3

Vol. 205. Ajith Abraham, Aboul-Ella Hassanien, and Václav Snášel (Eds.)
Foundations of Computational Intelligence Volume 5, 2009
ISBN 978-3-642-01535-9

Vol. 206. Ajith Abraham, Aboul-Ella Hassanien, André Ponce de Leon F. de Carvalho, and Václav Snášel (Eds.)
Foundations of Computational Intelligence Volume 6, 2009
ISBN 978-3-642-01090-3

Vol. 207. Santo Fortunato, Giuseppe Mangioni, Ronaldo Menezes, and Vincenzo Nicosia (Eds.)
Complex Networks, 2009
ISBN 978-3-642-01205-1

Vol. 208. Roger Lee, Gongzu Hu, and Huaikou Miao (Eds.)
Computer and Information Science 2009, 2009
ISBN 978-3-642-01208-2

Vol. 209. Roger Lee and Naohiro Ishii (Eds.)
Software Engineering, Artificial Intelligence, Networking and Parallel/Distributed Computing, 2009
ISBN 978-3-642-01202-0

Vol. 210. Andrew Lewis, Sanaz Mostaghim, and Marcus Randall (Eds.)
Biologically-Inspired Optimisation Methods, 2009
ISBN 978-3-642-01261-7

Vol. 211. Godfrey C. Onwubolu (Ed.)
Hybrid Self-Organizing Modeling Systems, 2009
ISBN 978-3-642-01529-8

Vol. 212. Viktor M. Kureychik, Sergey P. Malyukov, Vladimir V. Kureychik, and Alexander S. Malyoukov
Genetic Algorithms for Applied CAD Problems, 2009
ISBN 978-3-540-85280-3

Vol. 213. Stefano Cagnoni (Ed.)
Evolutionary Image Analysis and Signal Processing, 2009
ISBN 978-3-642-01635-6

Vol. 214. Been-Chian Chien and Tzung-Pei Hong (Eds.)
Opportunities and Challenges for Next-Generation Applied Intelligence, 2009
ISBN 978-3-540-92813-3

Vol. 215. Habib M. Ammari
Opportunities and Challenges of Connected k-Covered Wireless Sensor Networks, 2009
ISBN 978-3-642-01876-3

Vol. 216. Matthew Taylor
Transfer in Reinforcement Learning Domains, 2009
ISBN 978-3-642-01881-7

Vol. 217. Horia-Nicolai Teodorescu, Junzo Watada, and Lakhmi C. Jain (Eds.)
Intelligent Systems and Technologies, 2009
ISBN 978-3-642-01884-8

Horia-Nicolai Teodorescu, Junzo Watada,
and Lakhmi C. Jain (Eds.)

Intelligent Systems and Technologies

Methods and Applications

Springer

Prorector Prof. Horia-Nicolai Teodorescu
Universitatea Tehnica "Gheorghe Asachi"
Rectorat
B-dul Prof. D. Mangeron nr. 67
700050 Iasi
Romania
E-mail: hteodor@etc.tuiasi.ro

Prof. Junzo Watada
Graduate School of Information
Production and Systems (IPS)
Waseda University
2-7 Hibikino, Wakamatsuku
Kitakyushu, Fukuoka 808-0135
Japan
E-mail: junzow@osb.att.ne.jp

Prof. Dr. Lakhmi C. Jain
Professor of Knowledge-Based Engineering
University of South Australia
Adelaide City
The Mawson Lakes, SA 5095
Australia
E-mail: Lakhmi.Jain@unisa.edu.au

ISBN 978-3-642-10187-8 e-ISBN 978-3-642-01885-5

DOI 10.1007/978-3-642-01885-5

Studies in Computational Intelligence ISSN 1860949X

Typeset & Cover Design: Scientific Publishing Services Pvt. Ltd., Chennai, India.

Printed in acid-free paper

9 8 7 6 5 4 3 2 1

springer.com

Preface

The extent of the use of Artificial Intelligence (AI) into real-life systems, tools, equipments, and appliances in recent years is astonishing. AI is no more a curious term and a toy for the benefit of mathematicians and their closed collaborators, but a necessary addition to products to keep them competitive and responding to the expectations of the experts and of the general public users. The developments in the field are so fast that the definitions proposed for AI are almost too limited and outdated. All fields, from avionics to medicine and from pharmaceutics and food industry to car manufacturing require intelligent systems.

This book is primarily based on contributions to the Fifth European Conference on Intelligent Systems and Technologies, but all contributions have been significantly expanded and modified to compose self-consistent chapters. All contributions derived from the conference papers have been extended by the authors and more chapters have been added, specifically submitted for this book.

The book includes chapters on some of the recent developments in theoretical, methodological, and applicative sides of intelligent systems. The blend of fundamental theory, methods, and practical results would not satisfy every reader, but offers a consistent overview of the field for most readers.

In a field striving with new concepts, methods and tools domain like of intelligent systems, a new book is almost always useful to show new developments and applications, or to summarize and clarify recent developments. Unlike a journal, where new developments are emphasized, a volume is expected to bring more mature, better systematized, better interconnected treatment of the topics.

While mainly rooted in a conference, this volume is conceived independently. Out of about 100 papers proposed in the conference, only about 13% have been selected as suitable for further development as chapters in this volume. The criteria used in selecting the contributions for this volume have been the importance of the topic in the current stage of intelligent systems development, the scientific and applicative merit of the original form of the proposed manuscript, and the quality, clarity and systematic treatment of the presentation. Some manuscripts with excellent scientific quality but not fitting the purpose of this volume have been proposed for journal publication. The conference has its own volume of proceedings which does not overlap with the present book.

The volume is organized in three parts. The first one is an introduction to the intelligent systems. The second part is devoted to Methods and Tools used in designing intelligent systems. The third part includes the applications of intelligent systems.

While in some of our previous volumes we emphasized on specific tools and fields of applications, like neuro-fuzzy systems in medicine [1], soft computing in human related sciences [2], intelligent systems and technologies in rehabilitation engineering [3] and in one volume on the hardware for intelligent systems [4]. In this volume, Web architectures and Web applications, and new directions like DNA computing and bioinformatics are somewhat favored.

We acknowledge with thanks all authors and reviewers for their contributions. Thanks are due to the Springer-Verlag and team for their excellent work in the production of this volume.

Horia -Nicolai Teodorescu
Junzo Watada
Lakhmi C. Jain

References

[1] Teodorescu, H.N., Kandel, A. and Jain, L.C. (Editors), Fuzzy and Neuro-fuzzy Systems in Medicine, CRC Press USA, 1998.
[2] Teodorescu, H.N., Kandel, A. and Jain, L.C. (Editors), Soft Computing Techniques in Human Related Science, CRC Press USA, 1999.
[3] Teodorescu, H.N. and Jain, L.C. (Editors), Intelligent Systems and Technologies in Rehabilitation Engineering, CRC Press USA, 2001.
[4] Teodorescu, H.N., Jain, L.C. and Kandel, A. (Editors), Hardware Implementation of Intelligent Systems, Springer-Verlag, Germany, 2001.

Contents

Part III: Applications

Part I
Introduction

1
Advances in Intelligent Methodologies and Techniques

Lakhmi C. Jain[1] and Chee Peng Lim[2]

[1] School of Electrical & Information Engineering
University of South Australia, Australia
[2] School of Electrical & Electronic Engineering
University of Science Malaysia, Malaysia

Abstract. This chapter introduces a number of intelligent methodologies and techniques stemmed from Artificial Intelligence (AI). An overview of various intelligent models arisen from expert systems, artificial neural networks, fuzzy logic, genetic algorithms, decision trees, and agent technologies is presented. Application examples of these intelligent models in various domains are also presented. Then, the contribution of each chapter included in this book is described. A summary of concluding remarks is presented at the end of the chapter.

1 Introduction

The development of the electronic computer in 1940s has revolutionalized the technologies for storage and processing of voluminous data and information. A variety of computing methodologies and technologies has emerged as a result of the availability and accessibility of various kinds of data and information in electronic form. One of the advancements that was made possible by the computer is the creation of machines with intelligent behaviours. Subsequently, the linkage between human intelligence and machines received much attention of researchers. The term Artificial Intelligence (AI) was first coined in 1956 at the Dartmouth conference; and the Turing Test, which aimed to determine the possibility of creating machines with true intelligence, was the first serious proposal in the philosophy of AI (Russell and Norvig, 2003) [41]. Since then, theories and applications of AI started to proliferate and to have impacts on technological developments in various fields, and finally to enter human life. A snapshot on history of AI is described in Buchanan (2005) [8].

There are a variety of definitions of AI. They include "Artificial intelligence is the science of making machines do things that would require intelligence if done by men" (Minsky, 1968) [30]; "The goal of work in artificial intelligence is to build machines that perform tasks normally requiring human intelligence", (Nilsson, 1971) [32]; "Artificial intelligence is the study of how to make computers do things at which, at the moment, people are better" (Rich, 1983) [39]; "Artificial intelligence is concerned with the attempt to develop complex computer programs that will be capable of performing difficult cognitive tasks" (Eysenck, 1990) [16];

H.-N. Teodorescu, J. Watada, and L.C. Jain (Eds.): Intel. Sys. and Tech., SCI 217, pp. 3–28.
springerlink.com

"Artificial intelligence may be defined as the branch of computer science that is concerned with the automation of intelligent behaviour" (Luger, 2002) [23].

While AI is a multi-faceted research field, two of the fundamental challenges that are often faced by AI researchers are: creating machines that could efficiently solve problems, and making machines that could learn by themselves. Nevertheless, there is no clear boundary from one AI methodology or technique to another, nor there is an established unifying framework or paradigm that encompasses all AI research.

The organisation of this chapter is as follows. In section 2, an overview of different intelligent paradigms under the umbrella of AI is presented. The applicability of these intelligent models to tackling complex, real-world problems are discussed. The contribution of each chapter in this book is described in section 3. Section 4 gives some concluding remark of this chapter.

2 Intelligent Models and Applications

The main aim of this section is to share and disseminate a small fraction of commonly available AI methodologies and techniques for the development of intelligent machines and systems, as well as their application domains. The methodologies and techniques covered are by no means a comprehensive treatment of all the domains spanned by AI research.

2.1 Expert Systems

Expert systems (ESs) are software programs that attempt to solve problems in a particular field using knowledge extracted from human experts and encoded in the if-then format. According to Doran (1988) [13], "an expert system is a computer program which uses non-numerical domain-specific knowledge to solve problems with a competence comparable with that of human experts". Typically, an expert system comprises three main modules: a knowledge base that contains if-then rules elucidated from human experts; an inference engine that reasons about the information in the knowledge base using forward and backward chaining techniques for reaching conclusions; a user interface that asks questions and solicits responses from the users, as well as presents answers with explanations to the users.

One of the fruitful areas that ESs have demonstrated successful implementation is the medical domain. MYCIN (Shortliffe, 1976) [44] was one of the earliest ES-based medical decision support tool that aimed to manage and provide therapy advice for treating patients who suffered from bacterial infections of the blood. MYCIN was able to provide natural language explanations from its knowledge base for clinicians to understand how the system made its conclusions. Other ESs for medical applications include PIP (Pauker *et. al.*, 1976) [36] for renal diseases; pathology (Nathwani et al., 1997) [31], CASNET (Weiss *et al.*, 1978) [47] for glaucoma management; and INTERNIST (Miller *et. al.*, 1982) [28] for general internal medicine. On the other hand, PROSPECTOR, a pioneering ES for geological applications, appears to be the first "computerized advisor" who has successfully determined the location of mineralization (Campbell et al., 1982).

2.2 Artificial Neural Networks

Artificial Neural Networks (ANNs) are some kind of massively parallel computing models with a large number of interconnected simple processors (known as neurons) that are able to adapt themselves to data samples. Research in ANNs stems from the operation of the nervous system in the brain. McCulloch and Pitts (McCulloch & Pitts, 1943) [27] were the pioneers who initiated mathematical modeling of artificial neurons. To date, some of the popular ANNs models include the Multi-Layer Perceptron (MLP) network (Rumelhart et al., 1986) [40], Radial Basis Function (RBF) network (Broomhead & Lowe, 1988) [7], Self-Organizing Map (SOM) network (Kohonen, 1984) [22].

ANNs are data driven self-organizing learning models which can acquire knowledge from the data samples without any explicit specification of functional or distributional form for the underlying model (Zhang, 2000) [49]. ANNs have several advantages that make them reliable tools for tackling modeling, estimation, prediction, and classification problems. They are self-adaptive nonlinear models, which makes them flexible in modeling real-world complex relations. The MLP and RBF networks have been employed for classification of Middle Cerebral Artery (MCA) stenosis in diabetes patients based on Transcranial Doppler signals (Ergün *et. al.*, 2004) [15]. The SOM and MLP networks have also been used for estimating the efficiency of an activated sludge process in a wastewater treatment plant (Grieu *et. al.*, 2006) [19]. The SOM is applied to analyze the relationship of process variables, while the MLP is used as an estimation tool to predict the process efficiency.

2.3 Fuzzy Logic

The theory of fuzzy logic and fuzzy set was introduced by Zadeh (1965) [48]. Fuzzy systems are designed based on fuzzy set theory to process ambiguous, imprecise data and/or information expressed using human-like linguistic variables such as “many”, “often”, “few”, and “sometimes”. This fuzziness feature is useful in many real-world situations whereby it is difficult to categorize a data sample or a piece of information exactly into a specific class. This allows the degree by which the data and/or information is present or absent to be measured in a fuzzy manner. An inference mechanism using a set of fuzzy *if-then* rules for reasoning is applied in fuzzy systems.

Fuzzy logic is useful for tackling a variety of problems that require human-like reasoning and inference capabilities. It is widely used in control applications. Indeed, fuzzy logic provides an alternative design methodology to tackle non-linear control problems. It is able to improve the controller performance, simplify the implementation, as well as reduce the development time and costs. Application examples of fuzzy logic include the control of the isomerized hop pellets production (Alvarez, *et. al.*, 1999) [3]; parking capability of a car-type mobile robot (Chang and Li, 2002) [11]; a robotic manipulator (Chalhoub and Bazzi, 2004) [10]; obstacle avoidance of an autonomous vehicle (Lilly, 2007) [24]; as well as frequency control of a variable-speed wind generator (El Makodem *et. al.*, 2009) [14].

2.4 Evolutionary Computation

Evolutionary computation techniques are a collection of algorithms based on the evolution of a population towards a solution to a certain problem. These techniques include Genetic Algorithms (GAs), Genetic Programming and Evolutionary Algorithms. First proposed by Holland (1975) [20], GAs are adaptive, robust algorithms based on the principle of evolution, heredity, and natural selection. They are population-based algorithms for which any communication and interaction are performed within the population. They are also regarded as search algorithms for which they are capable of exploring a space using heuristics inspired by natural evolution.

A typical GA procedure is as follows. Given a fitness function to be optimized, a set of candidate solutions is randomly generated. The candidate solutions are assessed by the fitness function. Based on the fitness values, fitter candidate solutions have a better chance to be selected (i.e., survival of the fittest) and to go through the crossover (recombination) and mutation operators. The crossover operator is used to produce new offspring based on two or more parent solutions. The mutation operator is applied to the new offspring to inject diversity into the new population. The procedure selection, recombination, and mutation repeats from one generation of population to another until a terminating criterion is satisfied.

In comparison with traditional methodologies, GAs are global search techniques that are able to escape from being trapped in a local minimum. As such, GA provides good solutions for undertaking problems that require search and optimization. For examples, GAs are applied to large-scale layout design (Tam, 1992) [45]; topology optimization of trusses (Ohsaki, 1995) [33]; design optimization of electrical machines (Fuat Üler, 1995) [18]; the travelling salesman problem (Potvin, 1996) [37]; the shortest path routing problem (Ahn and Ramakrishna, 2002) [2]; as well as embedded microcontroller optimization (Mininno et. al., 2008) [29].

2.5 Decision Trees

Decision trees are commonly used as classification and decision support tools for choosing several courses of action. As a type of machine learning technique, a decision tree is essentially a map of the reasoning process in which a tree-like graph is constructed to explore options and investigate the possible outcomes of choosing the options. The reasoning process starts from a root node, transverses along the branches tagged with decision nodes, and terminates in a leaf node. A test on the attribute is conducted at each decision node, with each possible outcome resulting in a branch. Each branch leads to either another decision node or a leaf node (Markov and Larose, 2007) [26].

Decision trees are popular in the operation research domain such as evaluating between different strategies by taking into account the constraints in resources. Two well-known decision trees are CART (Classification and Regression Tree) (Breiman et. al., 1984) [6] and C4.5 (Quinlan, 1993) [38]. CART is a non-parametric technique that handles either categorical attributes (for classification) or numeric attributes (for regression). An application of CART is in constructing predictive models of blending of different types of coals to make coke for the blast furnace operation (Vasko et. al., 2005) [46]. The C4.5 and CART models have also been applied to building models pertaining to rating determinants of European insurance firms (Florez-Lopez, 2007) [17].

2.6 Agent Technology

Agents are software entities that possess the properties of autonomy, sociability, reactivity, and proactiveness (Jennings and Wooldridge, 1998) [21]. There are a lot of definitions for agents. According to Balogh et. al. (2000) [4], agents are defined as "sophisticated computer programs that act autonomously on behalf of their users, across open distributed environments, to solve a growing number of complex problems". This definition indicates that agents are capable of making decisions and performing tasks autonomously. Indeed, intelligent agents are flexible to changing environments and changing goals, and their capabilities have been recognized as 'the next significant breakthrough in software development' (Sargent, 1992) [42], and 'the new revolution in software' (Ovum, 1994) [34].

A pool of agents can be linked together to form a Multi-Agent System (MAS). To perform a task, the agents in an MAS interact according to some pre-defined reasoning model. One of the earliest models is the Beliefs, Desires, Intentions (BDI) model (Bratman, 1987) [5]. Beliefs represent an agent's understanding of the external world; desires are the goals that the agent wants to achieve; whereas intentions are the plans the agent uses to realize its desires.

Agent-based technologies have been employed to solve problems across many different areas. An MAS system called YAMS (Yet Another Manufacturing System) is proposed for manufacturing control (Parunak, 1987) [35]. It aims to efficiently manage the production processes in workcells grouped into a flexible manufacturing system. Maes (1994) [25] describes the use of the agent technologies in information management, whereby agents are used to filter electronic mails on behalf of the users, as well as to sort news articles retrieved from the Internet. A survey on development methodologies, negotiation technologies, and trust building mechanisms of agent technology in the area of electronic commerce is described in Sierra (2002) [43]. Chen and Wasson (2005) [12] discuss how agents can support students and instructors in distributed collaborative-learning environments. How intelligent agents can be incorporated into the FLE (Future Learning Environment), a Web-based groupware for computer-supported collaborative learning, is discussed. In surveillance applications, Aguilar-Ponce *et. al.* (2007) [1] use agent technology on wireless visual sensors that are scattered across an area to detect and track objects of interests and their movements.

3 Chapters Included in This Book

The book includes a sample of most recent research on the theoretical foundation and practical applications of intelligent systems. The book is divided into three parts. Part one provides an introduction to the book. Part two is on methods and tools used in designing intelligent systems. It includes six chapters. Chapter two is on a fuzzy density analysis of subgroups using oligonucleotides. The authors present a situation related to the industrial and organisational relationships between employees using fuzzy values. Chapter three is on the evolution of cooperating classification rules with an archiving strategy to underpin collaboration. The authors have validated their novel algorithm on two real world problems.

Chapter four is on designing agents with dynamic capability. The idea is to provide autonomous capabilities to agent supervisors within agent teams without the need to re-instantiate that team. A concept demonstrator is reported to validate the scheme reported in the chapter. Chapter five is on localized versus locality preserving representation methods in face recognition tasks. The authors have presented four localized representation methods and two manifold learning procedures and compared them in terms of recognition accuracy for several face processing tasks. It is demonstrated that the relative performance ranking of the methods is highly task dependent, and varies significantly upon the distance metric used. Chapter six is on the invariance properties of recurrent neural networks. The authors present the criteria for testing the set invariance. The scheme is formulated for two types of time dependent sets.

The third part of our book is on the applications of intelligent systems. It includes ten chapters. Chapter seven is on the application of soft computing paradigms in solving bioinformatics problems. The protein structure comparison problem is solved using a number of techniques. Chapter eight is on transforming an interactive expert code into a statefull service and a multicore-enabled system. The author has demonstrated the technique on a ten year old expert system for solving initial value problems for ordinary differential equations.

Chapter nine is on Paradigmatic Morphology and Subjectivity Mark-up in the RoWordNet Lexical Ontology. The authors state that the lexical ontology was developed for English language, but currently there are more than 50 similar projects for languages all over the world. RoWordNet is one of the largest lexical ontologies available today. It is sense-aligned to the Princeton WordNet 2.0 and the SUMO&MILO concept definitions have been translated into Romanian. Chapter presents the current status of the RoWordNet and enhancement of the knowledge encoded into it. Chapter ten presents special cases of relative object qualification using the AMONG operator. The authors have proposed the models and their evaluation schemes.

Chapter eleven is on the Development of a Speaker Tracking Module for Multi-Party Human-Computer Dialogue. The authors present a scheme for speaker tracking in multi-party human-computer dialogue. The technique is validated on a virtual librarian dialogue application, in Romanian language, and exhibits good runtime performance. Chapter twelve is on fuzzy interpolative control for passive greenhouses. The authors have proposed fuzzy interpolative control in their design. Chapter thirteen presents a GPS safety system for airports. The system is able to locate several objects simultaneously in real time and thus able to avoid accidents on the runway.

Chapter fourteen is on Exploring the use of 3D Collaborative Interfaces for E-learning. The author examines the use of Three Dimensional onscreen Graphical User Interfaces to stimulate users which can be combined with multi-user and synchronous communication techniques to facilitate meaningful interaction. The system called Collaborative Learning Environments with Virtual Reality is presented. Chapter fifteen presents an overview of Open Projects in Contemporary eLearning. The authors review several e-learning platforms and discuss the importance of open source e-learning platforms and analyze the total costs of implementation/educational output ratio. The final chapter is on software platforms for archaeological inventory and management system.

4 Summary

This chapter has presented an overview of various AI-based paradigms for designing and developing intelligent machines and systems that can be useful in solving problems in our daily activities. The methodologies and techniques discussed include expert systems, artificial neural networks, fuzzy logic, genetic algorithms, decision trees, and agent-based technology. Applicability of these intelligent models to various domains including medical, industrial, control, optimization, operation research, e-commerce, and education, has been highlighted. It is envisaged that AI-based methodologies and techniques will eventually enter our life and help solve complex real-world problems faced in our daily activities.

5 Resources

Following is a sample of additional resources on intelligent systems and technologies.

5.1 Journals

- International Journal of Knowledge-Based intelligent Engineering systems, IOS Press, The Netherlands.
  http://www.kesinternational.org/journal/
- International Journal of Hybrid Intelligent Systems, IOS Press, The Netherlands.
  http://www.iospress.nl/html/14485869.html
- Intelligent Decision Technologies: An International Journal, IOS Press, The Netherlands.
  http://www.iospress.nl/html/18724981.html
- IEEE Intelligent Systems, IEEE Press, USA.
  www.computer.org/intelligent/
- IEEE Transactions on Neural Networks.
- IEEE Transactions on Evolutionary Computing.
- IEEE Transactions on Fuzzy Systems.
- IEEE Computational Intelligence Magazine.
- Neural Computing and applications, Springer.
- Neurocomputing, Elsevier.

- International Journal of Intelligent and Fuzzy Systems, IOS Press, The Netherlands.
- Fuzzy Optimization and Decision Making, Kluwer.
- AI Magazine, USA
  www.aaai.org/

5.2 Special Issue of Journals

- Jain, L.C., Lim, C.P. and Nguyen, N.T. (Guest Editors), Recent Advances in Intelligent Paradigms Fusion and Their Applications, International Journal of Hybrid Intelligent Systems, Volume 5, Issue 3, 2008.
- Lim, C.P., Jain, L.C., Nguyen, N.T. and Balas, V. (Guest Editors), Advances in Computational Intelligence Paradigms and Applications, An International Journal on Fuzzy Optimization and Decision Making, Kluwer Academic Publisher, Volume 7, Number 3, 2008.
- Nguyen, N.T., Lim, C.P., Jain, L.C. and Balas, V.E. (Guest Editors), Theoretical Advances and Applications of Intelligent Paradigms, Journal of Intelligent and Fuzzy Systems, IOS Press, Volume 19, Issue 6, 2008.
- Abraham, A., Jarvis, D., Jarvis, J. and Jain, L.C. (Guest Editors), Special issue on Innovations in agents: An International Journal on Multiagent and Grid Systems, IOS Press, Volume 4, Issue 4, 2008.
- Ghosh, A., Seiffert, U. and Jain, L.C. (Guest Editors), Evolutionary Computation in Bioinformatics, Journal of Intelligent and Fuzzy Systems, IOS Press, The Netherlands, Volume 18, Number 6, 2007.
- Abraham, A., Smith, K., Jain, R. and Jain, L.C. (Guest Editors), Network and Information Security: A Computational Intelligence Approach, Journal of Network and Computer Applications, Elsevier Publishers, Volume 30, Issue 1, 2007.
- Palade, V. and Jain, L.C. (Guest Editors), Practical Applications of Neural Networks, Journal of Neural Computing and Applications, Springer, Germany, Volume 14, No. 2, 2005.
- Abraham, A. and Jain, L.C. (Guest Editors), Computational Intelligence on the Internet, Journal of Network and Computer Applications, Elsevier Publishers, Volume 28, Number 2, 2005.
- Abraham, A., Thomas, J., Sanyal, S. and Jain, L.C. (Guest Editors), Information Assurance and Security, Journal of Universal Computer Science, Volume 11, Issue 1, 2005.
- Abraham, A. and Jain, L.C. (Guest Editors), Optimal Knowledge Mining, Journal of Fuzzy Optimization and Decision Making, Kluwer Academic Publishers, Volume 3, Number 2, 2004.
- Palade, V. and Jain, L.C. (Guest Editors), Engineering Applications of Computational Intelligence, Journal of Intelligent and Fuzzy systems, IOS Press, Volume 15, Number 3, 2004.

- Alahakoon, D., Abraham, A. and Jain, L.C. (Guest Editors), Neural Networks for Enhanced Intelligence, Neural Computing and Applications, Springer, UK, Volume 13, No. 2, June 2004.
- Abraham, A., Jonkar, I., Barakova, E., Jain, R. and Jain, L.C. (Guest Editors), Special issue on Hybrid Neurocomputing, Neurocomputing, Elsevier, The Netherlands, Volume 13, No. 2, June 2004.
- Abraham, A. and Jain, L.C. (Guest Editors), Knowledge Engineering in an Intelligent Environment, Journal of Intelligent and Fuzzy Systems, IOS Press, The Netherlands, Volume 14, Number 3, 2003.
- Jain, L.C. (Guest Editor), Fusion of Neural Nets, Fuzzy Systems and Genetic Algorithms in Industrial Applications, IEEE Transactions on Industrial Electronics, USA, Volume 46, Number 6, December 1999.
- De Silva, C. and Jain, L.C. (Guest Editors), Intelligent Electronic Systems, Engineering Applications of Artificial Intelligence, Pergamon Press, USA, Volume 11, Number 1, January 1998.
- Jain, L.C. (Guest Editor), Intelligent Systems: Design and Applications - 2, Journal of Network and Computer Applications, Elsevier, **Vol. 2**, April 1996.
- Jain, L.C. (Guest Editor), Intelligent Systems: Design and Applications - 1, Journal of Network and Computer Applications, Elsevier, **Vol.1**, January, 1996.

5.3 Conferences

- KES International Conference Series
  www.kesinternational.org/
- AAAI Conference on Artificial Intelligence
  www.aaai.org/aaai08.php
- European Conferences on Artificial Intelligence (ECAI)

5.4 Conference Proceedings

- *Håkansson*, A., Nguyen, N.T., Hartung, R., Howlett, R.J. and Jain, L.C. (Editors), Agents and Multi-Agents Systems: Technologies and Applications, Lecture Notes in Artificial Intelligence, Springer-Verlag, Germany, 2009, in press.
- Nakamatsu, K., Phillips-Wren, G., Jain, L.C. and Howlett, R.J. (Editors), New Advances in Intelligent Decision Technologies, Springer-Verlag, 2009, in press.
- Nguyen, N.T., Jo, G.S., Howlett, R.J. and Jain, L.C. (Editors), Agents and Multi-Agents Systems: Technologies and Applications, Lecture Notes in Artificial Intelligence LNAI 4953, Springer-Verlag, Germany, 2008.
- Lovrek, I., Howlett, R.J. and Jain, L.C. (Editors), Knowledge-Based and Intelligent Information and Engineering Systems, Lecture Notes in Artificial Intelligence, Volume 1, LNAI 5177, Springer-Verlag, Germany, 2008.

- Lovrek, I., Howlett, R.J. and Jain, L.C. (Editors), Knowledge-Based and Intelligent Information and Engineering Systems, Lecture Notes in Artificial Intelligence, Volume 2, LNAI 5178, Springer-Verlag, Germany, 2008.
- Lovrek, I., Howlett, R.J. and Jain, L.C. (Editors), Knowledge-Based and Intelligent Information and Engineering Systems, Lecture Notes in Artificial Intelligence, Volume 3, LNAI 5179, Springer-Verlag, Germany, 2008.
- Pan, J. S., Niu, X.M., Huang, H. C. and Jain, L.C., Intelligent Information Hiding and Multimedia Signal Processing, IEEE Computer Society Press, USA, 2008.
- Jain, L.C., Lingras, P., Klusch, M., Lu, J., Zhang, C., Cercone, N. and Cao, L. Proceedings of the 2008 IEEE/WIC/ACM International Conference on Web Intelligence, WI 2008, IEEE Computer Society, USA, 2008.
- Jain, L.C., Gini, M., Faltings, B.B., Terano, T., Zhang, C., Cercone, N. and Cao, L. Proceedings of the 2008 IEEE/WIC/ACM International Conference on Intelligent Agent Technology, IAT 2008, IEEE Computer Society, USA, 2008.
- Apolloni, B., Howlett, R.J. and Jain, L.C. (Editors), Knowledge-Based Intelligent Information and Engineering Systems, Lecture Notes in Artificial Intelligence, Volume 1, LNAI 4692, KES 2007, Springer-Verlag, Germany, 2007.
- Apolloni, B., Howlett, R.J.and Jain, L.C. (Editors), Knowledge-Based Intelligent Information and Engineering Systems, Lecture Notes in Artificial Intelligence, Volume 2, LNAI 4693, , KES 2007, Springer-Verlag, Germany, 2007.
- Apolloni, B., Howlett, R.J. and Jain, L.C. (Editors), Knowledge-Based Intelligent Information and Engineering Systems, Lecture Notes in Artificial Intelligence, Volume 3, LNAI 4694, KES 2007, Springer-Verlag, Germany, 2007.
- Nguyen, N.T., Grzech, A., Howlett, R.J. and Jain, L.C., Agents and Multi-Agents Systems: Technologies and Applications, Lecture Notes in artificial Intelligence, LNAI 4696, Springer-Verlag, Germany, 2007.
- Howlett, R.P., Gabrys, B. and Jain, L.C. (Editors), Knowledge-Based Intelligent Information and Engineering Systems, Lecture Notes in Artificial Intelligence, KES 2006, Springer-Verlag, Germany, LNAI 4251, 2006.
- Howlett, R.P., Gabrys, B. and Jain, L.C. (Editors), Knowledge-Based Intelligent Information and Engineering Systems, Lecture Notes in Artificial Intelligence, KES 2006, Springer-Verlag, Germany, LNAI 4252, 2006.
- Howlett, R.P., Gabrys, B. and Jain, L.C. (Editors), Knowledge-Based Intelligent Information and Engineering Systems, Lecture Notes in Artificial Intelligence, KES 2006, Springer-Verlag, Germany, LNAI 4253, 2006.

- Liao, B.-H., Pan, J.-S., Jain, L.C., Liao, M., Noda, H. and Ho, A.T.S., Intelligent Information Hiding and Multimedia Signal Processing, IEEE Computer Society Press, USA, 2007. ISBN: 0-7695-2994-1.
- Khosla, R., Howlett, R.P., and Jain, L.C. (Editors), Knowledge-Based Intelligent Information and Engineering Systems, Lecture Notes in Artificial Intelligence, KES 2005, Springer-Verlag, Germany, LNAI 3682, 2005.
- Skowron, A., Barthes, P., Jain, L.C., Sun, R., Mahoudeaux, P., Liu, J. and Zhong, N.(Editors), Proceedings of the 2005 IEEE/WIC/ACM International Conference on Intelligent Agent Technology, Compiegne, France, IEEE Computer Society Press, USA, 2005.
- Khosla, R., Howlett, R.P., and Jain, L.C. (Editors), Knowledge-Based Intelligent Information and Engineering Systems, Lecture Notes in Artificial Intelligence, KES 2005, Springer-Verlag, Germany, LNAI 3683, 2005.
- Khosla, R., Howlett, R.P., and Jain, L.C. (Editors), Knowledge-Based Intelligent Information and Engineering Systems, Lecture Notes in Artificial Intelligence, KES 2005, Springer-Verlag, Germany, LNAI 3684, 2005.
- Khosla, R., Howlett, R.P., and Jain, L.C. (Editors), Knowledge-Based Intelligent Information and Engineering Systems, Lecture Notes in Artificial Intelligence, KES 2005, Springer-Verlag, Germany, LNAI 3685, 2005.
- Negoita, M., Howlett, R.P., and Jain, L.C. (Editors), Knowledge-Based Intelligent Engineering Systems, KES 2004, Lecture Notes in Artificial Intelligence, LNAI. 3213, Springer, 2004
- Negoita, M., Howlett, R.P., and Jain, L.C. (Editors), Knowledge-Based Intelligent Engineering Systems, KES 2004, Lecture Notes in Artificial Intelligence, LNAI 3214, Springer, 2004
- Negoita, M., Howlett, R.P., and Jain, L.C. (Editors), Knowledge-Based Intelligent Engineering Systems, KES 2004, Lecture Notes in Artificial Intelligence, LNAI 3215, Springer, 2004
- Palade, V., Howlett, R.P., and Jain, L.C. (Editors), Knowledge-Based Intelligent Engineering Systems, Lecture Notes in Artificial Intelligence, LNAI 2773, Springer, 2003
- Palade, V., Howlett, R.P., and Jain, L.C. (Editors), Knowledge-Based Intelligent Engineering Systems, Lecture Notes in Artificial Intelligence, LNAI 2774, Springer, 2003
- Damiani, E., Howlett, R.P., Jain, L.C. and Ichalkaranje, N. (Editors), Proceedings of the Fifth International Conference on Knowledge-Based Intelligent Engineering Systems, **Volume 1**, IOS Press, The Netherlands, 2002.
- Damiani, E., Howlett, R.P., Jain, L.C. and Ichalkaranje, N. (Editors), Proceedings of the Fifth International Conference on Knowledge-Based Intelligent Engineering Systems, **Volume 2**, IOS Press, The Netherlands, 2002.
- Baba, N., Jain, L.C. and Howlett, R.P. (Editors), Proceedings of the Fifth International Conference on Knowledge-Based Intelligent Engineering Systems (KES'2001), **Volume 1,** IOS Press, The Netherlands, 2001.
- Baba, N., Jain, L.C. and Howlett, R.P. (Editors), Proceedings of the Fifth International Conference on Knowledge-Based Intelligent Engineering Systems (KES'2001), **Volume 2,** IOS Press, The Netherlands, 2001.

- Howlett, R.P. and Jain, L.C.(Editors), Proceedings of the Fourth International Conference on Knowledge-Based Intelligent Engineering Systems, IEEE Press, USA, 2000. **Volume 1.**
- Howlett, R.P. and Jain, L.C.(Editors), Proceedings of the Fourth International Conference on Knowledge-Based Intelligent Engineering Systems, IEEE Press, USA, 2000. **Volume 2**.
- Jain, L.C.(Editor), Proceedings of the Third International Conference on Knowledge-Based Intelligent Engineering Systems, IEEE Press, USA, 1999.
- Jain, L.C. and Jain, R.K. (Editors), Proceedings of the Second International Conference on Knowledge-Based Intelligent Engineering Systems, **Volume 1**, IEEE Press, USA, 1998.
- Jain, L.C. and Jain, R.K. (Editors), Proceedings of the Second International Conference on Knowledge-Based Intelligent Engineering Systems, **Volume 2**, IEEE Press, USA, 1998.
- Jain, L.C. and Jain, R.K. (Editors), Proceedings of the Second International Conference on Knowledge-Based Intelligent Engineering Systems, **Volume 3**, IEEE Press, USA, 1998.
- Jain, L.C. (Editor), Proceedings of the First International Conference on Knowledge-Based Intelligent Engineering Systems, **Volume 1**, IEEE Press, USA, 1997.
- Jain, L.C. (Editor), Proceedings of the First International Conference on Knowledge-Based Intelligent Engineering Systems, **Volume 2**, IEEE Press, USA, 1997.
- Narasimhan, V.L., and Jain, L.C. (Editors), The Proceedings of the Australian and New Zealand Conference on Intelligent Information Systems, IEEE Press, USA, 1996.
- Jain, L.C. (Editor), Electronic Technology Directions Towards 2000, ETD2000, **Volume 1**, IEEE Computer Society Press, USA, May 1995.
- Jain, L.C. (Editor), Electronic Technology Directions Towards 2000, ETD2000, **Volume 2**, IEEE Computer Society Press, USA, May 1995

5.5 Book Series

5.5.1 Advanced Intelligence and Knowledge Processing, Springer-Verlag, Germany: www.springer.com/series/4738

- Zudilova-Seinstra, E. et al. (Editors), Trends in Interactive Visualization, Springer-Verlag, London, 2009.
- Monekosso, D., et al. (Editors), Intelligent Environments, Springer-Verlag, London, 2009.
- Chli, M. and de Wilde, P., Convergence and Knowledge Processing in Multi-agent Systems, Springer-Verlag, London, 2009
- Chein, M. and Mugnier, M.L., Graph-based Knowledge Representation, Springer-Verlag, London, 2009.

- Narahari, Y., et al., Game Theoretic Problems in Network Economics and Mechanism Design Solutions, Springer-Verlag, London, 2009.
- Zarri, G.P., Representation and Management of Narrative Information, Springer-Verlag, London, 2009.
- Hu, C., et al. (Editors), Knowledge Processing with Interval and Soft Computing, Springer-Verlag, London, 2008.
- Simovici, D.A. and Djeraba, C., Mathematical Tools for Data Mining, Springer-Verlag, London, 2008.
- Okada, A., et al. (Editors), Knowledge Cartography, Springer-Verlag, London, 2008.
- Nguyen, N.T., Advanced Methods for Inconsistent Knowledge Management, Springer-Verlag, London, 2008.
- Meisels, A., Distributed Search by Constrained Agents, Springer-Verlag, London, 2008.
- Camastra, F. and Vinciarelli, A., Machine Learning for audio, Image, and Video Analysis, Springer-Verlag, London, 2008.
- Kornai, A., Mathematical Linguistics, Springer-Verlag, London, 2008.
- Prokopenko, M. (Editor), Advances in Applied Self-Organising Systems, Springer-Verlag, London, 2008.
- Scharl, A., Environmental Online Communication, Springer-Verlag, London, 2007.
- Pierre, S. (Editor), E-Learning Networked Environments and Architectures, Springer-Verlag, London, 2007
- Karny, M. (Editor), Optimized Bayesian Dynamic Advising, Springer-Verlag, London, 2006.
- Liu, S. and Lin, Y., Grey Information: Theory and Practical Applications, Springer-Verlag, London, 2006.
- Maloof, M.A. (Editor), Machine Learning and Data Mining for Computer Security, Springer-Verlag, London, 2006.
- Wang, J.T.L., et al. (Editors), Data Mining in Bioinformatics, Springer-Verlag, London, 2005.
- Grana, M., et al. (Editors), Information Processing with Evolutionary Algorithms, Springer-Verlag, London, 2005.
- Fyfe, C., Hebbian Learning and Negative Feedback Networks,, Springer-Verlag, London, 2005.
- Chen-Burger, Y. and Robertson, D., Automatic Business Modelling,, Springer-Verlag, London, 2005.
- Husmeier, D., et.al. (Editors), Probabilistic Modelling in Bioinformatics and Medical Informatics, Springer-Verlag, London, 2005.
- Tan, K.C., et al., Multiobjective Evolutionary Algorithms and Applications, Springer-Verlag, London, 2005.
- Bandyopadhyay, S., et. al. (Editors), Advanced Methods for Knowledge Discovery from Complex Data, Springer-Verlag, London, 2005.
- Stuckenschmidt, H. and Harmelen, F.V., Information Sharing on the Semantic Web, Springer-Verlag, London, 2005.

- Abraham, A., Jain, L.C. and Goldberg, R., Evolutionary Multiobjective Optimization, Springer-Verlag, London, 2005.
- Gomez-Perez, et al., Ontological Engineering, Springer-Verlag, London, 2004.
- Zhang, S., et. al., Knowledge Discovery in Multiple Databases, Springer-Verlag, London, 2004.
- Ko, C.C., Creating Web-based Laboratories, Springer-Verlag, London, 2004.
- Mentzas, G., et al., Knowledge Asset Management, Springer-Verlag, London, 2003.
- Vazirgiannis, M., et al., Uncertainty Handling and Quality Assessment in Data Mining, Springer-Verlag, London, 2003.

5.5.2 Advanced Information Processing, Springer-Verlag, Germany

- Harris, C., Hong, X. and Gan, Q., Adaptive Modelling, Estimation and Fusion from Data, Springer-Verlag, Germany, 2002.
- Ohsawa, Y. and McBurney, P. (Editors), Chance Discovery, Springer-Verlag, Germany, 2003.
- Deen, S.M. (Editor), Agent-Based Manufacturing, Springer-Verlag, Germany, 2003.
- Gasós J. and Hirsch B., e-Business Applications, Springer-Verlag, Germany, 2003.
- Chen, S.H. and Wang, P.P. (Editors), Computational Intelligence in Economics and Finance, Springer-Verlag, Germany, 2004.
- Liu, J. and Daneshmend, L., Spatial reasoning and Planning, Springer-Verlag, Germany, 2004.
- Wang, L. and Fu, X., Data Mining with Computational Intelligence, Springer-Verlag, 2005.
- Ishibuchi, H., Nakashima, T. and Nii, M., Classification and Modeling with Linguistic Information Granules, Springer-Verlag, Germany, 2005.

5.5.3 Computational Intelligence and Its Applications Series, IGI Publishing, USA, http://www.igi-pub.com/bookseries/details.asp?id=5

- Chen, S.-H., Jain, L.C. and Tai, C.-C. (Editors), Computational Economics: A Perspective from Computational Intelligence, IGI Publishing, 2006.
- Begg, R. and Palaniswami, M., Computational Intelligence for Movement Sciences, IGI Publishing, 2006.
- Fulcher, J. (Editor), Advances in Applied Intelligences, IGI Publishing, 2006.
- Zhang, D., et al., Biometric Image Discrimination technologies, IGI Publishing, 2006.

5.5.4 Knowledge-Based Intelligent Engineering Systems Series, IOS Press, The Netherlands, http://www.kesinternational.org/bookseries.php

- Velasquez, J.D. and Palade, V., Adaptive Web Sites, IOS Press, Vol. 170, 2008.
- Zha, X.F. and Howlett, R.J., Integrated Intelligent System for Engineering Design, Vol. 149, IOS Press, 2006.
- Phillips-Wren, G. and Jain, L.C. (Editors), Intelligent Decision Support Systems in Agent-Mediated Environments, IOS Press, The Netherlands, Volume 115, 2005.
- Nakamatsu, K. and Abe, J.M., Advances in Logic Based Intelligent Systems, IOS Press, The Netherlands, Volume 132, 2005.
- Abraham, A., Koppen, M. and Franke, K. (Editors), Design and Applications of Hybrid Intelligent Systems, IOS Press, The Netherlands, Volume 104, 2003.
- Turchetti, C., Stochastic Models of Neural Networks, IOS Press, The Netherlands, Volume 102, 2004
- Wang, K., Intelligent Condition Monitoring and Diagnosis Systems, IOS Press, The Netherlands, Volume 93, 2003.
- Abraham, A., et al. (Editors), Soft Computing Systems, IOS Press, The Netherlands, Volume 87, 2002.
- Lee, R.S.T. and Liu, J.H.K., Invariant Object Recognition based on Elastic Graph Matching, IOS Press, The Netherlands, Volume 86, 2003.
- Loia, V. (Editor), Soft Computing Agents, IOS Press, The Netherlands, Volume 83, 2002.
- Motoda, H., Active Mining, IOS Press, The Netherlands, Volume 79. 2002.
- Namatame, A., et al. (Editors), Agent-Based Approaches in Economic and Social Complex Systems, IOS Press, The Netherlands, Volume 72, 2002.

5.5.5 The CRC Press International Series on Computational Intelligence, The CRC Press, USA, http://www.crcpress.com/shopping_cart/products/product_series.asp?id=&series=747975&parent_id=&sku=1965&isbn=9780849319655&pc

- Teodorescu, H.N. Kandel, A. and Jain, L.C. (Editors), Intelligent systems and Technologies in Rehabilitation Engineering, CRC Press USA, 2001.
- Jain, L.C. and Fanelli, A.M. (Editors), Recent Advances in Artificial Neural Networks: Design and Applications, CRC Press, USA, 2000.
- Medsker, L., and Jain, L.C. (Editors) Recurrent Neural Networks: Design and Applications, CRC Press, USA, 2000.
- Jain, L.C., Halici, U., Hayashi, I., Lee, S.B. and Tsutsui, S. (Editors) Intelligent Biometric Techniques in Fingerprint and Face Recognition, CRC Press, USA, 2000.

- Jain, L.C. (Editor), Evolution of Engineering and Information Systems, CRC Press USA, 2000.
- Dumitrescu, D., Lazzerini, B., Jain, L.C. and Dumitrescu, A., Evolutionary Computation, CRC Press USA, 2000.
- Dumitrescu, D., Lazzerini, B., Jain, L.C., Fuzzy Sets and their Applications to Clustering and Training, CRC Press USA, 2000.
- Jain, L.C. and De Silva, C.W. (Editors), Intelligent Adaptive Control, CRC Press USA, 1999.
- Jain, L.C. and Martin, N.M. (Editors), Fusion of Neural Networks, Fuzzy Logic and Evolutionary Computing and their Applications, CRC Press USA, 1999.
- Jain, L.C. and Lazzerini, B., (Editors), Knowledge-Based Intelligent Techniques in Character Recognition, CRC Press USA, 1999.
- Teodorescu, H.N. Kandel, A. and Jain, L.C. (Editors), Soft Computing Techniques in Human Related Science, CRC Press USA, 1999.
- Jain, L.C. and Vemuri, R. (Editors), Industrial Applications of Neural Networks, CRC Press USA, 1998.
- Jain, L.C., Johnson, R.P., Takefuji, Y. and Zadeh, L.A. (Editors), Knowledge-Based Intelligent Techniques in Industry, CRC Press USA, 1998.
- Teodorescu, H.N., Kandel, A. and Jain, L.C. (Editors), Fuzzy and Neurofuzzy Systems in Medicine, CRC Press USA, 1998.

5.5.6 International Series on Natural and Artificial Intelligence, AKI http://www.innoknowledge.com

- Apolloni, B., et al, Algorithmic Inference in Machine Learning, Advanced Knowledge International, Australia, 2006.
- Lee, R.S.T., Advanced Paradigms in Artificial Intelligence, Advanced Knowledge International, Australia, 2005.
- Katarzyniak, R, Ontologies and Soft Methods in Knowledge Management, Advanced Knowledge International, Australia, 2005.
- Abe, A. and Ohsawa, Y. (Editors), Readings in Chance Discovery, Advanced Knowledge International, Australia, 2005.
- Kesheng Wang, K., Applied Computational Intelligence in Intelligent Manufacturing Systems, Advanced Knowledge International, Australia, 2005.
- Murase, K., Jain, L.C., Sekiyama, K. and Asakura, T. (Editors), Proceedings of the Fourth International Symposium on Human and Artificial Intelligence Systems, University of Fukui, Japan, Advanced Knowledge International, Australia, 2004.
- Nguyen N.T., (Editor) Intelligent Technologies for inconsistent Knowledge Processing, Advanced Knowledge International, Australia, 2004.
- Andrysek, J., et al., (Editors), Multiple Participant Decision Making, Advanced Knowledge International, Australia, 2004.

- Matsuda, K., Personal Agent-Oriented Virtual Society, Advanced Knowledge International, Australia, 2004.
- Ichimura, T. and Yoshida, K. (Editors), Knowledge-Based Intelligent Systems for Healthcare, Advanced Knowledge International, Australia, 2004.
- Murase, K., and Asakura, T. (Editors), Dynamic Systems Approach for Embodiment and Sociality: From Ecological Psychology to Robotics, Advanced Knowledge International, Australia, 2003
- Jain, R., et al. (Editors), Innovations in Knowledge Engineering, Advanced Knowledge International, Australia, 2003.
- Graziella, T., Jain, L.C., Innovations in Decision Support Systems, Advanced Knowledge International, Australia, 2003.
- Galitsky, B., Natural Language Question Answering System Technique of Semantic Headers, Advanced Knowledge International, Australia, 2003.
- Guiasu, S., Relative Logic for Intelligence-Based Systems, Advanced Knowledge International, Australia, 2003.

5.5.7 Series on Innovative Intelligence, World Scientific: http://www.worldscientific.com/

- Jain, L.C., Howlett, R.J., Ichalkaranje, N., and Tonfoni, G. (Editors), Virtual Environments for Teaching and Learning, World Scientific Publishing Company Singapore, Volume 1, 2002.
- Jain, L.C., Ichalkaranje, N. and Tonfoni, G. (Editors), Advances in Intelligent Systems for Defence, World Scientific Publishing Company Singapore, Volume 2, 2002.
- Howlett, R., Ichalkaranje, N., Jain, L.C. and Tonfoni, G. (Editors), Internet-Based Intelligent Information Processing, World Scientific Publishing Company Singapore, Volume 3, 2002.
- Zaknich, A., Neural Nets for Intelligent Signal Processing, World Scientific Publishing Company Singapore, Volume 4, 2003.
- Hirose, A., Complex Valued Neural Networks, World Scientific Publishing Company Singapore, Volume 5, 2003.
- Shapiro, A.F. and Jain, L.C. (Editors), Intelligent and Other Computational Intelligence Techniques in Insurance, World Scientific Publishing Company Singapore, Volume 6, 2003.
- Pan, J-S., Huang, H.-C. and Jain, L.C. (Editors), Intelligent Watermarking Techniques, World Scientific Publishing Company Singapore, Volume 7, 2004.
- Hasebrook, J. and Maurer, H.A., Learning Support Systems for Organisational Learning, World Scientific Publishing Company Singapore, Volume 8, 2004.

5.6 Books

- Jain, L.C. and Nguyen, N.T. (Editors), Knowledge Processing and Decision Making in Agent-Based Systems, Springer-Verlag, Germany, 2009.
- Tolk, A. and Jain, L.C. (Editors), Complex Systems in Knowledge-based Environments, Springer-Verlag, Germany, 2009.
- Nguyen, N.T. and Jain, L.C. (Editors), Intelligent Agents in the Evolution of Web and Applications, Springer-Verlag, Germany, 2009.
- Jarvis, J., Ronnquist, R, Jarvis, D. and Jain, L.C., Holonic Execution: A BDI Approach, Springer-Verlag, 2008.
- Jain, L.C., Sato, M., Virvou, M., Tsihrintzis, G., Balas, V. and Abeynayake, C. (Editors), Computational Intelligence Paradigms: Volume 1 – Innovative Applications, Springer-Verlag, 2008.
- Phillips-Wren, G., Ichalkaranje, N. And Jain, L.C. (Editors), Intelligent Decision Making-An AI-Based Approach, Springer-Verlag, 2008.
- Fulcher, J. and Jain, L.C. (Editors), Computational Intelligence: A Compendium, Springer-Verlag, 2008.
- Sordo, M., Vaidya, S. and Jain, L.C.(Editors), Advanced Computational Intelligence Paradigms in **Healthcare 3**, Springer-Verlag, 2008.
- Virvou, M. And Jain, L.C.(Editors) , Intelligent Interactive Systems in Knowledge-Based Environments, Springer-Verlag, 2008.
- Sommerer, C., Jain, L.C. and Mignonneau, L. (Editors), The Art and Science of Interface and Interaction Design, Volume 1, Springer-Verlag, 2008.
- Nayak, R., Ichalkaranje, N. and Jain, L.C. (Editors), Evolution of the Web in Artificial Intelligence Environments, Springer-Verlag, 2008.
- Tsihrintzis, G. and Jain, L.C. (Editors), Multimedia Services in Intelligent Environments, Springer-Verlag, 2008.
- Tsihrintzis, G., Virvou, M., Howlett, R.J. and Jain, L.C. (Editors), New Directions in Intelligent Interactive Multimedia, Springer-Verlag, 2008.
- Holmes, D. and Jain, L.C. (Editors), Innovations in Bayesian Networks, Springer-Verlag, Germany, 2008.
- Magnenat-Thalmann, Jain, L.C. and Ichalkaranje, N., New Advances in Virtual Humans, Springer-Verlag, Germany, 2008.
- Jain, L.C., Palade, V. and Srinivasan, D. (Editors), Advances in Evolutionary Computing for System Design, Springer-Verlag, 2007.
- Baba, N., Handa, H. and Jain, L.C. (Editors), Advanced Intelligent Paradigms in Computer Games, Springer-Verlag, 2007.
- Chahl, J.S., Jain, L.C., Mizutani, A. and Sato-Ilic, M. (Editors), Innovations in Intelligent Machines 1, Springer-Verlag, 2007.
- Zharkova, V. and Jain, L.C. (Editors), Artificial Intelligence in Recognition and Classification of Astrophysical and Medical Images, Springer-Verlag, 2007.
- Pan, J-S., Huang, H.-C., Jain, L.C. and Fang, W.-C. (Editors), Intelligent Multimedia Data Hiding, Springer-Verlag, 2007.

- Yoshida, H., Jain, A., Ichalkaranje, A., Jain, L.C. and Ichalkaranje, N. (Editors), Advanced Computational Intelligence Paradigms in **Healthcare 1**, Springer-Verlag, 2007.
- Vaidya, S., Jain, L.C. and Yoshida, H. (Editors), Advanced Computational Intelligence Paradigms in **Healthcare 2**, Springer-Verlag, 2007.
- Jain, L.C, Tedman, R. and Tedman, D. (Editors), Evolution of Teaching and Learning in Intelligent Environment, Springer-Verlag, 2007.
- Sato, M. and Jain, L.C., Innovations in Fuzzy Clustering, Springer-Verlag, 2006.
- Patnaik, S., Jain, L.C., Tzafestas, S.G., Resconi, G. and Konar, A. (Editors), Innovations in Robot Mobility and Control, Springer-Verlag, 2006.
- Apolloni, B., Ghosh, A., Alpaslan, F., Jain, L.C. and Patnaik, S. (Editors), Machine Learning and Robot Perception, Springer-Verlag, 2006.
- Palade, V., Bocaniala, C.D. and Jain, L.C. (Editors), Computational Intelligence in Fault Diagnosis, Springer-Verlag, 2006.
- Holmes, D. and Jain, L.C. (Editors), Innovations in Machine Learning, Springer-Verlag, 2006.
- Ichalkaranje, N., Ichalkaranje, A. and Jain, L.C. (Editors), Intelligent Paradigms for Assistive and Preventive Healthcare, Springer-Verlag, 2006.
- Seiffert, U., Jain, L.C. and Schweizer, P. (Editors), Bioinformatics Using Computational Intelligence Paradigms, Springer-Verlag, 2005.
- Ghosh, A. and Jain, L.C. (Editors), Evolutionary Computation in Data Mining, Springer-Verlag, Germany, 2005.
- Phillips-Wren, G. and Jain, L.C.(Editors), Intelligent Decision Support Systems in Agent-Mediated Environments, IOS Press, The Netherlands, 2005.
- Silverman, B., Jain, A., Ichalkaranje, A. and Jain, L.C. (Editors), Intelligent Paradigms in Healthcare Enterprises, Springer-Verlag, Germany, 2005.
- Ghaoui, C., Jain, M., Bannore, V., and Jain, L.C. (Editors), Knowledge-Based Virtual Education, Springer-Verlag, Germany, 2005.
- Pal, N. and Jain, L.C.(Editors), Advanced Techniques in Knowledge Discovery and Data Mining, Springer-Verlag, London, 2005.
- Khosla, R., Ichalkaranje, N. and Jain, L.C.(Editors), Design of Intelligent Multi-Agent Systems, Springer-Verlag, Germany, 2005.
- Abraham, A., Jain, L.C. and van der Zwaag, B.(Editors), Innovations in Intelligent Systems, Springer-Verlag, Germany, 2004.
- Tonfoni, G. and Jain, L.C., Visualizing Document Processing, Mouton De Gruyter, Germany, 2004.
- Fulcher, J. and Jain, L.C.(Editors), Applied Intelligent Systems, Springer-Verlag, Germany, 2004.
- Damiani, E., Jain, L.C. and Madravio, M. (Editors), Soft Computing in Software Engineering, Springer-Verlag, Germany, 2004.

- Resconi, G. and Jain, L.C., Intelligent Agents: Theory and Applications, Springer-Verlag, Germany, 2004.
- Abraham, A., Jain, L.C. and Kacprzyk, J. (Editors), Recent Advances in Intelligent Paradigms and Applications, Springer-Verlag, Germany, 2003.
- Tonfoni, G. and Jain, L.C., The Art and Science of Documentation Management, Intellect, UK, 2003.
- Seiffert, U. and Jain, L.C. (Editors), Self-Organising Neural Networks, Springer-Verlag, Germany, 2002.
- Jain, L.C., Howlett, R.J., Ichalkaranje, N., and Tonfoni, G. (Editors), Virtual Environments for Teaching and Learning, World Scientific Publishing Company Singapore, 2002.
- Schmitt, M. Teodorescu, H.-N., Jain, A., Jain, A., Jain, S. and Jain, L.C. (Editors), Computational Intelligence Processing in Medical Diagnosis, Springer- Verlag, 2002.
- Jain, L.C. and Kacprzyk, J. (Editors), New Learning Paradigms in Soft Computing, Springer-Verlag, Germany, 2002.
- Jain, L.C., Chen, Z. and Ichalkaranje, N. (Editors), Intelligent Agents and Their Applications, Springer-Verlag, Germany, 2002.
- Jain, L.C. and De Wilde, P. (Editors), Practical Applications of Computational Intelligence Techniques, Kluwer Academic Publishers, USA, 2001.
- Howlett, R.J. and Jain, L.C. (Editors), Radial Basis Function Networks 1, Springer-Verlag, Germany, 2001.
- Howlett, R.J. and Jain, L.C. (Editors), Radial Basis Function Networks 2, Springer-Verlag, Germany, 2001.
- Teodorescu, H.N., Jain, L.C. and Kandel, A. (Editors), Hardware Implementation of Intelligent Systems, Springer-Verlag, Germany, 2001.
- Baba, N. and Jain, L.C., Computational Intelligence in Games, Springer-Verlag, 2001.
- Jain, L.C., Lazzerini, B. and Halici, U. (Editors), Innovations in ART Neural Networks, Springer-Verlag, Germany, 2000.
- Jain, A., Jain, A., Jain, S. and Jain, L.C. (Editors), Artificial Intelligence Techniques in Breast Cancer Diagnosis and Prognosis, World Scientific Publishing Company, Singapore, 2000.
- Jain, L.C. (Editor), Innovative Teaching and Learning in Intelligent Environment, Springer-Verlag, 2000.
- Jain, L.C. and Fukuda, T. (Editors), Soft Computing for Intelligent Robotic Systems, Springer-Verlag, Germany, 1998.
- Jain, L.C. (Editor), Soft Computing Techniques in Knowledge-Based Intelligent Engineering Systems, Springer-Verlag, Germany, 1997.
- Sato, M., Sato, S. and Jain, L.C., Fuzzy Clustering Models and Applications, Springer-Verlag, Germany, 1997.
- Jain, L.C. and Jain, R.K. (Editors), Hybrid Intelligent Engineering Systems, World Scientific Publishing Company, Singapore, 1997.

- Vonk, E., Jain, L.C. and Johnson, R.P., Automatic Generation of Neural Networks Architecture Using Evolutionary Computing, World Scientific Publishing Company, Singapore, 1997.
- Van Rooij, A.J.F., Jain, L.C. and Jain, L.C., Neural Network Training Using Genetic Algorithms, World Scientific Publishing Company, Singapore, December 1996.

5.7 Book Chapters

- Jain, L.C., Lim, C.P. and Nguyen, N.T., Innovations in Knowledge Processing and Decision Making in Agent-Based Systems, Springer-Verlag, Germany, 2009, Chapter 1, pp. 1-18.
- Tweedale, J. and Jain, L.C., The Evolution of Intelligent Agents Within the World Wide Web, Springer-Verlag, Germany, 2009, Chapter 1, pp. 1-9.
- Tolk, A. and Jain, L.C., An Introduction to Complex Systems in Knowledge-based Environments, Volume 168, SCI Series, Springer-Verlag, Germany, 2009, pp. 1-5
- Pedrycz, W., Ichalkaranje, N., Phillips-Wren, G., and Jain, L.C., Introduction to Computational Intelligence for Decision Making, Springer-Verlag, 2008, pp. 75-93, Chapter 3.
- Tweedale, J., Ichalkaranje, N., Sioutis, C., Urlings, P. and Jain, L.C., Future Directions: Building a Decision Making Framework using Agent Teams, Springer-Verlag, 2008, pp. 381-402, Chapter 14.
- Virvou, M. and Jain, L.C., Intelligent Interactive Systems in Knowledge-Based Environments: An Introduction, Springer-Verlag, 2008, pp. 1-8, Chapter 1.
- Tsihrintzis, G. and Jain, L.C., An Introduction to Multimedia Services in Intelligent Environments, Springer-Verlag, pp. 1-10, 2008, Chapter 1.
- Jain, L.C. and Lim, C.P., An Introduction to Computational Intelligence Paradigms, Springer-Verlag, pp. 1-15, 2008 Chapter 1.
- Nayak, R. and Jain, L.C., An Introduction to the Evolution of the Web in an Artificial Intelligence Environment, pp. 1-15, 2008, Chapter 1.
- Nayak, R. and Jain, L.C., Innovations in Web Applications using Artificial Intelligence Paradigms, pp. 17-40, 2008, Chapter 2.
- Zharkova, V.V. and Jain, L.C., Introduction to Recognition and Classification in Medical and Astrophysical Images, Springer-Verlag, 2007, pp. 1-18, Chapter 1.
- Yoshida, H., Vaidya, S. and Jain, L.C., Introduction to Computational Intelligence in Healthcare, Springer-Verlag, 2007, pp. 1-4, Chapter 1.
- Huang, H.C., Pan, J.S., Fang, W.C. and Jain, L.C., An Introduction to Intelligent Multimedia Data Hiding, Springer-Verlag, 2007, pp. 1-10, Chapter 1.
- Jain, L.C., et al., Intelligent Machines :An Introduction, Springer-Verlag, 2007, pp. 1-9, Chapter 1.
- Jain, L.C., et al., Introduction to Evolutionary Computing in System Design, Springer-Verlag, 2007, pp. 1-9, Chapter 1.

- Jain, L.C., et al., Evolutionary Neuro-Fuzzy Systems and Applications, Springer-Verlag, 2007, pp. 11-45, Chapter 1.
- Do, Q.V, Lozo, P. and Jain, L.C., Vision-Based Autonomous Robot Navigation, in Innovations in Robot Mobility and Control, Springer-Verlag, 2006, pp. 65-103, Chapter 2.
- Tran, C., Abraham, A. and Jain, L., Soft Computing Paradigms and Regression Trees in Decision Support Systems, in Advances in Applied Artificial Intelligence, Idea Group Publishing, 2006, pp. 1-28, Chapter 1.
- Jarvis, B., Jarvis, D. and Jain, L., Teams in Multi-Agent Systems, in IFIP International Federation for Information Processing, Vol. **228**, Intelligent Information Processing III, Springer, 2006, pp. 1-10, Chapter 1.
- Abraham, A. and Jain, L.C., Evolutionary Multiobjective Optimization, Springer-Verlag, 2005, pp. 1-6, Chapter 1.
- Sisman-Yilmaz, N.A., Alpaslan, F. and Jain, L.C., Fuzzy Multivariate Auto-Regression Method and its Application, in Applied Intelligent Systems, Springer, 2004, pp. 281-300.
- Lozo, P., Westmacott, J., Do, Q., Jain, L.C. and Wu, L., Selective Attention ART and Object Recognition, in Applied Intelligent Systems, Springer, 2004, pp. 301-320.
- Wang, F., Jain, L.C. and Pan, J., Genetic Watermarking on Spatial Domain, in Intelligent Watermarking Techniques, World Scientific, 2004, pp. 481-514, Chapter 17.
- Wang, F., Jain, L.C. and Pan, J., Watermark Embedding System based on Visual Cryptography, in Intelligent Watermarking Techniques, World Scientific, 2004, pp. 377-394, Chapter 13.
- Sioutis, C., Urlings, P., Tweedale, J., Ichalkaranje, N., Forming Human-Agent Teams within Hostile Environments, in Applied Intelligent Systems, Springer-Verlag, 2004, pp. 255-279.
- Jain, L.C. and Chen, Z., Industry, Artificial Intelligence In, in Encyclopedia of Information Systems, Elsevier Science, USA, 2003, pp. 583-597.
- Jain, L.C. and Konar, A., An Introduction to Computational Intelligence Paradigms, in Practical Applications of Computational Intelligence Techniques, Springer, 2001, pp. 1-38.
- Tedman, D. and Jain, L.C., An Introduction to Innovative Teaching and Learning, in Teaching and Learning, Springer, 2000, pp. 1-30, Chapter 1.
- Filippidis, A., Russo, M. and Jain, L.C., Novel Extension of ART2 in Surface Landmine Detection, Springer-Verlag, 2000, pp.1-25, Chapter 1.
- Jain, L.C. and Lazzerini, B., An Introduction to Handwritten Character and Word Recognition, in Knowledge-Based Intelligent Techniques in Character Recognition, CRC Press, 1999, 3-16.
- Filippidis, A., Jain, L.C. and Martin, N.N., "Computational Intelligence Techniques in Landmine Detection," in Computing with Words in Information/Intelligent Systems 2, Edited by Zadeh, L. and Kacprzyk, J., Springer-Verlag, Germany, 1999, pp. 586-609.

- Halici, U., Jain, L.C. and Erol, A., Introduction to Fingerprint Recognition, in Intelligent Biometric Techniques in Fingerprint and Face Recognition, CRC Press, 1999, pp.3-34.
- Teodorescu, H.N., Kandel, A. and Jain, L., Fuzzy Logic and Neuro-Fuzzy Systems in Medicine: A historical Perspective, in Fuzzy and Neuro-Fuzzy Systems in Medicine, CRC Press, 1999, pp. 3-16.
- Jain, L.C. and Vemuri, R., An Introduction to Intelligent Systems, in Hybrid Intelligent Engineering Systems, World Scientific, 1997, pp. 1-10, Chapter 1.
- Karr, C. and Jain, L.C., Genetic Learning in Fuzzy Control, in Hybrid Intelligent Engineering Systems, World Scientific, 1997, pp. 69-101, Chapter 4.
- Karr, C. and Jain, L.C., Cases in Geno- Fuzzy Control, in Hybrid Intelligent Engineering Systems, World Scientific, 1997, pp. 103-132, Chapter 5.
- Katayama, R., Kuwata, K. and Jain, L.C., Fusion Technology of Neuro, Fuzzy, GA and Chaos Theory and Applications, in Hybrid Intelligent Engineering Systems, World Scientific, 1997, pp. 167-186, Chapter 7.
- Jain, L.C., Medsker, L.R. and Carr, C., Knowledge-Based Intelligent Systems, in Soft Computing Techniques in Knowledge-Based Intelligent Systems, Springer-Verlag, 1997, pp. 3-14, Chapter 1.
- Babri, H., Chen, L., Saratchandran, P., Mital, D.P., Jain, R.K., Johnson, R.P. and Jain, L.C., Neural Networks Paradigms, in Soft Computing Techniques in Knowledge-Based Intelligent Systems, Springer-Verlag, 1997, pp. 15-43, Chapter 2.
- Jain, L.C., Tikk, D. and Koczy, L.T., Fuzzy Logic in Engineering, in Soft Computing Techniques in Knowledge-Based Intelligent Systems, Springer-Verlag, 1997, pp. 44-70, Chapter 3.
- Tanaka, T. and Jain, L.C., Analogue/Digital Circuit Representation for Design and Trouble Shooting in Intelligent Environment, in Soft Computing Techniques in Knowledge-Based Intelligent Systems, Springer-Verlag, 1997, pp. 227-258, Chapter 7.
- Jain, L.C., Hybrid Intelligent System Design Using Neural Network, Fuzzy Logic and Genetic Algorithms - Part I, Cognizant Communication Corporation USA, 1996, pp. 200-220, Chapter 9.
- Jain, L.C., Hybrid Intelligent System Applications in Engineering using Neural Network and Fuzzy Logic - Part II, Cognizant communication Corporation USA,1996, pp. 221-245, Chapter 10.
- Jain, L.C., Introduction to Knowledge-Based Systems, Electronic Technology Directions to the Year 2000, IEEE Computer Society Press USA, 1995, pp. 17-27, Chapter 1.
- Jain, L.C. and Allen, G.N., Introduction to Artificial Neural Networks, Electronic Technology Directions to the Year 2000, IEEE Computer Society Press USA, 1995, pp. 36-62, Chapter 2.
- Jain, L.C. and Karr, C.L., Introduction to Fuzzy Systems, Electronic Technology Directions to the Year 2000, IEEE Computer Society Press USA, 1995, pp. 93-103, Chapter 3.

- Jain, L.C. and Karr, C.L., Introduction to Evolutionary Computing Techniques, Electronic Technology Directions to the Year 2000, IEEE Computer Society Press USA, 1995, pp. 121-127, Chapter 4.
- Sato, M., Jain, L.C. and Takagi, H., Electronic Design and Automation, Electronic Technology Directions to the Year 2000, IEEE Computer Society Press USA, 1995, pp. 402-430, Chapter 9.
- Furuhashi, T., Takagi, H. and Jain, L.C., Intelligent Systems using Artificial Neural Networks, fuzzy Logic and Genetic Algorithms in Industry, Electronic Technology Directions to the Year 2000, IEEE Computer Society Press USA, 1995, pp. 485-4

References

[1] Aguilar-Ponce, R., Kumara, A., Tecpanecatl-Xihuitla, J.L., Bayoumia, M.: A network of sensor-based framework for automated visual surveillance. Jouranl of Network and Computer Applications 30, 1244–1271 (2007)

[2] Ahn, C.W., Ramakrishna, R.S.: A genetic algorithm for shortest path routing problem and the sizing of populations. IEEE Trans. on Evolutionary Computation 6, 566–579 (2002)

[3] Alvarez, E., Cancela, M.A., Correa, J.M., Navaza, J.M., Riverol, C.: Fuzzy logic control for the isomerized hop pellets production. Journal of Food Engineering 39, 145–150 (1999)

[4] Balogh, Z., Laclavik, M., Hluchy, L.: Multi-agent system for negotiation and decision support. In: Proceedings of Fourth International Scientific Conference Electronic Computers and Informatics, pp. 264–270 (2000)

[5] Bratman, M.E.: Intention, Plans, and Practical Reason. Harvard University Press (1987)

[6] Breiman, L., Friedman, J., Olshen, R., Stone, C.: Classification and Regression Trees. Wadsworth and Brooks (1984)

[7] Broomhead, D.S., Lowe, D.: Multivariate functional interpolation and adaptive networks. Complex Systems 2, 321–355 (1988)

[8] Buchanan, B.G.: A (very) brief history of artificial intelligence. AI Magezine 26, 53–60 (2005)

[9] Campbell, A.N., Hollister, V.F., Duda, R.O., Hart, P.E.: Recognition of a hidden mineral deposit by an artificial intelligence program. Science 217, 927–929 (1982)

[10] Chalhoub, N.G., Bazzi, B.A.: Fuzzy logic control for an integrated system of a micromanipulator with a single flexible beam. Journal of Vibration and Control 10, 755–776 (2004)

[11] Chang, S.J., Li, T.H.S.: Design and implementation of fuzzy parallel-parking control for a car-type mobile robot. Journal of Intelligent and Robotic Systems 34, 175–194 (2002)

[12] Chen, W., Wasson, B.: Intelligent Agents Supporting Distributed Collaborative Learning. In: Lin, F.O. (ed.) Designing Distributed Learning Environments with Intelligent Software Agents, pp. 33–66. Information Science Publishing (2005)

[13] Doran, J.: Expert systems and archaeology: What lies ahead? In: Ruggles, Rahtz (eds.) Computer and Quantitative Methods in Archaeology. BAR International Series, pp. 235–241 (1988)

[14] El Makodem, M., Courtecuisse, V., Saudemont, C., Robyns, B., Deuse, J.: Fuzzy logic supervisor-based primary frequency control experiments of a variable-speed wind generator. IEEE Trans. on Power Systems 24, 407–417 (2009)

[15] Ergün, U., Barýþçý, N., Ozan, A.T., Serhatlýoðlu, S., Oğur, E., Hardalaç, F., Güler, I.: Classificaiton of MCA stenosis in diabetes by MLP and RBF neural network. Journal of Medical Systems 28, 475--487 (2004)

[16] Eysenck, M.W.: Artificial Intelligence. In: Eysenck, M.W. (ed.) The Blackwell Dictionary of Cognitive Psychology. Basil Blackwell, Malden (1990)

[17] Florez-Lopez, R.: Modelling of insurers' rating determinants: an application of machine learning techniques and statistical models. European Journal of Operational Research 183, 1488–1512 (2007)

[18] Fuat Üler, G., Mohamed, O.A., Koh, C.S.: Design optimization of electrical machines using genetic algorithms. IEEE Transactions on Magnetics 31, 2008–2011 (1995)

[19] Grieu, S., Thiery, F., Traore, A., Nguyen, T.P., Barreau, M., Polit, M.: KSOM and MLP neural networks for on-line estimating the efficiency of an activated sludge process. Chemical Engineering Journal 116, 1–11 (2006)

[20] Holland, J.H.: Adaptation in Natural and Artificial Systems. The University of Michigan Press (1975)

[21] Jennings, N., Wooldridge, M.J.: Applications of Intelligent Agents. In: Jennings, N., Wooldridge, M.J. (eds.) Agent Technology: Foundations, Applications, and Markets, pp. 3–48. Springer, Heidelberg (1998)

[22] Kohonen, T.: Self-Organization and Associative Memory. Springer, Heidelberg (1984)

[23] Luger, G.F.: Artificial Intelligence: Structures and Strategies for Complex Problem Solving, 4th edn. Pearson Education Ltd., London (2002)

[24] Lilly, J.H.: Evolution of a negative-rule fuzzy obstacle avoidance controller for an autonomous vehicle. IEEE Trans. on Fuzzy Systems 15, 718–728 (2007)

[25] Maes, P.: Agents that reduce work and information overload. Communications of the ACM 37, 31–40 (1994)

[26] Markov, Z., Larose, D.T.: Data Mining the Web: Uncovering Patterns in Web Content, Structure, and Usage. Wiley-Interscience, Hoboken (2007)

[27] McCulloch, W.S., Pitts, W.: A logical calculus of the ideas immanent in nervous activity. Bulletin of Mathematical Biophysics 5, 115–133 (1943)

[28] Miller, R.A., Pople, H.E., Myers, J.D.: INTERNIST- 1: An experimental computer-based diagnostic consultant for general internal medicine. The New England Journal of Medicine 307, 468–476 (1982)

[29] Mininno, E., Cupertino, F., Naso, D.: Real-valued compact genetic algorithms for embedded microcontroller optimization. IEEE Trans. on Evolutionary Computation 12, 203–219 (2008)

[30] Minsky, M.L.: Semantic Information Processing. MIT Press, Cambridge (1968)

[31] Nathwani, B., Clarke, K., Lincoln, T., Berard, C., Taylor, C., Ng, K., Patil, R., Pike, M., Azen, S.: Evaluation of an expert system on lymph node pathology. Human Pathology 28, 1097–1110 (1997)

[32] Nilsson, N.J.: Problem-Solving Methods in Artificial Intelligence. McGraw-Hill, New York (1971)

[33] Ohsaki, M.: Genetic algorithm for topology optimization of trusses. Computers and Structures 57, 219–225 (1995)

[34] Ovum Report, Intelligent agents: the new revolution in software (1994)

[35] Parunak, H.V.D.: Manufacturing experience with the contract net. In: Huhns, M.N. (ed.) Distributed AI. Morgan Kaufmann, San Francisco (1987)
[36] Pauker, S.G., Gorry, G.A., Kassirer, J.P., Schwartz, W.B.: Toward the simulation of clinical cognition: Taking a present illness by computer. The American Journal Medicine 60, 981–995 (1976)
[37] Potvin, J.Y.: Genetic algorithms for the travelling salesman problem. Annals of Operations Research 63, 337–370 (1996)
[38] Quinlan, J.R.: C4.5 Programs for Machine Learning. Morgan Kaufmann, San Francisco (1993)
[39] Rich, E.: Artificial Intelligence. McGraw-Hill, New York (1983)
[40] Rumelhart, D.E., Hinton, G.E., William, R.J.: Learning internal representation by error propagation. In: Rumelhart, D.E., McLelland, J.L. (eds.) Parallel Distributed Processing, I, pp. 318–362. MIT Press, Cambridge (1986)
[41] Russell, S.J., Norvig, P.: Artificial Intelligence: A Modern Approach, 2nd edn. Prentice Hall, Englewood Cliffs (2003)
[42] Sargent, P.: Back to school for a brand new ABC. The Guardian, p. 28 (March 12, 1992)
[43] Sierra, C.: Agent-mediated electronic commerce. Autonomous Agents and Multi-Agent Systems 9, 285–301 (2004)
[44] Shortliffe, E.H.: Computer-based medical consultation: MYCIN. Elsevier/North Holland, Amsterdam (1976)
[45] Tam, K.Y.: Genetic algorithsm, function optimization, and facility layout design. European Journal of Operational Research 63, 322–346 (1992)
[46] Vasko, F.J., Newhart, D.D., Strauss, A.D.: Coal blending models for optimum cokemaking and blast furnace operation. Journal of Operational Research Society 56, 235–243 (2005)
[47] Weiss, S.M., Kulikowski, C.A., Amarel, S., Safir, A.: A model-based method for computer-aided medical decision making. Artificial Intelligence 11, 145–172 (1978)
[48] Zadeh, L.A.: Fuzzy Sets. Information and Control 8, 338–353 (1965)
[49] Zhang, G.P.: Neural networks for classification: a survey. IEEE Trans. on Systems, Man, and Cybernetics, Part C: Applications and Reviews 30, 451–462 (2000)

Part II
Methods and Tools

A Fuzzy Density Analysis of Subgroups by Means of DNA Oligonucleotides

Ikno Kim and Junzo Watada

Graduate School of Information, Production and Systems, Waseda University
octoberkim@akane.waseda.jp, watada@waseda.jp

Abstract. In complicated industrial and organizational relationships between employees or workers, it is difficult to offer good opportunities for their psychological and skill growth, since our progressive information and industrial societies have created many menial tasks. Redesigning subgroups in a personnel network for work rotation is a method that organizes employees appropriately to address these types of problems. In this article, we focus on a fuzzy density analysis of subgroups where employees are connected via their relationships with fuzzy values. However, it becomes extremely hard to rearrange those employees when there are vast numbers of them, meaning it is an NP-hard problem. In the personnel network, all the possible cohesive subgroups can be detected by making the best use of DNA oligonucleotides, which is also applied as a method by which to rearrange employees via fuzzy values based on the results of a fuzzy density analysis.

1 Introduction

Specialized menial tasks and occupations typically emerge in advanced industrial societies and environments and generally have an effect on employees or workers. To workers, these menial tasks rarely offer opportunities for achievement or satisfaction. Personnel managers employ a variety of methods to improve tasks and to enhance qualities of work life for reducing those problems. Especially, human relationships in business situations influence the achievements of employees and their organizations as indicated by Watada *et al.* in 1998 [18] and Toyoura *et al.* in 2004 [17].

Therefore, qualitatively understanding the relationships between employees requires competent rating of human relationships in business situations. One useful approach is known as work rotation, which has employees move and transfer from their present tasks to other new tasks [9]. These rotations can break the monotony of highly specialized and menial tasks while developing different skills and capabilities. The organization usually benefits from having employees who can take on several tasks rather than only one task. A variety of tasks and jobs easily improves the employee's self-image, and provides both personal and organizational psychological growth.

H.-N. Teodorescu, J. Watada, and L.C. Jain (Eds.): Intel. Sys. and Tech., SCI 217, pp. 31–45.
springerlink.com

The main problem is that if personnel managers must have knowledge of close interpersonal relationships between employees in a large personnel network, it can be intractable to properly redesign all the subgroups among employees because all the cliques and components of employees are hard to determine by conventional methods. Further, detecting the maximum clique of employees becomes NP-hard. The majority of personnel managers redesign subgroups based on the frequency data of employee relationships in business trading and transactions. To address this problem, DNA oligonucleotides are adapted to employee relationships with fuzzy values to solve these rearranging problems.

Since L. Adleman [1] discovered molecular computation in 1994, the attention on DNA computing has been almost entirely from either the computer science or biotechnology fields. On the other hand, we have found that DNA computing can be a useful tool for a variety of management problems, thus we have shown that efficient solutions are obtained by DNA computing. The efficiency of DNA computing is examined by comparing and analyzing the personnel network in this article. The objective of this article is to provide weighted evaluations of human relationships among employees as a fuzzy graph and to find fuzzy cliques in the organization between employees [15]. Also, we propose a way to apply DNA computing to human resource management that is a part of the engineering management field, and to measure the efficiency of DNA computing for this personnel management problem.

The structure of the rest of this article consists of Section 2 that illustrates the basic concepts of graphs, fuzzy graphs, cliques and fuzzy cliques, Section 3 that gives the method of DNA computing and exemplifies a fuzzy clique problem, Section 4 that shows the redesigning of subgroups in the fuzzy density analysis, and Section 5 that finally gives our conclusions.

2 Analysis of Subgroups with Fuzzy Values

2.1 Roles of Cohesive Subgroups

In a personnel network, cohesive subgroups represent specific subsets of employees among whom there are relatively strong, dense, direct, frequent, intense, or positive ties. The relationships in cohesive groups enable employees to share or exchange their business information and tasks. Numerous direct contacts among all cohesive subgroups between employees, combined with few or null ties to outsiders, dispose a group toward a close interpersonal relationship in business. Examples of formal cohesive groups include the personnel, production, quality control, or finance departments in a company or a formal association.

2.2 Model Personnel Network

A model personnel network is selected as a highly skilled group and was made for providing thoughtful leadership and specialized support to the knowledge management for the company. The group was composed of employees who had technologically advanced degrees or extensive industry experience.

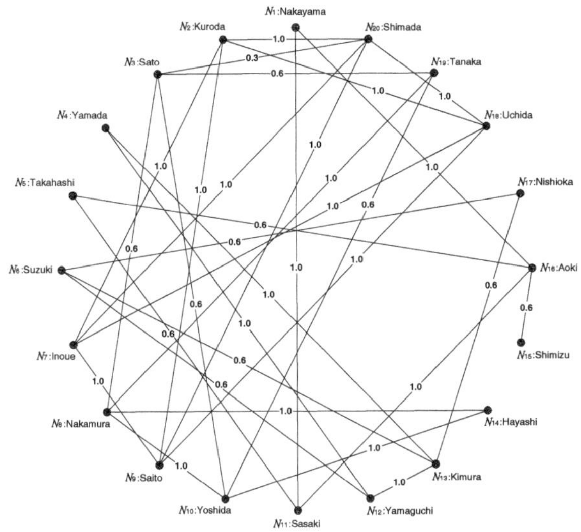

Node: N_i	Degree: $d(N_i)$	
N_1	2	Subgroup 1
N_2	4	
N_3	4	
N_4	2	
N_5	2	
N_6	3	Subgroup 2
N_7	4	
N_8	4	
N_9	4	
N_{10}	4	
N_{11}	3	Subgroup 3
N_{12}	3	
N_{13}	4	
N_{14}	2	
N_{15}	1	
N_{16}	4	Subgroup 4
N_{17}	2	
N_{18}	4	
N_{19}	3	
N_{20}	5	

Fig. 1. Example of a personnel network for employee relationships

The network for this model is given in Fig. 1, twenty actors of employees and thirty-two ties with fuzzy values. There are also the four circles that represent present subgroups, which represent the unorganized subgroups, and the connected lines that represent the relationships among the employees who mutually share their information through a close interpersonal relationship in business situations. In this case, cohesive subgroups should be created by determining each subgroup of all the cliques and components.

2.3 Fuzzy Graph in a Personnel Network [5], [10], [14], [20]

Let us denote an undirected graph with a set N_s of employees and a set E_s of connection lines is defined again to be relation $E_s \subseteq N_s \times N_s$ on a set N_s. A fuzzy relation μ: $N_s \times N_s \rightarrow [0, 1]$ is named a weighted graph or fuzzy graph, connection line $(x, y) \in N_s \times N_s$ has weight $\mu(x, y) \in [0, 1]$. Undirected graphs are considered for simplicity, i.e., the fuzzy relation is symmetric and all connection lines are regarded as unordered pairs of employees.

Definition 1. A fuzzy graph is denoted by $G = (\sigma, \mu)$, which is a pair of functions σ: $N_s \rightarrow [0, 1]$ and μ: $N_s \times N_s \rightarrow [0, 1]$, where for all x, y in N_s, then we obtain $\mu(x, y) \leq \sigma(x) \wedge \sigma(y)$.

Definition 2. A specific fuzzy graph $G_s = (\tau, \nu)$ is called a fuzzy subgraph of G if

$$\tau(x) \leq \sigma(x) \text{ for all } x \in N_s \tag{1}$$

and

$$\nu(x, y) \leq \mu(x, y) \text{ for all } x, y \in N_s . \quad (2)$$

Definition 3. An α-cut of a fuzzy graph which is included in σ and μ for $\alpha \in [0, 1]$, which are the fuzzy sets representing

$$\sigma_\alpha = \{x \in N_s \mid \sigma(x) \geq \alpha\} \quad (3)$$

and

$$\mu_\alpha = \{(x, y) \in N_s \times N_s \mid \mu(x, y) \geq \alpha\}, \quad (4)$$

where $\mu_\alpha \subseteq \sigma_\alpha \times \sigma_\alpha$ then $(\sigma_\alpha, \mu_\alpha)$ is a graph with the employee set σ_α and connection line set μ_α.

Definition 4. In a fuzzy graph, a path ρ is a sequence of distinct employees x_0, x_1, $x_2, \ldots, x_n$, such that $\mu(x_{i-1}, x_i) > 0$, $1 \leq i \leq n$; here $n \geq 0$ is called the length of ρ, then the strength of ρ is defined as

$$\wedge_{i=1}^{n} \mu(x_{i-1}, x_i) . \quad (5)$$

ρ is called a cycle if $x_0 = x_n$, and $n \geq 3$. A graph that has no cycles is called acyclic, or a forest; a connected forest is called a tree.

Definition 5. (1) if and only if $(supp(\sigma), supp(\mu))$ is a tree, then (σ, μ) is a tree; and (2) (σ, μ) is a fuzzy tree if and only if (σ, μ) has a fuzzy spanning subgraph (σ, ν) that is a tree, such that $\forall (u, v) \in supp(\mu) \setminus supp(\nu)$, $\mu(u, v) < \nu^\infty (u, v)$, meaning there is a path in (σ, ν) between u and v, whose strength is greater than $\mu(u, v)$.

Definition 6. (1) if and only if $(supp(\sigma), supp(\mu))$ is a cycle, then (σ, μ) is a cycle; and (2) (σ, μ) is a fuzzy cycle if and only if $(supp(\sigma), supp(\mu))$ is a cycle and $\nexists$ unique $(x, y) \in supp(\mu)$, such that $\mu(x, y) = \wedge \{\mu(u, v) \mid (u, v) \in supp(\mu)\}$.

2.4 Employees in Cliques and Components

A clique between employees is a useful starting point in cohesive subgroups for specifying the formal properties. The clique can be represented as well-specified mathematical properties, and captures much of the intuitive notion of cohesive subgroups in a personnel network. Also, the clique in a personnel network is often called as a maximal complete subgraph of three or more employees, and all of employees are clearly adjacent to each other.

Definition 7. Let G be a fuzzy graph on N_s and $G_s = (\tau, \nu)$ be a subgroup, which is induced by $T \subseteq N_s$, where T is a subset of N_s, then G_s is a clique if $(supp(\tau), supp(\nu))$ is a clique, and G_s is a fuzzy clique if G_s is a clique and every cycle is a fuzzy cycle in (τ, ν).

A graph $G = (N, E)$ is complete if all its nodes are pair-wise adjacent. The clique C in a graph G is a subset of nodes N such that the induced graph $G(C)$ is complete. The clique number of G is the size of the maximum number of employees in a clique is to find a clique of the maximum number cardinality in G [3], [16].

3 Biological Computation Approach

Many NP-complete problems are solved using heuristic and approximate methods instead of providing a complete solution or mathematical optimal approaches. The central reason comes from the huge computation time to solve such combinatorial problems by means of a conventional silicon computing based on the von Neumann architecture. To solve such NP-complete problems, this article provides alternative and innovative methods based on the best use of DNA oligonucleotides.

3.1 DNA Computing

DNA computing, also called molecular computation, is a new approach to massively parallel computation. DNA computing basically uses bio-molecules that constitute deoxyribonucleic acid (DNA), which consists of polymer chains, call DNA strands, that are composed of nucleotides adenine (A), guanine (G), cytosine (C) and thymine (T). Adenine always bonds with only thymine, while guanine always bonds with only cytosine. This phenomenon is called Watson-Crick complementarity as shown in Fig. 2.

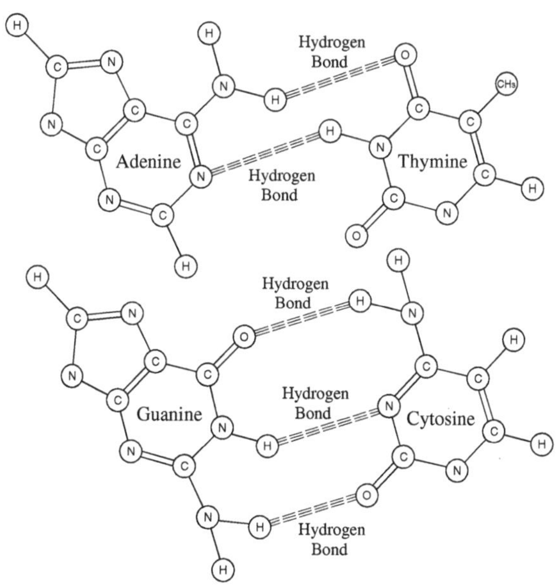

Fig. 2. Watson-Crick complementarity

DNA computing is sometimes called wet computation, which is based on the high ability of special molecule recognitions executed in the reaction among DNA molecules. L. Adleman reported on molecule computation in 1994, when he found that a DNA polymerase, which has an enzyme function of copying DNAs, is very similar to the function of a Turing machine. The DNA polymerase composes its complementary DNA molecules using a single strand helix of a DNA molecule as a mold. On the basis of this characteristic, if a huge amount of DNA molecules is mixed in a test tube, the reaction among the DNA molecules is pursued in parallel at the same time. Therefore, when a DNA molecule can express data or a program and the reaction among DNA molecules is executed, it is possible to realize super-parallel processing and a huge volume of memories comparing to present conventional electronic computers.

For example, we can realize 60,000 Tbytes of memories, if one string of a DNA molecule expresses one character. The total executing speed of DNA computing can outshine conventional electronic computers even if the execution time of one DNA molecule reaction is relatively slower than in conventional computers. DNA computing is appropriate to tackle such a problem as analysis of genome information and functional designing of DNA molecules.

3.2 Method of DNA Computing

The main idea behind DNA computing is to adopt a wet biological technique as an efficient computing vehicle where data are represented using a DNA strand itself. Even though a DNA reaction is slower than a silicon-based machine, the inherently parallel processing offered by the DNA process plays an important role. This parallelism of the DNA processing is of particular interest of NP-hard problems.

DNA computing has become a promising technique for solving NP-hard problems in various fields and applications. Real DNA capabilities can explored beyond the limitations of silicon machines. DNA computing has been applied to various fields such as nanotechnology [11], cable trench problems [7], combinatorial optimizations [8], [19], and so on.

As mentioned above, DNA molecules are used as information storage media. Usually, DNA oligonucleotides of about 8 to 20 base pairs (bp) are used to represent bits, and numerous methods have been developed to manipulate and evaluate them. In order to manipulate a wet technology into a computational approach, several techniques such as ligation, hybridization, polymerase chain reaction (PCR), gel electrophoresis, and restriction enzyme sites are used as computational operators for copying, sorting, and splitting or concatenating information in DNA molecules.

3.2.1 Encoding Scheme

In the DNA computing procedure, a main process is to encode each object of a focal problem into a DNA sequence. In this process, we will encode our data into DNA sequences according to our design. The correct design is essential in order to ensure we will get an optimal result. A wrong design may result in obtaining a wrong sequence after the ligation process.

3.2.2 Ligation and Hybridization

When DNA sequences are spoiled in a test tube using droppers, the DNA sequences recombine with each other in the test tube by means of enzyme reactions, as in Fig. 3. This process is called ligation. All the DNA sequences that we used in the experiment, with their complements, will be mixed together into one test tube. Normally, the oligonucleotide or DNA mixture is heated to 95 degrees centigrade and cooled to 20 degrees centigrade at 1 degree per minute for hybridization. The reaction is then subjected to a ligation. At the end of this process, a certain DNA sequence will be ligated together with another DNA sequence so as to produce a new DNA sequence.

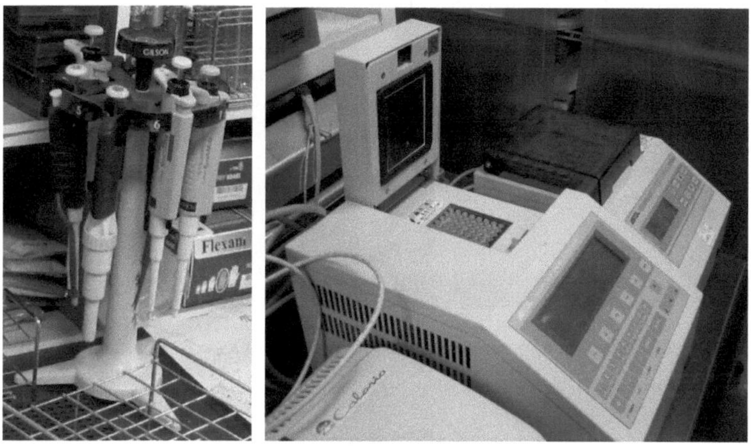

Fig. 3. Pipettes and a PCR machine taken at University of Occupational and Environmental Health Japan

3.2.3 Polymerase Chain Reaction (PCR)

PCR is a process that quickly amplifies the amount of specific molecules of DNA in a given solution using primer extension by polymerase. DNA polymerases perform several functions including the repair and duplication of DNA. Each cycle of the reaction doubles the quantity of this molecule, giving an exponential growth in the number of sequences.

3.2.4 Affinity Separation

The objective of an affinity separation process is to verify whether all the data have the same strands. This process permits single strands containing a given subsequence v to be filtered out from a heterogeneous pool of other sequences. After synthesizing strands complementary to v and attaching them to magnetic beads, the heterogeneous solution is passed over the beads. Those strands containing v anneal to the complementary sequence and are retained. Strands not containing v pass through without being retained.

Normally, in this process, a double-stranded DNA is incubated with the Watson-Crick complementarity of data that is conjugated to magnetic beads. Only single-stranded DNA molecules are retained as the sequences of data are annealed to the bond, meaning the process is repeated.

3.2.5 Gel Electrophoresis

Gel electrophoresis is an important technique for sorting DNA strands by their size [6]. Electrophoresis enables charged molecules to move in an electric field as shown in Fig. 4. Basically, DNA molecules carry a negative charge, so when we put them in an electrical field, they tend to migrate towards a positive pole. Since DNA molecules have the same charge per unit length, they all migrate with the same force in an electrophoresis process. Smaller molecules, therefore, migrate faster through the gel, thus we can sort them according to its size. At the end the resulting DNA is photographed as shown in Fig. 5.

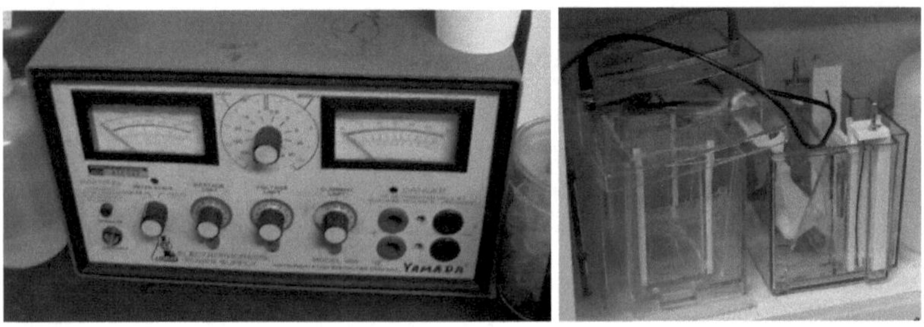

Fig. 4. Gel electrophoresis apparatus taken at University of Occupational and Environmental Health Japan

Fig. 5. Digital camera machine taken at University of Occupational and Environmental Health Japan

3.3 Comparison with Program

A DNA computing technique employs completely different tactics when they allocate an independent letter code such as ATCG, GTAC or CAAC, to each of the samples. DNA sequences corresponded to a number of possible combinations are prepared. After they are hybridized in super parallel, the remaining DNA fragments are amplified to obtain an answer sequence and the procedure is carried out only once.

DNA computing creates all different feasible solutions at once. This is the main benefit to using DNA computing in solving complex problems. This is known as parallel processing. Humans and most electronic computers must solve problems, step by step, known as linear processing. DNA itself provides the added benefits of being a cheap, energy-efficient resource [2].

(1) Separate (T, s): The operation separates given set T into set, +(T, s), of characters including character string s and set, -(T, s) of character strings without character string s. This operation corresponds to abstract experiment on DNA molecules in a test tube.
(2) Mix: The operation mixes sets T_1 and T_2 into union set $T_1 \cup T_2$. This operation corresponds to mix test tubes T_1 and T_2.
(3) Detect (T): Detect (T) returns YES if T is not empty and NO if T is empty in a test tube. The operation corresponds to the experimental treatment that detects the existence of DNA molecules by electrophoretical fluorescent method.
(4) Amplify (T): The operation corresponds to create multi-sets T_1 and T_2 with the same contents as given set T. This amplifying experimental treatment amplifies the amount of molecules using PCR, where this treatment corresponds to experimental treatment to amplify the amount of molecules.

The Watson-Crick complementarity is essential to realize the above mentioned separate operation. That is, it is possible to separate a partial string of characters "ad" so that a DNA sequence complementary to the DNA denoting "ad" is marked, input into a test tube, hybridized to form a double strand helix of DNA and abstracted. This nature also enables us to randomly create a set of character strings according to some rules.

In Adleman's model, a set of character strings is computed according to a program denoted using the four kinds of instructions mentioned above. Using this computation an NP-complete problem can be solved by an algorithm based on the production-detection method PCR. DNA computing is a computing method to solve real-world problems using this nature.

3.4 Algorithm to Find Cliques and Components

Redesigning cohesive subgroups in a personnel network can define specific graph theoretic properties that should be satisfied in order to identify a subset of specific employees. For examining a set of personnel network data in cohesive subgroups, finding collections of employees who have relatively strong ties is important and becomes visible by displaying functions or rearrangements.

	N_1	N_2	N_3	N_4	N_5	N_6	N_7	N_8	N_9	N_{10}	N_{11}	N_{12}	N_{13}	N_{14}	N_{15}	N_{16}	N_{17}	N_{18}	N_{19}	N_{20}
N_1	**1**	0	0	0	0	0	0	0	0	0	1.0	0	0	0	0	1.0	0	0	0	0
N_2	0	**1**	0	0	0	0	1.0	0	1.0	0	0	0	0	0	0	0	0	1.0	0	1.0
N_3	0	0	**1**	0	0	0	0	0.6	0	0.6	0	0	0	0	0	0	0	0	0.6	0.3
N_4	0	0	0	**1**	0	0	0	0	0	0	0	1.0	1.0	0	0	0	0	0	0	0
N_5	0	0	0	0	**1**	0	0	0	0	0	0.6	0	0	0	0	0.6	0	0	0	0
N_6	0	0	0	0	0	**1**	0	0	0	0	0	0.6	0.6	0	0	0	0.6	0	0	0
N_7	0	1.0	0	0	0	0	**1**	0	1.0	0	0	0	0	0	0	0	0	1.0	0	1.0
N_8	0	0	0.6	0	0	0	0	**1**	0	1.0	0	0	0	1.0	0	0	0	0	1.0	0
N_9	0	1.0	0	0	0	0	1.0	0	**1**	0	0	0	0	0	0	0	0	1.0	0	1.0
N_{10}	0	0	0.6	0	0	0	0	1.0	0	**1**	0	0	0	1.0	0	0	0	0	0.6	0
N_{11}	1.0	0	0	0	0.6	0	0	0	0	0	**1**	0	0	0	0	1.0	0	0	0	0
N_{12}	0	0	0	1.0	0	0.6	0	0	0	0	0	**1**	1.0	0	0	0	0	0	0	0
N_{13}	0	0	0	1.0	0	0.6	0	0	0	0	0	1.0	**1**	0	0	0	0.6	0	0	0
N_{14}	0	0	0	0	0	0	0	1.0	0	1.0	0	0	0	**1**	0	0	0	0	0	0
N_{15}	0	0	0	0	0	0	0	0	0	0	0	0	0	0	**1**	0.6	0	0	0	0
N_{16}	1.0	0	0	0	0.6	0	0	0	0	0	1.0	0	0	0	0.6	**1**	0	0	0	0
N_{17}	0	0	0	0	0	0.6	0	0	0	0	0	0	0.6	0	0	0	**1**	0	0	0
N_{18}	0	1.0	0	0	0	0	1.0	0	1.0	0	0	0	0	0	0	0	0	**1**	0	1.0
N_{19}	0	0	0.6	0	0	0	0	1.0	0	0.6	0	0	0	0	0	0	0	0	**1**	0
N_{20}	0	1.0	0.3	0	0	0	1.0	0	1.0	0	0	0	0	0	0	0	0	1.0	0	**1**

Fig. 6. Example of a socio-matrix for the personnel network

A socio-matrix [4] is a matrix that represents employee relationships and is the most important procedure to find all the cliques and components among employees. In the socio-matrix, a systematic way for ordering rows and columns reveals the subgroup structure of the personnel network. Fig. 6 shows a model socio-matrix with the rows and columns, employees who have ties and close interpersonal business relationships with fuzzy membership grades. The socio-matrix of size $m \times m$ becomes 20 rows and 20 columns for the model personnel network as shown in Fig. 1. There is a row and column for each employee, and the rows and columns are labeled 1, 2,…, 20. $x_{i,j}$ denotes the value of the tie from employee i to employee j, and $x_{i,j}$ records which pairs of nodes are adjacent. Nodes N_i and N_j are adjacent with some grade $x_{i,j}$, and if nodes N_i and N_j are not adjacent, then $x_{i,j} = 0$. In addition, an edge between two nodes is either present or absent. If an edge is present, it goes both from N_i to N_j and from N_j to N_i. The total sum of rows and columns represents the level of the connections between edges, which is divided by 2, thus the number of edges is denoted as L. For the model personnel network, the number of edges L becomes 32.

This socio-matrix makes the best use of this DNA application, and we designed a new algorithm based on the algorithm of the maximal clique problem solution that was proposed by Ouyang *et al.* [12], the method to find the maximum clique. However, we designed a new algorithm that finds all the 1-cliques and components, as well as the maximum clique of employees for cohesive subgroups in the personnel network.

Step 1: A graph has n nodes, each possible clique represented by a binary number, of which 1 is in the clique, and 0 is vice versa. Also, if the total is 1, 1 is in the independent line.

Step 2: All the possible combinations between encoded employee DNA fragments are created from 2^n using PCR and parallel overlap assembly (POA).

Step 3: In this graph, called the complementary graph, are all edges missing in the original graph. Any two nodes connected in the complementary connection are connected in the invalid connection. Remove those nodes of cliques and independent lines that contain the invalid connections, corresponding to $x_{i,j} = 0$ in Fig. 6.

Step 4: The remaining data pool is sorted to select the DNA sequence from 2 bits of value 1 to 20 bits of the value 1 in existence. Find the nodes connected between the possible cliques, and distinguish those connected cliques from other cliques.

Step 5: Find the connection lines in each subset of all the α-cut groups, remove all the connection lines that have less strength than the selected α-cut group, and construct all the cliques of employees in each α-cut group of all the employees in the personnel network.

3.5 Experimental Studies and Results

In the experimental studies, an α-cut is taken at $\alpha = 0.3$, then computing the personnel network is done to solve the rigid clique problem.

The DNA sequences basically are designed in the form of double-stranded DNA. Each node of the DNA sequence in a binary number is composed of two sequences that are a position sequence E_i and a value sequence N_i. The position sequences are used for connecting each node of the DNA sequence, and the value sequence is used for distinguishing whether those position sequences contain that node or not. For the given twenty employees in the model personnel network, the value sections N_1 to N_{20} are prepared sandwiched sequentially between the position sections E_1 to E_{21}. E_i was set with the length of 10 bp. N_i was set with the length of 0 bp if the value is 1 and 6 bp if the value is 0. Each DNA oligonucleotide consists of two different position motifs and contains its own restriction enzyme site. Each of these two different single-stranded DNA sequences stick to each other to become a double-stranded DNA based on the sticking operation [13]. Using a gel electrophoresis apparatus, we repeat selecting the shortest DNA strands that correspond to all the possible cliques of employees, and the cliques of employees are shown in the picture of gel electrophoresis and positioned on the same lines, meaning those employees are connected in each component.

The clique of largest size is represented by the shortest length of DNA. The maximum clique of nodes consists of five employees N_2, N_7, N_9, N_{18}, and N_{20}. The second largest clique of nodes consists of four employees N_3, N_8, N_{10}, and N_{19}. There are also six possible cliques of nodes that are all connected by three employees, and there are also the two independent lines. In addition, all the employees are determined to be divided into three components in the personnel network while the possible cliques and components of employees are mutually connected together.

4 Fuzzy Density Analysis

Many different types of cliques and components in a huge and complicated personnel network should be determined and rearranged in order to be able to promote cohesion. Fig. 7 shows the redesigned personnel network with strength 0.3, and Fig. 8 shows the redesigned personnel network with strength 1.0 based on all the DNA experiment results. To prove the efficiency of the redesigned personnel network using DNA computing, we show the differences between the redesigned personnel network and the previous personnel network. Therefore, we calculated the inclusiveness and the density of the three different personnel networks, which correspond to the previous personnel network, the new personnel network with strength 0.3, and the new personnel network with strength 1.0 as shown in Tables 1, 2 and 3. As Fig. 8 shows, the fuzzy clique problem can be solved by applying a finite number of an easy solving method.

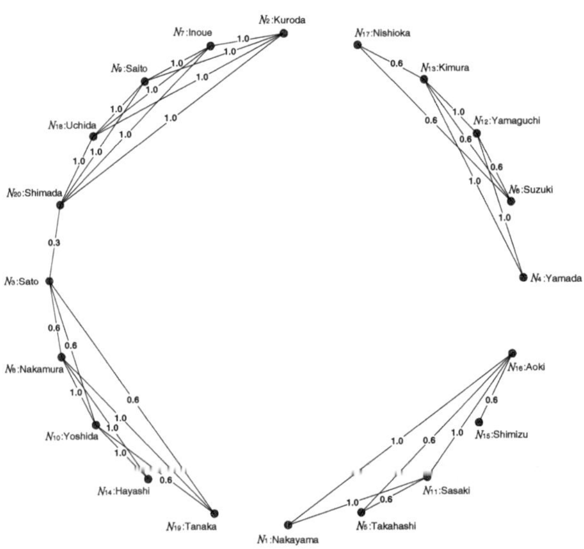

Node: N_i	Degree: $d(N_i)$	
N_2	4	New Subgroup 1
N_7	4	
N_9	4	
N_{18}	4	
N_{20}	5	
N_3	4	
N_8	4	
N_{10}	4	
N_{14}	2	
N_{19}	3	
N_1	2	New Subgroup 2
N_5	2	
N_{11}	3	
N_{15}	1	
N_{16}	4	
N_4	2	New Subgroup 3
N_6	3	
N_{12}	3	
N_{13}	4	
N_{17}	2	

Fig. 7. Redesigned personnel network for employee relationships with strength 0.3

Table 1. Inclusiveness and density comparisons for the previous personnel network

	Subgroup 1	Subgroup 2	Subgroup 3	Subgroup 4
No. of Connected Nodes	0	4	2	2
Inclusiveness	0	0.80	0.40	0.40
Sum of Degrees	0	4	2	2
No. of Edges	0	2	1	1
Density	0	0.20	0.10	0.10

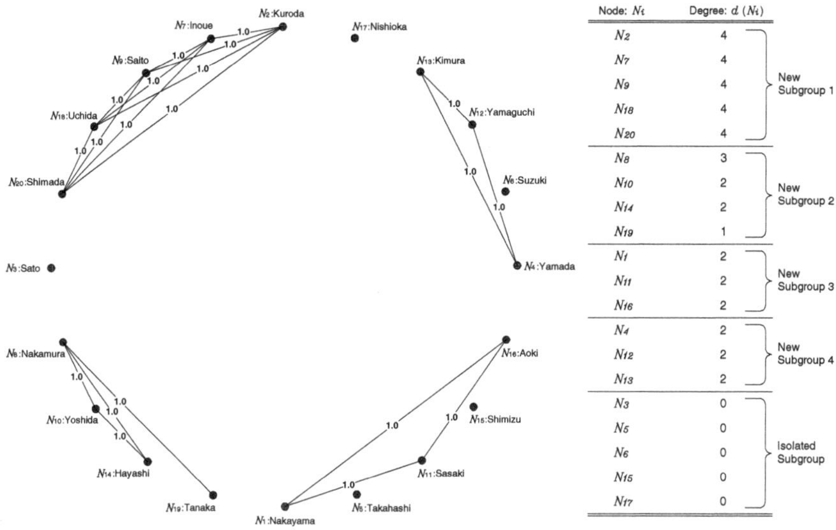

Fig. 8. Redesigned personnel network for employee relationships with strength 1.0

Table 2. Inclusiveness and density comparisons for the redesigned personnel network with strength 0.3

	New Subgroup 1	New Subgroup 2	New Subgroup 3
No. of Connected Nodes	10	5	5
Inclusiveness	1.00	1.00	1.00
Sum of Degrees	38	12	14
No. of Edges	19	6	7
Density	0.42	0.60	0.70

Table 3. Inclusiveness and density comparisons for the redesigned personnel network with strength 1.0

	New Subgroup 1	New Subgroup 2	N. S. 3	N. S. 4	Isolated Subgroup
No. of Connected Nodes	5	4	3	3	0
Inclusiveness	1.00	1.00	1.00	1.00	0
Sum of Degrees	20	8	6	6	0
No. of Edges	10	4	3	3	0
Density	1.00	0.67	1.00	1.00	0

5 Conclusions

The subgroups were redesigned more efficiently and became cohesive subgroups as shown in the two redesigned personnel networks under the consideration of a fuzzy graph. DNA computing can be a massively parallel computation executed to provide the results of the redesigned personnel network for an efficient work rotation, and the concepts of the fuzzy graph made possible to analyze the redesigned personnel network in different relational strengths between the employees.

DNA computing with the fuzzy concepts provides various ideas on personnel management problems or other management problems to overcome the limitation of electronic computations. Moreover, the fuzzy concepts are important in order to analyze the employee relationships with grades for real situations. It is necessary in the future to investigate a new type of DNA computing that is clearly adapted to fuzzy graphs to develop fuzzy DNA computing for dealing with uncertainties in scientific or other related data.

References

[1] Adleman, L.: Molecular computation of solutions to combinatorial problems. Science 266(11), 1021–1024 (1994)

[2] Amos, M., Paun, G., Rozenberg, G., Salomaa, A.: Topics in the theory of DNA computing. Theoretical Computer Science 287(1), 3–38 (2002)

[3] Bomze, I.M., Pelillo, M., Stix, V.: Approximating the maximum weight clique using replicator dynamics. IEEE Trans. Neural Networks 11(6), 1228–1241 (2000)

[4] Carrington, P.J., Scott, J., Wasserman, S.: Models and methods in social network analysis, pp. 77–97. Cambridge University Press, Cambridge (2005)

[5] Dubois, D., Prade, H.: Fuzzy sets and systems, theory and applications, pp. 19–80. Academic Press, New York (1980)

[6] Hartl, D., Jones, E.: Essential genetics: a genomics perspective, 3rd edn., pp. 210–242. Jones and Bartlett Publishers, Inc. (2005)

[7] Jeng, D.J.-F., Kim, I., Watada, J.: Bio-inspired evolutionary method for cable trench problem. International Journal of Innovative Computing, Information and Control 3(1), 111–118 (2006)

[8] Kim, I., Jeng, D.J.-F., Watada, J., Pedrycz, W.: Molecular computation applied to clustering problems with a statistical method. In: The 4th International Symposium on Management Engineering, ISME 2007, Kitakyushu, Japan, Proceedings, pp. R08-1-R08-8 (2007)

[9] Luthans, F.: Organizational behavior, 10th edn., pp. 478–508. McGraw-Hill International Edition, New York (2005)

[10] Nair, P.S., Cheng, S.-C.: Cliques and fuzzy cliques in fuzzy graphs. In: IFSA World Congress and 20th NAFIPS International Conference, Proceedings, pp. 2277–2280 (2001)

[11] van Noort, D., Landweber, L.F.: Towards a re-programmable DNA computer. In: Chen, J., Reif, J.H. (eds.) DNA 2003. LNCS, vol. 2943, pp. 190–196. Springer, Heidelberg (2004)

[12] Ouyang, Q., Kaplan, P.D., Liu, S., Libacher, A.: DNA solution of the maximal clique problem. Science 278(5337), 446–449 (1997)

[13] Kari, L., Păun, G., Rozenberg, G., Salomaa, A., Yu, S.: DNA computing, sticker systems, and universality. Acta Informatica 35(5), 401–420 (1998)

[14] Pedrycz, W.: Shadowed sets: representing and presenting fuzzy sets. IEEE Trans. on Systems, Man, and Cybernetics, part B 28(1), 103–109 (1998)
[15] Rosenfeld, A.: Fuzzy graphs, Fuzzy Sets and Their Applications. In: Zadeh, L.A., Fu, K.S., Shimura, M. (eds.), pp. 77–95. Academic Press, New York (1975)
[16] Stix, V.: Finding all maximal cliques in dynamic graphs. Computation Optimization and Applications 27(2), 173–186 (2004)
[17] Toyoura, Y., Watada, J., Yabuuchi, Y., Ikegame, H., Sato, S., Watanabe, K., Tohyama, M.: Fuzzy regression analysis of software bug structure. Central European Journal of Operations Research 12(1), 13–23 (2004)
[18] Watada, J., Tanaka, T., Arredondo, A.R.: Analysis of safety from macro-ergonomics approach. Japanese Journal of Ergonomics, Japan Ergonomics Society 34(6), 333–339 (1998)
[19] Watada, J.: DNA computing and its application. In: Fulcher, J., Jain, L.C. (eds.) Computational Intelligence: A Compendium, pp. 1065–1086. Springer, Heidelberg (2008)
[20] Zadeh, L.A.: Fuzzy sets. Information and Control 8(3), 338–353 (1965)

3

Evolution of Cooperating Classification Rules with an Archiving Strategy to Underpin Collaboration

Catalin Stoean and Ruxandra Stoean

University of Craiova, Faculty of Mathematics and Computer Science, A. I. Cuza, 13, 200585 Craiova, Romania
{catalin.stoean,ruxandra.stoean}@inf.ucv.ro
http://inf.ucv.ro/{~cstoean, ~rstoean}

Abstract. Individuals encoding potential rules to model an actual partition of samples into categories may be evolved by means of several well-known evolutionary classification techniques. Nevertheless, since a canonical evolutionary algorithm progresses towards one (global or local) optimum, some special construction or certain additional method are designed and attached to the classifier in order to maintain several basins of attraction of the different prospective rules. With the aim of offering a simpler option to these complex approaches and with an inspiration from the state-of-the-art cooperative coevolutionary algorithms, this chapter presents a novel classification tool, where rules for each class are evolved by a distinct population. Prototypes evolve simultaneously while they collaborate towards the goal of a good separation, in terms of performance and generalization ability. A supplementary archiving mechanism, which preserves a variety of the best evolved rules and eventually yields a thorough and diverse rule set, increases the forecasting precision of proposed technique. The novel algorithm is tested against two real-world decision problems regarding tumor diagnosis and obtained results demonstrate the initial presumption.

1 Introduction

Classification may well be regarded as a pursuit to deliver an accurate and general collection of rules that are able to record the patterns that trigger a decision-making process. Among the many powerful paradigms that have addressed this key field of data mining, evolutionary algorithms (EAs) have successfully established themselves as a flexible and robust choice to acquire optimal decision rules.

As discrimination between patterns of similar configurations, while of divergent kinds, is an intricate task and EAs naturally lead to a homogeneous set of solutions, the establishment and the preservation of diversity within the collection of evolved rules is an essential aspect, which adds complexity to existing evolutionary classifiers.

The aim of this chapter is hence to put forward a novel evolutionary classification technique, deriving from the major field of cooperative coevolutionary algorithms (CCEAs), which has proven to be a simpler and viable alternative. Proposed

H.-N. Teodorescu, J. Watada, and L.C. Jain (Eds.): Intel. Sys. and Tech., SCI 217, pp. 47–65.
springerlink.com

approach accomplishes the collaboration between evolving rule populations of different outcomes towards improvement in prediction accuracy. A potential architecture considers the final decision suite as to contain only one or more randomly selected rules for each outcome of the classification task. The construction is enhanced by an archiving mechanism, which preserves the best performing combinations during each evolutionary cycle and displays a complete and varied output rule set. The theoretical assumptions are validated against two practical problems related to tumor diagnosis.

The chapter is structured as follows. Section 2 brings a formal definition of a classification instance, while section 3 reviews the most common EA classifiers. Section 4 introduces the CCA framework and details on the particular mechanisms and concepts. Section 5 brings forward proposed classification engine within CCA, while section 6 sustains the superior version of the archiving policy. The developed training and test environments are described in all features. The experimental evidence in section 7 comes in support of the ability of the novel methodology to manage real-world test cases; the diagnosis of breast cancer and the early prediction of hepatic cancer are explored as application.

2 Classification: A Perspective

Classification can assume different characterizations, however this chapter regards it from the general point of view of pattern recognition. Given $\{(x_i, y_i)\}_{i=1,2,\ldots,m}$, a training set where every $x_i \in R^n$ represents a data sample (values that correspond to a sequence of attributes or indicators) and each $y_i \in \{1, 2, \ldots, p\}$ represents a class (outcome, decision attribute), a classification task consists in learning the optimal mapping that minimizes the discrepancy between the actual classes of data samples and the ones produced by the learning machine. Subsequently, the learnt patterns are confronted with each of the test data samples, without an a priori knowledge of their real classes. The predicted outcome is then compared with the given class: If the two are identical for a certain sample, then the sample is considered to be correctly classified. The percentage of correctly labeled test data is reported as the classification accuracy of the constructed learning machine.

The data are split into the training set consisting of a higher number of samples and a test set that contains the rest of the data. The training and test sets are disjoint. In present discussion, the samples that form the training set are chosen in a random manner from the entire specific data set.

The aim of a classification technique may be further conceived as to stepwise learn a set of rules that model the training set as good as possible. When the learning stage is finished, the obtained rules are applied on previously unseen samples within the test set.

3 Evolutionary Approaches to Classification

Apart from the hybridization with non-evolutionary specialized classification techniques, such as fuzzy sets, neural networks or decision trees, the evolutionary computation community has targeted classification through the development of special standalone EAs for the particular task.

On a broader sense, an evolutionary classification technique is concerned with the discovery of IF-THEN rules that reproduce the correspondence between the given samples and corresponding classes. Given an initial set of training samples, the system learns the patterns, i.e. evolves the classification rules, which are then expected to predict the class of new examples.

Remark 1. An IF-THEN rule is imagined as a first-order logic implication where the condition part is made of the conjunction of attributes and the conclusion part is represented by the class.

There are two state-of-the-art approaches to evolutionary classification techniques. The first direction is represented by De Jong's classifier [1], [2], an evolutionary system which considers an individual to represent an entire set of rules, constructed by means of disjunctions and conjunctions of attributes. Rule sets are evolved using a canonical EA and the best individual from all generations represents the solution of the classification problem. The opposite related approach is Holland's classifier system [3], [1]. Here, each individual encodes only one IF-THEN rule and the entire population represents the rule set. Thus, detection and maintenance of multiple solutions (rules) in a multiple subpopulations environment is required. As a canonical EA cannot evolve non-homogeneous individuals, Holland's approach suggested doubling the EA by a credit assignment system that would assign positive credit to rules that cooperate and negative credit to the opposite.

Another standard method is characterized by a genetic programming approach to rule discovery [4], [5]. The internal nodes of the individual encode mathematical functions (e.g. AND, OR, +, -, *, <, =) while the leaf nodes refer the attributes. Given a certain individual, the output of the tree is computed and, if it is greater than a given threshold, a certain outcome of the classification task is predicted.

If discussion revolves around those classification evolutionary models that are specifically for coadapted components, then we must refer the abovementioned Holland's classifier system [HolML86] and the REGAL system [6], where stimulus-response rules in conjunctive form were evolved by EAs. In Holland's system, cooperation is achieved through a bucket brigade algorithm that awards rules for collaboration and penalizes them otherwise. In the REGAL classifier, problem decomposition is performed by a selection operator, complete solutions are found by choosing best rules from each component, a seeding operator maintains diversity and fitness of individuals within one component depends on their consistency with the negative samples and on their simplicity.

However, the existing evolutionary classification techniques have quite intricate engines and thus their application is not always straightforward: they use complex credit assignment systems that penalize or reward good rules, as well as complicated schemas of the entire system. To the best of our knowledge, there has been no attempt in utilizing the cooperative coevolution of simple IF-THEN rules to classification.

4 Cooperative Coevolution – Prerequisites

The chapter will tackle classification from the viewpoint of cooperative coevolution. However, this paradigm belongs to the wider field of coevolutionary algorithms.

4.1 Artificial Coevolution

According to the Darwinian principles, an individual evolves through the interaction with the environment. However, a significant segment of its surroundings is, in fact, represented by other individuals. As a consequence, evolution actually implies coevolution. This interactive process may assume collaboration towards the achievement of a specific mutual purpose, or, on the contrary, competition for the common resources in the spirit of the survival of the fittest. Accordingly, two kinds of artificial coevolutionary systems exist: cooperative and competitive, respectively.

In CCEA, collaborations between two or more individuals are necessary in order to evaluate one complete potential solution, while in the competitive approach, the evaluation of an individual is determined by a set of competitions between the current individual and several others.

Coevolutionary algorithms bring an interesting angle of perception upon evolution, as they promote a different manner of fitness evaluation of a candidate solution, which takes into account its relation to the other surrounding individuals. In addition, the coevolutionary evaluation is continuously altered throughout the existence of an individual as a result of various tests.

4.2 The Mechanics

CCEAs imply a decomposition of a candidate solution of the problem to be solved into a number of components [7], [8], [9]. Each of these parts is subsequently attributed to a population (species) of an EA. The species evolve independently (although concurrently), while interactions between populations appear only at the moment when fitness is computed. Each individual of a species stands for a part of the solution, therefore, a candidate for each component in turn cannot be evaluated separately from the complementary ones. Hence, when the fitness of an individual is assessed, collaborators from each of the remaining populations are selected in order to form a complete solution. The performance of the established solution is measured and returned as the fitness evaluation of the considered individual.

Evolution is thus directed by the collaboration between species towards the joint goal of assembling a near optimal solution to the problem.

The general outline of a canonical CCEA is given by the algorithm in Fig. 1. The evolutionary process starts with the initialization of each population. In order to evaluate the initial fitness of each individual, a random selection of collaborators from each of the other populations is performed and obtained solutions are measured and scored accordingly. After this starting phase, each population is evolved through a canonical EA. Subsequently, the evaluation of a member of one species is performed through its fusion to individuals of the complementary population, which are at this point selected through a certain strategy.

Require: A search/optimization problem
Ensure: A complete solution to the problem given by CCEA
Begin
$t = 0$;
For each component s **do**
    Randomly initialize $P_s(t)$;
End For
For each component s **do**
    Evaluate $P_s(t)$;
End For
While stop condition = false **do**
    $t = t + 1$;
    For each component s **do**
        Select $P_s(t)$ from $P_s(t - 1)$;
        Apply genetic operators to $P_s(t)$;
        Evaluate $P_s(t)$;
    End For
End While
End

Fig. 1. Schema of a CCEA

4.3 The Parameters

The main issue within CCEAs concerns the choice of collaborators. As a result, there are three attributes (parameters) that control this option, whose values have to be properly decided.

1. *Collaborator selection pressure* refers to the manner in which individuals are chosen from each of the complementary populations with the purpose of forming complete solutions to the problem; it must be decided whether to pick the best individual according to its previous fitness score, take a random individual or use a classic selection scheme.
2. *Collaboration pool size* represents the number of collaborators that are selected from each population.
3. *Collaboration credit assignment* decides the way to compute the fitness of the current individual. This attribute appears solely in the case when the *collaboration pool size* is higher than one. In this situation, the evaluation of an individual consists of several collaborations. Since every such collaboration has its personal score for the objective function, these multiple values must be somehow encapsulated into a single quality value. There are three methods for deciding the final assignment:

    a. *Optimistic* - the fitness of the current individual is the value of its best collaboration.

b. *Hedge* - the average value of its collaborations is returned as the fitness score.
c. *Pessimistic* - the value of its worst collaboration is assigned to the considered individual.

The value for the collaboration pool size parameter will be further on denoted by *cps*. Algorithm in Fig. 2 demonstrates the modality of evaluation of an individual *c* with respect to the three referred attributes. It is presumed that it is dealt with a maximization problem.

Require: An individual *c* from a population
Ensure: The *evaluation* of individual c
Begin
$t = 0$;
For each $i = 1, 2, ..., cps$ **do**
    Select one collaborator d_j, $j = 1, 2, ...,$ *number of species* from each population different from that of *c*;
    Form a complete potential solution;
    Compute the fitness f_i of the solution in the terms of the objective criterion;
End For
If *Collaboration credit assignment* = *Optimistic* **then**

$$evaluation \leftarrow \max_{i=1}^{cps}(f_i);$$

Else
    If *Collaboration credit assignment* = *Pessimistic* **then**

$$evaluation \leftarrow \min_{i=1}^{cps}(f_i);$$

    Else

$$evaluation \leftarrow \underset{i=1}{\overset{cps}{avg}}(f_i);$$

    End If
End If
End

Fig. 2. Fitness evaluation within CCEAs

In order to evaluate an individual c from a certain population, a number of complete potential solutions are formed according to the chosen collaboration pool size. In order to aggregate a solution, collaborators from each population different from that of *c* are selected through a certain strategy (collaboration selection pressure). Each solution is evaluated according to the objective function of the current problem. Once all candidate solutions are gathered and assessed, the preferred type for the collaboration credit assignment decides the value that will be returned as the performance of individual *c*.

4.4 The Application

CCEA was introduced as an alternative evolutionary approach to function optimization [7]. For this task, one considers as many populations as the number of variables of the function, i.e. each variable represents a component of the solution vector and is separately treated using some type of EA. Several functions with multiple local optima and one global optimum were considered and the CCEA proved to be effective [7], [9].

The CCEA technique has been recently successfully applied to develop a rule-based control system for agents; two species were considered, each consisting of a population of rule sets for a class of behaviors [10].

5 Cooperative Coevolution Approach to Classification

Within aimed CCEA approach to classification [11], [12], [13], the output has been imagined as to be represented by a set of rules that contains at least one prototype for each class. The decomposition of each potential problem solution into components is performed by assigning to each population the task of building the rule(s) for one certain class. Thus, the number of species equals the number of outcomes of the classification problem.

A rule is considered to be a first order logic entity in conjunctive form (1), where $a_1, a_2, , ..., a_n$ are the attributes of every sample, $v_1, v_2, , ..., v_n$ are values in the definition domain of each indicator, while $k \in \{1, 2, ..., p\}$ denotes the class.

$$\text{if } (a_1 = v_1) \text{ and } (a_2 = v_2) \text{ and ... and } (a_n = v_n) \text{ then class } k. \tag{1}$$

5.1 Training Stage: The EA Behind

Recall the training data set $\{(x_i, y_i)\}_{i=1,2,...,m}$, where $x_i \in R^n$ and $y_i \in \{1, 2, ..., p\}$. As the task of the CCEA technique is to build p rules, one for each class, p populations are considered, each with the purpose of evolving one of the p individuals.

Representation. Each individual (or rule) c in every population follows the same encoding as a sample from the data set, i.e. it contains values for the corresponding attributes, $c = (c_1, c_2, ..., c_n)$. As already stated, individuals represent simple IF-THEN rules having the condition part in the attributes space and the conclusion in the classes space. Within the CCEA approach to classification, an individual will not however encode the class, as all individuals within a population have the same outcome.

Initialization. The values for the genes of all individuals are randomly initialized following a uniform distribution in the definition intervals of the corresponding attributes in the data set.

In case the considered data set is normalized, the values for the genes of the individuals are initialized in the [0, 1] interval, again following a uniform distribution.

```
Require: A rule c from a population
Ensure: The evaluation (accuracy) of rule c
Begin
For i = 1 to cps do
    correct_i = 0;
    select a random collaborating rule from each population different
    from that of c according to the collaborator selection pressure pa-
    rameter;
    For each sample s in the training set do
        find the rule r from the set of all collaborators that is closest to s;
        found class for s = class of r;
        If found class of s = real class of s then
            correct_i = correct_i + 1
        End If
    End For
End For
If optimistic then
    success = max_{i=1}^{p}(correct_i)
Else
    If pessimistic then
        success = min_{i=1}^{p}(correct_i)
    Else
        success = avg_{i=1}^{p}(correct_i)
accuracy = (100 · success) / (number of training samples)
    End If
End If
End
```

Fig. 3. Fitness evaluation of a rule *c* using either optimistic, pessimistic, or hedge collaboration credit assignment within CCEA for classification

Fitness Evaluation. In order to measure the quality of a rule, this has to be integrated into a complete set of rules which is subsequently applied to the training set. The obtained accuracy reflects the quality of the initial rule. Obviously, the value of the accuracy very much depends on the other rules that are selected in order to form a complete collection: For a more objective assessment of its quality by means of the prediction value, the rule is tested within several different sets of rules, i.e. various values for *cps* are considered.

For evaluating an individual from a certain population – that is a rule of a certain outcome – a collaborating rule from each of the other populations is selected according to the collaborator selection pressure choice for a number of times equal to *cps*. Every time, the set of rules is applied to the entire training collection. Obtained accuracy represents the evaluation of the current individual. The fitness of an individual *c* may be given by the best of the *cps* acquired accuracies (optimistic assignment), by the worst one of them (pessimistic assignment) or by the average of all *cps* accuracies (hedge assignment). Algorithm in Fig. 3 describes the way evaluation takes place in these cases.

Require: A rule *c* from a population
Ensure: The evaluation (*accuracy*) of rule *c*
Begin
For each sample s in the training set **do**
    Set the score for each possible outcome of *s* to 0;
End For
For $i = 1$ to *cps* **do**
    select a random collaborating rule from each population different from that of *c* according to the collaborator selection pressure parameter;
    For each sample *s* in the training set **do**
        find the rule *r* from the set of all collaborators that is closest to *s*;
        found class for *s* = class of *r*;
        increase the score of found class for *s* by one;
    End For
End For
success = 0;
For each sample *s* in the training set **do**
    If the real class of *s* equals the class with the highest score for *s* **then**
        s is correctly classified;
        success = *success* + 1;
    End If
End For

$$accuracy = \frac{100 \cdot success}{number\ of\ training\ samples}$$

End

Fig. 4. Fitness evaluation of a rule *c* using a collaboration credit assignment based on scores within CCEA for classification

In addition to the classical CCEA types, we propose a novel assignment (Fig. 4). Departing from the current individual, for each sample *s* in the training set, multiple sets of rules are formed and applied in order to predict its class. All rules within a set have different outcomes. Scores are computed for sample *s*, for each of the possible outcomes, in the following manner: When a rule set is applied to a sample, a certain outcome is established for it. The score of that outcome is increased by unity. Each of the *cps* sets of rules is applied to *s*. Finally, the class of *s* is concluded to be the class that obtains the highest score. The fitness of the considered rule is computed as the correctly labeled cases over the total number.

Independently of the chosen algorithm for evaluating rule performance, the distance between individuals and samples has to be computed when it is decided which rule is closer to each example in the training set. In the conducted experiments, normalized Manhattan was adopted for distance calculation (2). However, other measures may be as well employed, depending of the considered problem. Note that the distance does not depend on the class of the sample/individual.

$$d(c,x_i)=\sum_{j=1}^{n}\frac{|c_j-x_{ij}|}{b_j-a_j}. \tag{2}$$

Recall that $x_i = (x_{i1}, x_{i2}, ..., x_{in})$ stands for a sample from the training set, while $c = (c_1, c_2, ..., c_n)$ represents an individual (or rule). a_j and b_j denote the lower and upper bounds of the j-th attribute. As usually the values for the attributes belong to different intervals, the distance measure has to refer to their bounds. Obviously, if data is normalized, the denominator disappears as all attributes have their values between 0 and 1.

In both Fig. 3 and 4, the fitness of an individual is computed as the percent of correctly classified samples from the training set (variable *success* in the algorithms specifies the number of samples that were successfully labelled).

In the routine from Fig. 4, problematic situations may appear when, for a certain sample, there exist more classes that have the same maximum score. In this case, one class has to be decided and it was considered to choose the first one in the order of outcomes. As herein all combinations of rules count in the determination of accuracies, it might be stated that the new choice of assignment is closer to the classical hedge type.

Selection and Variation Operators. The selection operator presently discussed refers to the EA selection for reproduction within each population and not the CCEA one related to collaborators. Fitness proportional selection is employed, but any other selection scheme [14] may be successfully applied.

Intermediate recombination is used – having two randomly selected parents *p* and *q*, the value of a gene *i* of the offspring *o* is obtained according to (3).

$$o_i = p_i + r \cdot (q_i - p_i), \tag{3}$$

where *r* is a uniformly distributed random number over [0, 1]. The obtained offspring individual replaces the worst of its two parents.

Mutation with normal perturbation is considered for performed experiments – a statistically chosen gene *i* of an individual *p* is changed according to (4).

$$p_i = p_i + r \cdot (b_i - a_i) / ms, \tag{4}$$

where *r* is a random number with normal distribution, b_i and a_i are the upper and lower bounds of the *i*-th attribute in the data set and *ms* is the mutation strength parameter. As the domains for the values of the attributes in the data set have different ranges, the size of the interval for each indicator is again referred when the values of the genes are perturbed through mutation. In case the data set is normalized, the way the value of the gene i is modified changes to (5).

$$P_i = P_i + R \cdot ms. \tag{5}$$

No obstacle for using any other recombination or mutation operators [14] can be imagined.

Stop Condition. During experimentation, a fixed number of generations was set for the evolutionary process.

5.2 Cooperative Coevolution Parameters

In order to achieve the optimal configuration for the parameters of the CCEA classification approach, experiments were carried out as follows.

Concerning the *collaborator selection pressure* attribute, random selection had been used, on the one hand, and, on the other hand, a fitness proportional scheme had been employed.

All the three types of fitness *assignment* presented in algorithm from Fig. 3 together with the one based on scores had been tested.

As for the *collaboration pool size*, the number of collaborators had been varied in order to find the optimum balance between accuracy and runtime.

5.3 Test Stage: Rules Application

After the stop condition is reached, it is disposed of p populations of rules that had been evolved against the training set. In order to form a complete set of rules, an item from each population has to be chosen. Rules may be selected randomly, the best ones may be considered or a selection scheme may be used. In the last two cases, there are the final fitness evaluations of the individuals that are taken into account. However, it is not always the case that by selecting the local fittest rule from each population, their combination would obtain the best accuracy on the test set. Moreover, even if these best rules would give very good results on the training set, they may be in fact not general enough to be applied to previously unseen data.

Within the conducted experiments, it was considered that for a number of *cps* times, one rule from each population was randomly selected in order to form *cps* complete sets of rules. Each time, the rule set is applied to the test data in a similar manner to the performance calculation in algorithm from Fig. 4 and the final classification accuracy is acquired.

6 Enhancement through an Archiving Strategy

Although already competitive [11], [12], [13], the developed CCEA classifier can still be improved in what regards the production of multiple rules for each outcome of the decision problem. Since the regular EA that evolves every population conducts to the generation of rules around one global/local optimum, the technique must be endowed with a mechanism to conserve a diverse selection of rules.

It is in this respect that an archive population is created to coexist and update during the evolutionary loop. At the end of each generation, it seizes a fixed number of best collaborators from the current population and the previous archive. These fittest combinations are copied into the new archive such that there are no two rules alike.

Require: A k-class classification problem
Ensure: A rule set with multiple prototypes for each class
Begin
$t = 0$;
$P(t) = \emptyset$;
For each class k **do**
  Randomly initialize $P_k(t)$;
  $P(t) = P(t) \cup P_k(t)$;
End For
For each class k **do**
  Evaluate $P_k(t)$;
End For
Copy the best N collaborations from $P(t)$, based on previous evaluation scores and neglecting identical rules, to A_t;
While stop condition = false **do**
  $t = t + 1$;
  $P(t) = \emptyset$;
  For each class k **do**
    Select $P_k(t)$ from $P_k(t - 1)$;
    Apply genetic operators to $P_k(t)$;
    Evaluate $P_k(t)$;
    $P(t) = P(t) \cup P_k(t)$;
  End For
  Copy the best N collaborations from $P(t)$ and A_t to A_{t+1}
End While
End

Fig. 5. Schema of a CCEA classifier with an archive

The essence of the novel version of the CCEA classifier is depicted in Fig. 5. The steps fundamentally follow the principles of the initial approach. An external population of size $k{\cdot}N$ is additionally considered. It is initialized with the fittest rules, following from the first evaluation of the individuals in each population. The collaborators of each of these rules are also captured into the archive. Identical rules are ignored with the purpose of a diverse archive.

The evolutionary cycle is then entered and at the end of every generation, the best sets (with one rule per class) from the current population and archive are conserved to the archive of the next iteration.

After the termination condition is met, the archive contains a complete and varied decision rule set. For reason of performing a direct and objective comparison to the initial approach, it has been settled to maintain the same CCEA parameters, as well as the same selection, variation operators and evolutionary parameters for the archiving strategy. In the end of the algorithm, all the rules contained in the archive are applied to the test samples and the accuracy is reported.

The proposed enhanced approach brings a new parameter that refers to the size of the archive. In our experiments, its value had been varied from 2 up to 10. Given the manner in which we test the archive against the given data, i.e. apply all collected rules to the test samples, the new parameter cannot be very high both due to an increase in runtime and because a large archive does not necessary exhibit significantly better results, as noticed in the undertaken experiments.

7 Experiments

The purpose of experimentation is twofold. It is envisaged to validate the new classification approach against real-world decision tasks, as well as to check the assumption that an archive brings more diversity and subsequent better prediction to the CCEA classifier.

In all conducted experiments, for each parameter setting, the training set is formed of randomly picked samples and the test set contains the rest of the samples. In order to prove the stability of the approaches, each reported average result is obtained after 30 runs of the algorithm.

7.1 Test Problems

Experimentation has been carried on two real-world problems of tumor diagnosis, both having only two classes, one coming from the UCI Repository of Machine Learning Databases and the other collected from the University Hospital in Craiova, Romania. The two data sets are further on briefly presented.

Breast Cancer Data Set. The data set contains 699 observations on 9 discrete cytological factors. The factors that influence the diagnosis are: clump thickness, uniformity of cell size, uniformity of cell shape, marginal adhesion, single epithelial, cell size, bare nuclei, bland chromatin, normal nucleoli and mitoses. The objective is to identify whether a sample corresponds to either a benign or a malignant tumor. There are 16 missing values for the sixth attribute. In all experiments, they are replaced by the average value for that indicator.

Hepatic Cancer Early Diagnosis. Hepatocellular carcinoma (HCC) is the primary cancer of the liver that ranks fifth in frequency among all malignancies in the world. In patients with a higher suspicion of HCC, the best method of diagnosis involves a scan of the abdomen, but only at a high cost. A cheap and effective alternative consists in detecting small or subtle increases for serum enzymes levels. Consequently, based on a set of 14 significant serum enzymes, a group of 299 individuals and two possible outcomes (HCC and non-HCC), the aim is to provide an efficient computational means of checking the consistency of decision-making in the early detection of HCC at a significantly low expense.

7.2 Preexperimental Planning

Within preliminary experiments, we tested different settings for the coevolutionary parameters in order to verify their suitability for the classification problems. However, in this initial experimentation, we applied the technique only on the breast cancer data set.

We observed that there are no major differences between results obtained when different types of collaboration credit assignments are used.

As it was expected, when the collaboration pool size value is increased, the runtime of the algorithm also raises. As concerning the results, they also seem to be improved to some extent by the increase of the value for this parameter. The technique had been tested from one up to seven collaborators.

As regards the collaborator selection pressure parameter, we initially employed random selection which drove the coevolutionary process to very competitive results. We then chose the best individual from each population for collaborations, but, surprisingly, the obtained results were worse than in the case of a random selection pressure. The next step we took was that of using a selection scheme for choosing the collaborators. Proportional selection was employed and it proved to be efficient as results were slightly better than those obtained through random selection.

As concerns the enhanced technique with the archive preservation, we had noticed the necessity of checking whether the individuals that enter the archive are identical or not: Before performing this verification, we observed that, from the early stages of the evolution, the archive had been populated with copies of the same best combination of individuals. By allowing only different sets of individuals to become members of the archive, we assure a good diversity among the final set of rules of the external population. We noticed the promise of proposed method from the start when, even from the early generations, the application of the rules in the archive to the training set yielded good accuracies. This happens due to the fact that the archive contains the best found (up to the moment) suites of rules that match the known data and are, at the same time, different from each other.

7.3 Task

We want to evaluate whether the proposed approach produces viable results if compared to those obtained by other approaches and how appropriate parameters will be chosen. Secondly, we plan to evaluate if the CCEA classifier with an archive produces a viable learning machine when compared to the former method.

7.4 Setup

The values for all the parameters were manually tuned. Table 1 contains the values for both the parameters of the EA and the coevolutionary ones. The population size refers to the number of individuals from each of the considered species.

Table 1. Values for the evolutionary parameters of the CCEA classifiers with and without an archive

	Breast cancer	Hepatic tumor
Evolutionary Parameters		
Population size	100	100
Recombination probability	0.5	0.5
Mutation probability	0.6	0.6
Mutation strength	0.01	100
Number of generations	120	120
CCEA Parameters		
Collaboration pool size	3	5
Collaborator selection pressure	Proportional	Proportional
Collaboration credit assignment	Hedge	Pessimistic

30 runs were conducted for each data set; in every run, around 75% random cases were appointed to the training set and the remaining 25% went into test. Experiments showed the necessity for data normalization in breast cancer, while hepatic tumor diagnosis requires without.

The only difference between the values for the parameters used in relation to the two data sets can be remarked for the mutation strength indicator: That is due to the fact that one of the data set is normalized and the other is not. For the hepatic data set, the number that is used to mutate the gene values is computed by dividing the difference between the upper and lower bounds of each attribute to the value specified in the table as described in (4). Naturally, in order to have an objective comparison, the same parameters are used for both the CCEA classifier with and without an archive. As concerns the size of the archive, we tuned the value in search of the best configuration.

7.5 Results

For both data sets we varied the archive size from 2 (meaning 2 pairs of rules) up to 10. The output did not exhibit very important differences as a result of these variations, but only of less than one percent. Variances can very well be the result of setting a luckier or, on the contrary, unfortunate separation between the training and test sets. Note, however, that results reported for the CCEA classifier without an archive are obtained for precisely the same training/test configurations as in the case of the archive variant. Table 2 displays the results of both versions of CCEA classifier. The accuracies reported by the novel variant were obtained for an archive size equal to 9 in case of breast cancer, while for hepatic cancer the value was set to 5.

The outcome of experimentation argues in favor of the archiving strategy. The results for each problem were compared via a Wilcoxon rank-sum test. The p-values (see last column in Table 2) suggest to detect a more significant difference in the

Table 2. Accuracies of prediction of the CCEA classifier with and without an archive

Data set	Cooperative Classifier	Test accuracy in 30 runs (%)				p-value
		Average	Minimum	Maximum	Std. dev.	
Breast cancer	Without archive	94.42	89.14	98.85	2.26	0.05
	With archive	95.52	93.14	98.85	1.56	
Hepatic tumor	Without archive	90.1	86	95	2.67	0.31
	With archive	90.76	86	95	2.4	

case of breast cancer. Moreover, the absolute difference is not large for the hepatic data, although rendering the new approach a slight advantage. Nevertheless, it is more relevant for the breast cancer problem.

As concerns the runtime, while for an archive equal to 2 it takes approximately 8 seconds to evolve and test the rules, for an archive of 10, the runtime takes 11 seconds. The mentioned measurements were made for the breast cancer data set, they are the average result of 30 repeats and were performed on a Pentium IV with a 3 GHz processor and a RAM memory of 1 GB.

Table 3. Comparison to accuracies obtained by other techniques for the breast cancer data sets

Technique	Accuracy
CCEA with archive	**95.52**
Decision trees	94.2 – 95.6
Naïve Bayes	96.4
Linear discriminant analysis	96.0
Support vector machines [15]	97.2

While for the HCC data set there are no previous attempts to automatically classify the data, several other techniques had been tested against the breast cancer data set. However, their results (illustrated in Table 3) cannot be directly compared with the ones obtained by our model as they used 10-fold cross-validation and they removed the samples that had missing values. Except for the last result, all the other ones are from [16]. Nevertheless, accuracies from Table 3 indicate that the CCEA classification technique offers results comparable to other well-established techniques in the literature.

7.6 Observations

There is not a very strong dependence between the EA parameters and obtained results as very competitive accuracies are obtained for a large scale of their values. Concerning the CCEA parameters, changes within the collaboration credit assignment schemes do not bring important modifications to the final average results: The differences between various settings as regards the eventual accuracies do not overpass one percent.

As previously stated, the collaboration pool size parameter directly influences the runtime of the program that implements the approach; the final test accuracy is at the same time positively affected by the increase in this value, but a balance has to be established between runtime and accuracy. Another important observation is that, when a certain threshold for the collaboration pool size parameter is surpassed, no further gain in accuracy is reached. We generally achieved the best results when three collaborators were considered.

Obtained results indicate that the proposed type of collecting the suites of rules that performed best on the training set indeed leads to higher success than the initial CCEA classifier. The explanation could be that the archive already contains a collection of rules that are well-suited to the given data and so, no collaborations need to be selected in the test stage, as in the primary construction.

Moreover, during evolution, one rule can find collaborators and, in conjunction, obtains a very promising training accuracy; but then, when other collaborators are selected for the same individual, they jointly might output very poor results. The EA would, in such a case, encourage the modification of these individuals (through mutation and/or recombination) or, even more likely, their removal through the selection operator and, as a consequence, they would be lost. The archive however collects these promising combinations and saves them from becoming extinct.

The values for the standard deviations and, at the same time, the maximum and minimum accuracies point out the fact that the archiving strategy brings forward a more stable algorithm that gives more similar results, despite the fact that the training/test sets are randomly selected in each of the 30 repeats.

7.7 Discussion

The CCEA framework offers a complementary evolution of several prototypes of patterns in a classification task.

Diversity is secondarily increased through an additional archiving mechanism which has the advantage of conserving the promise of certain collaborations, throughout the generations. Some individuals meet at certain points and they collaborate in an excellent fashion by returning a very good accuracy on the training set. Only one of them is rewarded (the one that is evaluated) by being assigned a high fitness, but next time it is considered (in case it does not vanish because of selection or variation operators), it may not encounter proper collaborators and therefore provide a poor result. In conclusion, it will accordingly have a poor fitness and little chances of being selected further on.

The archive however collects these *good teams* of rules and conserves them over the generations. The reported results underline the fact that the archiving strategy indeed performs better than the standard CCEA approach for classification.

8 Conclusions and Future Work

An approach to classification based on cooperative coevolution is presented. The mechanism is very simple and yet, very efficient: Populations of rules with the same outcome are evolved and they cooperate during the evaluation process. The fitness assignment is designed to constrain the rules to resemble the samples in the training set with their exact outcome.

An archiving strategy for the CCEA for classification is then proposed in order to enhance accuracy. The mechanism assumes the conservation of the most promising sets of rules starting from the initial generation. The archive is updated with the best performing teams of rules from the population and the former archive after each iteration of the EA. The rules that form the archive are in the end of the algorithm applied together to the test set. Obtained results met the initial expectations as they proved to be more accurate than the ones acquired by appointing the primary cooperative classifier.

The archive approach was applied for the parameter configurations that generated the best results for the CCEA approach in its first construction. However, an automatic tuning mechanism for the parameters could lead to interesting results and correlations between parameters.

The future research in this direction includes the extension of the algorithm for application to a larger area of test cases. An a priori employment of a preprocessing technique to data, in order to substantially reduce them, could boost the results of CCEA for classification.

In conjunction with that (or by itself), we presume that employing a chunking technique in order to repeatedly pick only small parts from the training set could significantly improve runtime (maybe even the accuracy), especially when dealing with large data sets.

In the current version, the CCEA approach with an archive records only suites that contain one rule per class. Storing sets that allow several rules per outcome could make the classifier even more versatile and efficient.

Acknowledgements. The authors wish to thank Professors PhD MD Tudorel Ciurea and Adrian Saftoiu from the University of Medicine of Pharmacy, Craiova, Romania, for the data on hepatic cancer.

References

1. Michalewicz, Z.: Genetic Algorithms + Data Structures = Evolution Programs, 3rd edn. Springer, London (1996)
2. De Jong, K.A.: Genetic-algorithm-based learning. In: Machine Learning: An Artificial Intelligence Approach, vol. 3, pp. 611–638. Morgan Kaufmann Publishers Inc., San Francisco (1990)
3. Holland, J.H.: Escaping Brittleness: The possibilities of General Purpose Learning Algorithms Applied to Parallel Rule-Based Systems. Machine Learning 2, 593–623 (1986)
4. Freitas, A.A.: A Survey of Evolutionary Algorithms for Data Mining and Knowledge Discovery. In: Advances in Evolutionary Computation, pp. 819–845. Springer, New York (2002)
5. Freitas, A.A.: Data Mining and Knowledge Discovery with Evolutionary Algorithms. Springer, Heidelberg (2002)

6. Giordana, A., Saitta, L., Zini, F.: Learning Disjunctive Concepts by Means of Genetic Algorithms. In: Proceedings of the 11th International Conference on Machine Learning, pp. 96–104 (1994)
7. Potter, M.A., De Jong, K.A.: A Cooperative Coevolutionary Approach to Function Optimization. In: Davidor, Y., Männer, R., Schwefel, H.-P. (eds.) PPSN 1994. LNCS, vol. 866, pp. 249–257. Springer, Heidelberg (1994)
8. Potter, M.A., De Jong, K.A.: Cooperative Coevolution: An Architecture for Evolving Coadapted Subcomponents. Evolutionary Computation 8(1), 1–29 (2000)
9. Wiegand, P.R.: Analysis of Cooperative Coevolutionary Algorithms. PhD thesis. Department of Computer Science George Mason University, USA (2003)
10. Potter, M.A., Meeden, L.A., Schultz, A.C.: Heterogeneity in the Coevolved Behaviors of Mobile Robots: The Emergence of Specialists. In: Proceedings of the 17th International Conference on Artificial Intelligence, pp. 1337–1343 (2001)
11. Stoean, C., Stoean, R., Preuss, M., Dumitrescu, D.: Spam Filtering by Means of Cooperative Coevolution. In: Teodorescu, H.N. (ed.) 4th European Conference on Intelligent Systems and Technologies (ECIT 2006), Advances in Intelligent Systems and Technologies, Selected Papers, pp. 157–159. Performantica Press (2006)
12. Stoean, C., Preuss, M., Dumitrescu, D., Stoean, R.: Cooperative Evolution of Rules for Classification. In: IEEE Postproceedings Symbolic and Numeric Algorithms for Scientific Computing (SYNASC 2006), pp. 317–322 (2006)
13. Stoean, C., Stoean, R., Preuss, M., Dumitrescu, D.: Coevolution for Classification. Technical report, No. CI-239/08. Collaborative Research Center on Computational Intelligence University of Dortmund, Germany (2008)
14. Eiben, A.E., Smith, J.E.: Introduction to Evolutionary Computing. Springer, Heidelberg (2003)
15. Bennett, K.P., Blue, J.: A Support Vector Machine Approach to Decision Trees. Math Report, No. 97–100. Rensselaer Polytechnic Institute Troy, New York (1997)
16. Duch, W., Adamczak, R., Grabczewski, K., Zal, G.: Hybrid neural-global minimization method of logical rule extraction. Journal of Advanced Computational Intelligence 3(5), 348–356 (1999)

Dynamic Applications Using Multi-Agents Systems

Mohammad Khazab, Jeffrey Tweedale, and Lakhmi Jain

School of Electrical and Information Engineering,
Knowledge Based Intelligent Engineering Systems Centre,
University of South Australia, Mawson Lakes, SA 5095, Australia
Mohammad.Khazab@postgrads.unisa.edu.au,
{Jeffrey.Tweedale,Lakhmi.Jain}@unisa.edu.au

Abstract. The aim of this research is to efficiently provide reusable autonomous *capabilities* to agent supervisors within teams, without the need to re-instantiate agents representing specific capabilities. Agent teaming techniques have already been used to enhance the behaviour and flexibility of agent communication in the real world applications. The theory of this concept needs to be simulated in order to generate Measures Of Efficiency (MOE) and Measures Of Performance (MOP). A concept demonstrator uses persistent components that assemble at design-time called the Agent Factory Demonstrator (AFD). It has been developed to show how a Multi-Agent System (MAS) can dynamically create capabilities using a single agent supervisor. This architecture is also used to show how the changing composition of a team can be used to efficiently complete a variety of tasks using adaptive capabilities provided in a manner similar to a team of single agents. The simulator uses a Java Graphical User Interface (GUI) supported by an agent oriented design in order to autonomously coordinate MAS Teams which can be enhanced further by incorporating other MAS to dynamically improve communication and knowledge-sharing. Knowledge Interchange Format (KIF), Agent Communication Languages (ACL), Knowledge Query Manipulation Language (KQML), FIPA Agent Communication Languages (FIPA ACL) and Simple Object Access Protocol (SOAP) have already been revised in an attempt to create a universal communication model that adapts to transient agent teams dynamically. Complementary research on a variety of agent tools (specifically JACK, JADE and CIAgent) has also been conducted to adapt the lessons learned into the AFD to generate results worthy of further effort.

Keywords: Artificial Intelligence, Multi-Agent Systems, Agent Factory Demonstrator, Agent Communication Language.

1 Introduction

There are many definitions of an agent. The major reason for this variance is due to the exponential growth of diversity and functionality. Wooldridges' definition of weak and strong Agency [1] currently dominates most literature. The weaker notion defines the term agent as having ability to provide simple autonomy, sociability, reactivity or pro-activeness [2, 3], while the stronger notion is more descriptive and

H.-N. Teodorescu, J. Watada, and L.C. Jain (Eds.): Intel. Sys. and Tech., SCI 217, pp. 67–79.
springerlink.com

agent refers to computer systems that extend the above properties, as either abstract or personified concepts. It is quite common in Artificial Intelligence (AI) to characterise an agent using cognitive notions, such as knowledge, belief, intention, obligation [4]. An agent can be seen as either software or hardware components within a system capable of accomplishing the tasks on behalf of its source [5].

Automation is convenient and becoming a common place for dangerous or repetitive tasks. These applications are becoming so efficient that previous human input/control is becoming the bottleneck. The battle space is one field where automating tasks is deemed to be very important [6][1]. This supplementation can reduce the cost of operation and prevent loss of life. An autonomous, dynamic agent system can be used to minimize system and Human Machine Interface (HMI) disruptions. The Agent Factory Demonstrator (AFD) attempts to document these bottlenecks and recommend/implement solutions to these issues in an air environment. This research aims to develop an agent learning and teaming architecture suitable for controlling Unmanned Aerial Vehicles (UAVs). This introduction is followed by discussion of Agents in Section 2, Agent Commuincation Languages in Section 3, the concept demonstrator in Section 4, Human Computer Trust in Section 5, measuring trust in Section 6 and future efforts in Section 7.

2 Agents

Agents are considered as software or hardware system that can act on behalf of its user to accomplish a task [5]. They can be equipped with other capabilities such as learning, reasoning and mobility [7, 8]. System overheads could be minimized given the ability to context switch functionality and communication models using self discovery. A Multi-Agent System (MAS) framework[2] using an AI mixed with supervisors and functional agents that share the process of discovery and goal completion[3]. In a MAS team of agents, knowledge could ignore, be mis-informed, or force intelligent cooperation or collaboration within the group [10]. Agent collaboration and cooperation provide the ability of agents to work together in order to solve problems and achieve common goals and can voluntarily cooperative to problem solving. Autonomous agents have the ability to decide when they would interact with other agents if they have positive motivations [11]. This is the reason agents need to collaborate and negotiate with each other within the team to plan and take actions to solve problems based on the knowledge available [12].

The concept of knowledge is a collection of facts, principles, and related concepts. Knowledge representation is the key to any communication language and a fundamental issue in AI. The way knowledge is represented and expressed has

[1] Such as flying aircraft for surveillance purposes.

[2] Where the MAS is a group of agents or human and agents that are interacting with each other in order to achieve mutual goals [9, 10].

[3] MAS re-configurability refers to the ability of teams of agents to dynamically organize and form a new subgroup configured to map decomposed tasks to achieve the team goal. Factors of reconfigurability include the MAS environment (bandwidth, capabilities, sensors and processing abilities) and communication topologies.

to be meaningful so that the communicating entities can grasp the concept of the knowledge transmitted among them. This requires a good technique to represent knowledge. In computers symbols (numbers and characters) are used to store and manipulate the knowledge. There are different approaches for storing knowledge because there are different kinds of knowledge such as facts, rules, relationships, and so on. Some popular approaches for storing knowledge in computers include procedural, relational, and hierarchical representations [13].

3 Agent Communication

Austin was one of the pioneer researchers in the area of ACL and his work inspired other researches in this field [14]. The desired interaction[4] and cooperation of agents within MAS is not natively possible without some kind of Agent Communication Languages (ACL). The following text briefly describes Knowledge Interchange Format (KIF), Foundation of Intelligent Physical Agents (FIPA), Knowledge Query Manipulation Language (KQML), FIPA Agent Communication Languages (FIPA ACL) and our research is now exploring Simple Object Access Protocol (SOAP).

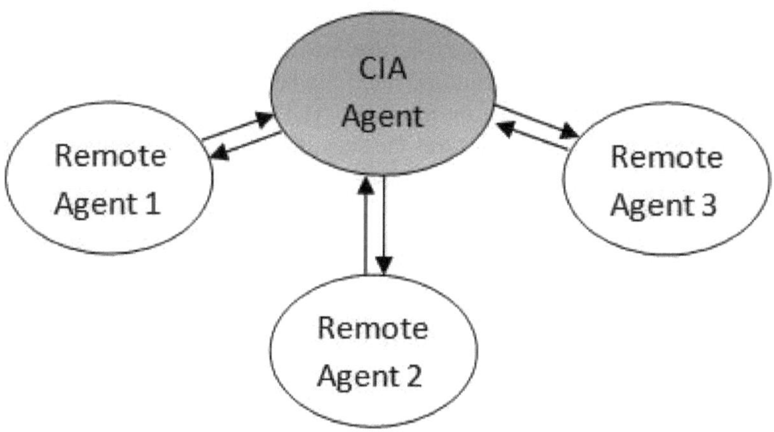

Fig. 1. The Concept of Inter-Agent Communications

ACLs were developed to handle complex semantics and transfer messages that describe a desired state[5]. They also enable the exchange of more complex objects such as shared plans, goals and experience [15], similar to Beliefs, Desires, Intentions (BDI).

[4] Interaction is generally achieved using duplex communications (like a telephone) in simplex mode (like an intercom) as shown in Figure 1.

[5] JATLite and the JAFMAS are agent communication architecture [12].

3.1 Knowledge Interchange Format

KIF is a language developed by the Defense Advanced Research Projects Agency (DARPA) Knowledge Sharing Effort (KSE) [15] group for interchanging knowledge between agents in order to improve interoperability in computing environment [16, 17]. KIF is not a programming language but can translate directly to and from other agent languages. The syntax consists of variables, constants, and operators. These variables can be of individual or sequenced types. Constants can be of all types of numbers, characters and strings. They can also indicate objects, relations, and functions over objects. There are also logical constants used to express facts in the form of Boolean expressions.

3.2 Knowledge Query Manipulation Language

KQML is a high-level, message-oriented communication language with three layers (content, message and communication [18]). It is based on speech act theory and consists of primitives which allow agents to communicate their attitudes. KQML is independent of the transport mechanism, content language, and ontology [15]. The syntax of KQML is based on the Lisp language.

3.3 FIPA Agent Communication Languages

The FIPA ACLs Semantic Language (SL) that uses a LISP-like encoding to embody concepts and actions. It is a quantified multi-modal logic with the BDI concepts [18] that extends previous work by the FIPA. The new specification is based on the BDI agent model [13] which stores the state of environment. Agents also have plans used to track sub-tasks being processed required to achieve its goal, updating the state of environment after completing each task. FIPA ACL is similar to which is also based on the speech act theory and their syntax is similar except for some reserved names of the primitives [15]. The messages in FIPA ACL are considered to be Communicative Acts and are the compulsory parameter of the message [12]. These transfer entities as primitives (like KQML) to share knowledge, negotiate and submit queries [13].

3.4 Simple Object Access Protocol

The latest ACL is an Extensible Markup Language (XML) based inter-application cross-platform communication mechanism derived from the JAVA programming language [19]. SOAP messages consist of two main elements (the envelope and body, with an optional header element) [20]. The envelope element contains attributes such as the encoding Style attributes that specifies the encoding style of the message. The body of message includes the information intended for recipient. This information should use the SOAP encoding rules. The header element can be used to add information about the message such as the return path for the response.

3.5 Decision Making

All the logic and decision processes required for agent's capabilities and actions is decision support. The decision support might be used to find a target or get to a point and so on. Agent teaming architecture divides the tasks into subtasks and allocates them to agents that have the capabilities to solve the problem. Agents in a MAS can be cooperating or competing with each other to achieve their goals. Multi-agent Planning specifies the actions for agents to perform [21]. Multi-body planning enables agents to find the possible joint plans to achieve the goal. The co-operation of agents requires them to agree on the same goal, which can be achieved by communication.

A BDI framework in a MAS architecture that operates in a constrained environment is an effective model for this research. Beliefs are the understandings of the agents from the environment, desires are the goals of the agent to achieve, and intentions are its plans or actions for achieving the goal [22]. The BDI model originated from philosophical theory of practical reasoning [23]. Practical reasoning involves two aspects of deliberation and reasoning (planning) that refer to decisions of what to be achieved and how to achieve it respectively. Based on the concept of BDI, agents are assumed to have beliefs and understanding about other agents and the state of environment. Agents also have plans for taking the required actions to achieve a particular goal and change the state of environment.

In the scenario of this research, agents were equipped with navigation skills to be able to patrol the virtual simulation environment. They were also capable of detecting objects at certain distance from them. They can realize the type of objects as well. The decision making feature enables them to use their rule-base to react accordingly based on the objects they reached to. For example, if the object they hit is an obstacle like a wall they would be using their collision avoidance rule and turn away, or if the object is the target they were searching for they would pick it up or send its position coordinates to supervisor agent. The decision support system is being extended to enable agents remember the actions they did that led to achieving their goal, e.g. the direction they took to reach the target. This information can be used by the agent or other agents to achieve the goal quicker the next time they encounter the same situation.

3.6 Dynamic Communication

In this simulation environment where multiple agents are running and other factors influence the virtual world dynamically, communication and cooperation are important for achieving the goal. When two agents reach certain distance apart they can start to communicate and share their knowledge about the environment so that the area is patrolled quicker and agents avoid passing the same route redundantly. A communication facilitator agent, or also referred to as supervisor agent, was used to facilitate the communications and interactions among other agents. The facilitator agent offers various useful communication and network services [24, 25]. When agents are created at run-time they inform the facilitator agent of their name, address, and capabilities or services they provide. The facilitator agent then registers

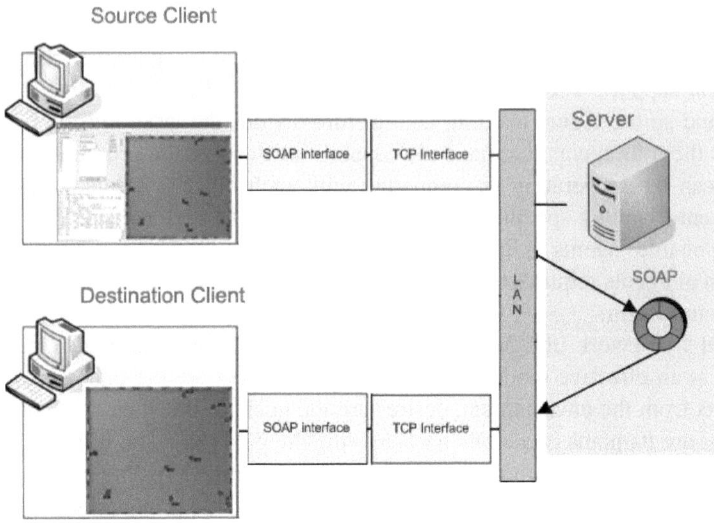

Fig. 2. Developing Distributed System using SOAP as communication protocol

them in its database based on this information, which then can be used to retrieve information to identify agents. The facilitator can also forward information between agents and acts as a translator between agents with different semantic and ontology in the knowledge content of their messages. The result of this was the achievement of a better teamwork by the MAS.

KQML messages were used to transmit information among agents. Data was structured into XML documents and transmitted between server-client applications via SOAP messages (Figure 2). This information includes messages transmitted between agents, data obtained from the simulation environment, and the requests and replies communicated between the server and client applications. Transmitting all these data at once can exceed the communication channel overhead and cease the system from its other operations and networking. Two styles of communication were considered in simulating the dynamic environment to handle the data load. One style is used when an event requires immediate attention, e.g. an agent finds a target. The other style deals with less frequent communication which provides agents with the slower update about the environment, e.g. an agent is moving from point A to B. Parsing the information into XML documents enabled client and server applications to extract and process only the related information that was required.

Dividing the tasks into subtasks also helped with the issue of data management. For example, few agents were used to listen for network connection and communication requests from client-side applications. The other agents were decomposed into subgroups, each conducting a different task such as finding a safe route to the target, moving obstacles, and sending the route coordinates to clients. The Supervisor agent also has an important role in coordinating these teams by annunciating record keeping of their capabilities and function status.

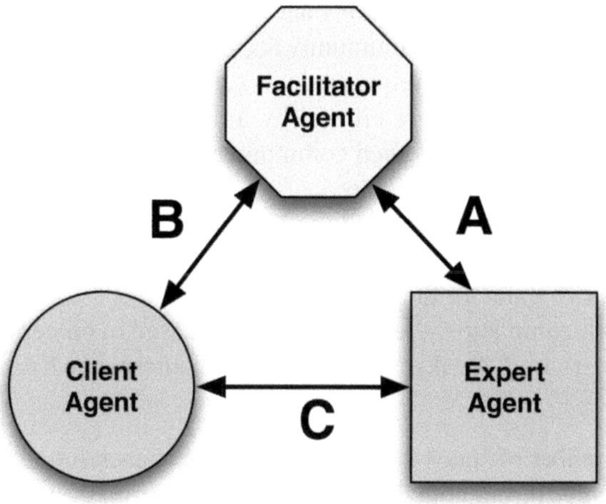

Fig. 3. Agent Factory Demonstrator Sequence

4 Agent Factory Demonstrator

The AFD is a simple MAS simulator using an agent factory design pattern to control its core functions. It employs a flexible communication language that is used to span multiple networks or systems and has the potential to improve the integration and inter-operability of large-scale distributed systems [25]. Context switching is achieved by way of a list of agent pointers using a poly-morphic abstract class to interface between each agent and the capability being transacted. This facilitator agent is a unique agent that has the list of all instantiated expert and client agents. When an expert agent is instantiated, it sends a message to the facilitator agent (registering itself as a performative component). The facilitator agent would add this capability to its functionality list (access to its publication stream is achieved via subscription). A client agent must approach the facilitator and seek access to the closest expert with a specific capability. The facilitator agent would register the client as a subscriber to that corresponding capability and its data stream. The facilitator agent can also keep track of the state of all expert agents to prevent *deadlock situations* using *semaphores* and *state variables*. Figure 3 displays the possible transaction between the facilitator, expert and client agents (There are many to many relationships and eventually each group of agents will be able to dynamically grow and shrink using present hysteresis parameters). The facilitator agent has previous knowledge of the client and expert or autonomously discovers their presence and availability. Free expert agents will be allocated to service client agents requests (after decomposing the goal into sub-task(s). If there are insufficient resources, some requests may be queued or additional experts instantiated on a priority/needs basis. The transactions occur in the order shown. This sequence of transaction is described as:

Start: When the program starts the Facilitator Agent (FA) is initiated. Separate Expert Agents (EAs) are also initiated as specified and selected by the user.

Step A: EAs communicate with the FA and advertise their expertise. FA registers and lists EAs in a particular community according to their expertise. EAs would then be ready for incoming requests.

Step B: Client Agents (CAs) are created by a user, scenario generator or sensor with a task to perform. CAs then communicates with the FA and are allocated suitable resources (EAs) as required.

Step C: The CA opens dialog with all the EAs and negotiates the accomplishment of each task. If EA can successfully fulfil CAs task then their conversation is finished and EA waits for the next CA to approach.

Cycle: Upon task completion, the CA would be released to process another task by the user. Otherwise CA asks FA to introduce an alternative EA to help fulfill its task.

There are a number of states that agents can have at each time [13]. The state of agents would change throughout their lifetime depending on the state and conditions they are at or the work they are doing. Some states are common among all agents while other states are specific for nominated agents. For example, some common states of agents includes Initiated, Active, Suspended, Uninitiated, Unknown. All agents start with initiated state and their state become active in the next step ready to do other tasks. Any agent can be suspended, disabled by changing its state. For

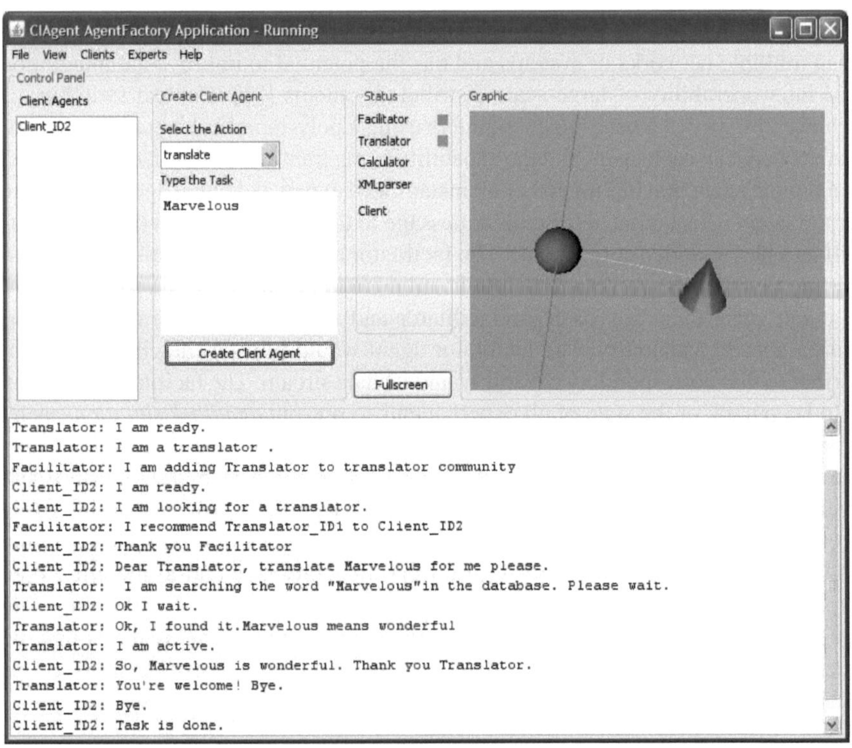

Fig. 4. Agent Factory Demonstrator Snap Shop

example, a client agent would be set to **PROCESSING** when one or more expert agent(s) is processing a sub-task. Alternatively, it may be set to **WAITING** prior to being allocated the resources required to process its goal. Other states are used to inform the facilitator about the state of each agent within its domain of influence. More sophisticated communications protocols are required to pass messages, commands/requests and disparate data sources between agents. For example the expert agents can change state, request tasking, process solutions and report faults. The client would do the reverse, where the facilitator agent communicates, negotiates and monitors its domain and external systems or clients. At present this research only uses a single communication model, however it deals with three capabilities as shown in Figure 4.

5 Human Computer Trust

Trust is increasingly becoming a factor in the relationship between humans and machines and in particular automated systems. It is this trust between a human operator and an automated system, that is commonly called human-computer trust (HCT). The trust between people is separated by the expectations people have about a process or object and therefore different approaches and may need to be considered when considering the trust of such systems.

Human-computer trust has been acknowledged as a limiting factor in delegating more responsibility to automated processes in many different areas, such as procurement, computing, on-line interactions and e-commerce, transportation and security. However, computers may also be perceived by some to be trustworthy due to the perception that they are autonomous and nonjudgmental [26–37].

One accepted definition of human-computer trust attributed to Madsen and Gregor states that:

> "trust is the extent to which a user is confident in, and willing to act on the basis of, the recommendations, actions, and decisions of [an automated] decision aid [38]".

Kelly et al. [39] considers trust to be an "enabler" to the introduction of new systems and therefore the ability to measure operator trust during real-time simulations is important in evaluating such systems. Muir crossed the taxonomies of Barber and Rempel to form a single two-dimensional taxonomy of human-computer trust. This shows that trust is an important intervening variable between automated systems, their use and performance [40].

6 Measuring Trust

There are too many ways of measuring trust. Two style continue to attract attention, which include [39]:

objectively by measures of operator performance (e.g. frequency, accuracy or speed in interaction) if the relationship between these measures and the automation could be unequivocally established; and

subjectively by asking the operators how they feel

The point is also made that the subjective measurements can be converted to objective measurements if the origin of the subjective ratings can be modeled.

6.1 Subjective Measures

Kelly et al. [39] uses the following scales which are considered the best method of measuring trust in the context of real-time simulations since they can be relatively easy to apply without being intrusive. The most interesting scales in the context of the current application are:

- the scale of "Human-Computer Trust", developed by Madsen and Gregor;
- a trust questionnaire, developed by Jian, Biznatz and Drury (employed in the context of military command and control decision-making);
- a questionnaire with a rating scale to determine operators views on the timeliness and appropriateness of adaptive computing aiding, developed by Taylor and Shadrake; and
- the report of studies on the human-electronic crew, by Haugh.

6.2 Objective Measures

Kelly et al. [39] suggest that the studies of Moray, in the context of supervisory control process, have shown that it is possible to develop empirical models of trust. Trust may be objectively measured by observing system physical properties, such as productivity output, selection of particular system functions and frequency of manual intervention. However the point is made that the approach may not be appropriate for ATC systems, and would require the establishment of the particular set of variables that are relevant.

A simple measure is to monitor whether or not an operator activates a particular automated system, given the assumption that if an operator chooses to use a tool they must have trust in it. However, problems with this measure include the possibility that although the operator trusts the system they consider they can perform the same task better without using it, or that even if they have the tool open, they may not be using it to make decisions. Kelly makes the comment that a more sophisticated measure is therefore required.

The development of a trust measure for the automated DSS developed as part of the current research should be addressed. Kelly et al. [39] suggest the use of a rating-scale based subjective measure as a simple and straightforward approach. The point is made, however, that a set of rating scales, each of which is used to measure a different dimension of trust, is probably more effective than a single trust rating scale. The dimensions considered to be the most appropriate for ATM automation were:

- reliability,
- accuracy,

- understanding,
- faith,
- liking,
- familiarity, and
- robustness.

Fuzzy set measures were also mentioned as being more appropriate rather than numerical scales.

7 Future Research Direction

This research demonstrates a MAS that provides autonomous *capabilities* using a facilitator agent to coordinate the teams actions, without the need to re-instantiate agents or capabilities. Agent teaming techniques have been used to enhance the behaviour and flexibility of agent communication using polymorphism. No Measures Of Efficiency (MOE) or Measures Of Performance (MOP) have been chosen as the system design needs to mature. The AFD model developed at the University of South Australia involves leading edge concepts intended to exploit the dynamic creation of agent capabilities with fast and efficient context switching. The simulator uses threads, programmed in Java, with a Graphical User Interface (GUI) that displays the state and results of all agents in the system. SOAP may be used in future versions of the demonstrator in an attempt to provide a universal translator approach for communication and tasking. Past research has already established architectures involving JACK, JADE and CIAgent cannot achieve these goals in isolation. Solutions to those lessons are being adapted for use in the AFD. One way of achieving better cooperation is by improving trust mechanisms. To achieve that, flexible methods of measuring trust must be established. Further work is required to mature this design, however the results so far are very encouraging.

Acknowledgements

Some of the material shown has already been presented by one or more authors in the First KES International Symposium on Intelligent Decision Technologies (IDT09) and the 10th International Conference on Knowledge Based Intelligent Information and Engineering Systems (KES06).

References

[1] Wooldridge, M., Jennings, N.R.: Theories, architectures, and languages: A survey, intelligent agents. In: Wooldridge, M.J., Jennings, N.R. (eds.) ECAI 1994 and ATAL 1994. LNCS (LNAI), vol. 890, pp. 1–39. Springer, Heidelberg (1995)

[2] Castelfranchi, C.: Guarantees for autonomy in cognitive agent architecture. In: Wooldridge, M., Jennings, N.R. (eds.) ECAI 1994 and ATAL 1994. LNCS, vol. 890, pp. 56–70. Springer, Heidelberg (1995)

[3] Genesereth, M.R., Ketchpel, S.P.: Software agents. Communications of the ACM 37(7), 48–53 (1994)
[4] Bratman, M.E.: Intentions Plans and Practical Reason. Center for the Study of Language and Information (1999)
[5] Nwana, H.S.: Software agents: An overview. In: McBurney, P. (ed.) The Knowledge Engineering Review, Cambridge Journals, Simon Parsons, City University of New York, USA, vol. 11(3), pp. 205–244 (1996)
[6] Finn, A., Kabacinski, K., Drake, S., Mason, K.: Design challenges for an autonomous cooperative of UAVs. In: Information Decision and Control (IDC 2007), Adelaide, DSTO, Australia, February 11-14 (2007)
[7] Chira, O., Chira, C., Roche, T., Tormey, D., Brennan, A.: An agent-based approach to knowledge management in distributed design. Journal of Intelligent Manufacturing, The Institution of Engineering and Technology 17(6), 737–750 (2006)
[8] Jennings, N., Wooldridge, M.: Software agents, vol. 42(1), pp. 17–20. IEEE Press, NY (1996)
[9] Panait, L., Luke, S.: Cooperative Multi-Agent learning: The state of the art. Autonomous Agents and Multi-Agent Systems 11(3), 387–434 (2005)
[10] Dudek, G., Jenkin, M., Milios, E., Wilkes, D.: Taxonomy for swarm robots. In: International Conference on Intelligent Robots and Systems 1993, IROS 1993, Yokohama, Japan, IEEE/RSJ, vol. 1, pp. 441–447. IEEE Press, Piscataway (1993)
[11] Wooldridge, M.: Verifying that agents implement a communication language. In: Proceedings Sixteenth National Conference on Artificial Intelligence (AAI 1999). Eleventh Innovative Applications of Artificial Intelligence Conference (IAAI 1999), Orlando, FL, USA, pp. 52–57 (1999)
[12] Tweedale, J., Jain, L.C.: The Evolution of Intelligent Agents within the World Wide Web. In: Nguyen, N., Jain, L.C. (eds.) Intelligent Agents in the Evolution of Web and Applications, pp. 1–9. Springer, Heidelberg (2009)
[13] Bigus, J.P., Bigus, J.: Constructing Intelligent Agents Using Java: Professional Developer's Guide, 2nd edn. Wiley, New York (2001)
[14] Austin, J.L.: How to Do Things with Words. University Press, Oxford (1962)
[15] Labrou, Y., Finin, T., Peng, Y.: The current landsscape in agent communication languages. IEEE Intelligent Systems 2 (1999)
[16] Finin, T., Labrou, Y., Mayfield, J.: Kqml as an agent communication language. In: Software Agents, p. 480. AAAI Press / The MIT Press (1997)
[17] Bradshaw, J.M.: Software Agents. AAAI, MIT Press (1997)
[18] Fasli, M.: Agent technology for e-commerce. John Wiley, Chichester (2007)
[19] Seely, S., Sharkey, K.: SOAP: Cross Platform Web Services Development Using XML. Pearson Education, London (2001)
[20] Graham, S., Davis, D., Simeonov, S., Daniels, G., Brittenham, P., Nakamura, Y., Fremantle, P., Koenig, D., Zentner, C.: Building Web services with Java: making sense of XML, SOAP, WSDL, and UDDI. Developer's Library (2002)
[21] Russell, S.J., Norvig, P.: Artificial intelligence: A Modern Approach. Prentice Hall/ Pearson Education, Inc., Upper Saddle River (2005)
[22] Wooldridge, M., Muller, J., Tambe, M.: Agent theories, architectures, and languages: a bibliography. In: IJCAI 1995 Workshop (ATAL) Proceedings. Intelligent Agents II. Agent Theories, Architectures, and Languages, pp. 408–431. IEEE Press, New York (1996)
[23] Bratman, M.E.: Intention, Plans, and Practical Reason. Harvard University Press, USA (1999)

[24] Patil, R., Fikes, R., Patel-Schneider, P., McKay, D.P., Finin, T., Gruber, T., Neches, R.: The darpa knowledge sharing effort: Progress report. In: Nebel, B. (ed.) Proceedings of the Third International Conference on Principles of Knowledge Representation And Reasoning. Morgan Kaufmann Publishers Inc., San Fransisco (1992)
[25] Finin, T., Fritzon, R., McKay, D., McEntire, R.: KQML as an Agent Communication Language. In: Adam, N., Bhargaa, B., Yesha, Y. (eds.) Proceeding of the 3rd international Conference on Information and Knowledge Managment (CIKM 1994), pp. 456–463. ACM Press, New York (1994)
[26] Shneiderman, B.: Designing trust into online experiences. Communications of the ACM 43(12), 57–59 (2000)
[27] Subramanian, K.R., Lee, S., Shiang, T.K., Sue, G.B.: Intelligent agent platform for procurement. In: Preceedings of the IEEE International Conference on Systems, Man, and Cybernetics (IEEE SMC 1999), vol. 3, pp. 107–112. IEEE, Los Alamitos (1999)
[28] Hoffman, R.R.: Whom (or what) do you (mis)trust?: Historical reflections on the psychology and sociology of information technology. In: Proceedings of the Fourth Annual Symposium on Human Interaction with Complex Systems, pp. 28–36 (1998)
[29] Rosenbloom, A.: Trusting technology: Introduction. Communications of the ACM 43(12), 31–32 (2000)
[30] Uslaner, E.M.: Trust online, trust offline. Communications of the ACM 47(4), 28–29 (2004)
[31] Gefen, D.: Reflections on the dimensions of trust and trustworthiness among online consumers. SIGMIS Database 33(3), 38–53 (2002)
[32] Schneidewind, N.F.: Reliability modeling for safety-critical software. IEEE Transactions on Reliability 46(1), 88–98 (1997)
[33] Cahill, V., Gray, E., Seigneur, J.M., Jensen, C.D., Chen, Y., Shand, B., Dimmock, N., Twigg, A., Bacon, J., English, C., Wagealla, W., Terzis, S., Nixon, P., Serugendo, G.D.M., Bryce, C., Carbone, M., Krukow, K., Nielson, M.: Using trust for secure collaboration in uncertain environments. IEEE Pervasive Computing 2(3), 52–61 (2003)
[34] Lindqvist, U., Olovsson, T., Jonsson, E.: An analysis of a secure system based on trusted components. In: Proceedings of the Eleventh Annual Conference on Computer Assurance Systems Integrity. Software Safety. Process Security, pp. 213–223. IEEE, Los Alamitos (1996)
[35] Oppliger, R.: Internet security enters the middle ages. Computer 28(10), 100–101 (1995)
[36] Wang, W., Zhu, Y., Li, B.: Self-managed heterogeneous certification in mobile ad hoc networks. In: Proceedings of the IEEE 58th Vehicular Technology Conference (VTC 2003-Fall), vol. 3, pp. 2137–2141. IEEE, Los Alamitos (2003)
[37] Wilson, W., Sachs, J., Wichers, D., Boucher, P.: MLS and trust issues at the user interface in MLS AISs. In: Proceeding of the Sixth Annual Computer Security Applications Conference, pp. 204–208 (1990)
[38] Madsen, M., Gregor, S.: Measuring human-computer trust. In: Proceedings of the Eleventh Australasian Conference on Information Systems, Brisbane (2000)
[39] Kelly, C., Boardman, M., Goillau, P., Jeannot, E.: Guidelines for trust in future atm systems: A literature review. Technical Report 030317-01, European Organisation for the Safety of Air Navigation, Naval Wepons Center, Chine Lake, CA (May 2003)
[40] Khasawneh, M.T., Bowling, S.R., Jiang, X., Gramopadhye, A.K., Melloy, B.J.: A model for predicting human trust in automated systems. In: Proceedings of the Eigth Annual International Conference of Industrial Engineering - Theory, Applications and Practice, Las Vegas, Nevada, USA, pp. 216–222 (2003)

5

Localized versus Locality Preserving Representation Methods in Face Recognition Tasks

Iulian B. Ciocoiu

Technical University of Iasi, Romania, Faculty of Electronics and Telecommunications,
Bd. Carol I, no. 11, Iasi, 700506, Romania
iciocoiu@etc.tuiasi.ro

Abstract. Four different localized representation methods and two manifold learning procedures are compared in terms of recognition accuracy for several face processing tasks. The techniques under investigation are: a) Non-negative Matrix Factorization (NMF); b) Local Non-negative Matrix Factorization (LNMF); c) Independent Components Analysis (ICA); d) NMF with sparse constraints (NMFsc); e) Locality Preserving Projections (Laplacianfaces); and f) Orthogonal Projection Reduction by Affinity (OPRA). A systematic comparative analysis is conducted in terms of distance metric used, number of selected features, and sources of variability on AR, Yale, and Olivetti face databases. Results indicate that the relative performance ranking of the methods is highly task dependent, and varies significantly upon the distance metric used.

1 Introduction

When thinking about defining our identity, we readily observe that we are not merely perceived as a set of credentials, but as a subtle, complex, variable mixture of distinct anatomical and behavioral features. Combining such features with automated signal processing techniques sets the grounds for the implementation of *biometric technologies*, largely used nowadays for recognition or verification purposes. Fingerprints, voice, face, iris, or the handshape may be preferred in security oriented tasks over passwords, PIN codes, or ID cards, since the former cannot be stolen, forgotten, or misplaced (although they could be accurately reproduced by *artificially* created data and still "fool" automatic recognition devices).

Human face plays a significant role in social life, not only for identification purposes, but also as a means of communicating strong feelings and emotional, mental, or health state. Our face recognition performances are quite impressive: in most cases, we may still correctly identify a person without having seen him for a long time, under poor illumination conditions, or despite the presence of eyeglasses, beard, or new haircut. This amazing robustness against adverse conditions has motivated an enormous research interest during the last decades and many innovative, flexible, scalable solutions aiming at replicating these performances

H.-N. Teodorescu, J. Watada, and L.C. Jain (Eds.): Intel. Sys. and Tech., SCI 217, pp. 81–103.
springerlink.com

have been reported. The commercial trends are also significant: the market is estimated to reach 5.5 billion $ at the end of 2010, with an annual growth rate of about 25% [1].

Face recognition is a difficult task due to the many sources of variability that are present in a realistic setting. These include among others the illumination level, pose, face expression, partial occlusion, and demographic features (age, gender, race). Inferring invariance against geometric transformations such as translation, in-plane rotation, or scale changes puts also a challenge. The block diagram of a generic automatic face recognition system is presented in Fig. 1, where we may identify the presence of two main modules, namely a feature extractor that should yield a proper face "signature", and a classifier that outputs the identity of the unknown person. Following specific terminology, the data used for designing and evaluating the performances of the automated biometric system is composed of "gallery" (training data) and "probe" (test data) sets, that are typically collected in controlled conditions.

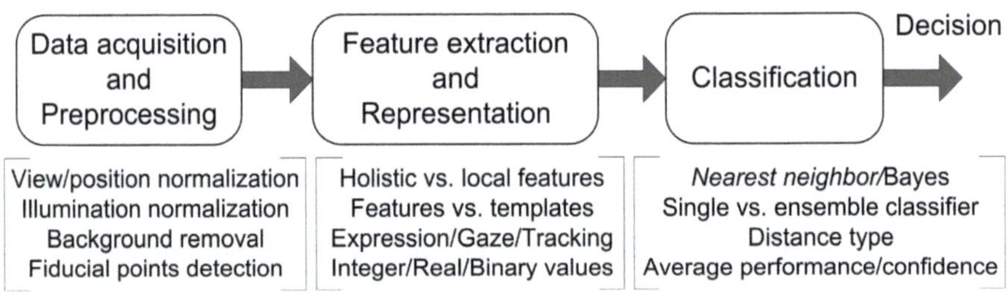

Fig. 1. Block diagram of a generic automated face recognition/verification system

Most of the feature extraction approaches may be classified into two categories [2]:

a) Template-based techniques, usually performing a projection of the original (high-dimensional) images onto lower dimensional subspaces spanned by specific basis vectors. Examples include Principal Components Analysis (PCA), Linear Discriminant Analysis (LDA), and their kernel-based variants. Eigenfaces [3] represent a *de facto* standard for this approach and, although superior solutions exist, still defines a performance reference against which any new method is compared.
b) Geometric feature-based techniques, relying on the identification of generic components of a face such as eyes, nose, mouth, and distances among them, followed by computation of specific local features. Elastic Graph Matching [4], active shape models [5], and Local Feature Analysis (LFA) [6] belong to this category of tools.

A number of recent surveys [7, 8] review modern trends in this area of research, including:

- kernel-type extensions of classical linear subspace projection methods such as Kernel PCA/LDA/ICA [9-12]

- holistic vs. component-based approaches [13, 14], compared in terms of stability to local deformations, lighting variations, and partial occlusion. The list is augmented by representation procedures using space localized basis images, some of which are described in the present contribution.
- the assumption that many real-world data lie near low-dimensional nonlinear manifolds exhibiting specific structure triggered the use of a significant set of *manifold learning* strategies in face oriented applications [15, 16], two of which are included in the present comparative analysis.

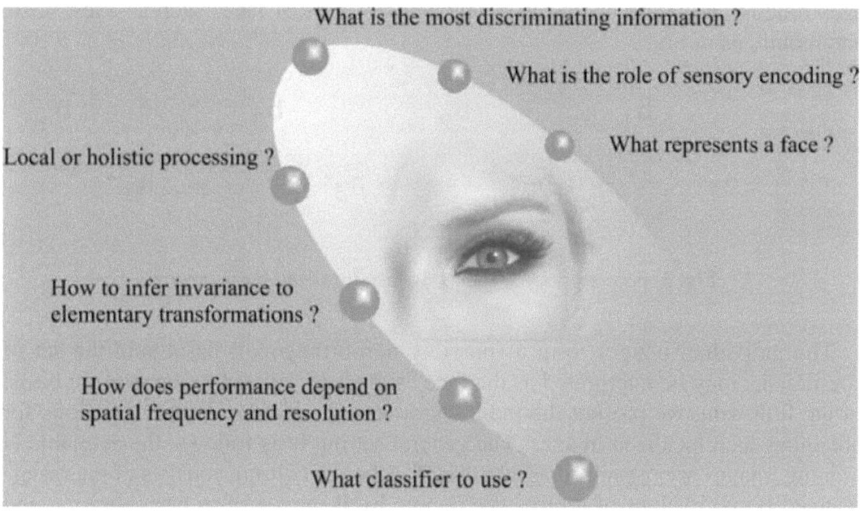

Fig. 2. Critical issues in face recognition/verification applications

Deeper understanding of the information content of face images is a key issue for designing automated systems able to cope with the performance targets and robustness requirements of commercial products, and may prove useful for other pattern recognition tasks as well. Some of the critical questions are indicated in Fig. 2, while several possible answers are given in the following paragraphs.

The present contribution focuses on a systematic comparative analysis of several distinct local feature extraction techniques and locality preserving projection methods, based on: the type of distance metric, the dimension of the feature vectors to be used for actual classification, the number of training images for each person in the database, the sources of face variability. The chapter is organized as follows: the theoretical grounds of the methods are described in Section 2. In Section 3 the experimental setup, including database description and preprocessing methods, is defined. Recognition performances are reported in Section 4, while discussion and conclusions are finally presented.

2 Localized and Locality Preserving Feature Extraction Techniques

From a taxonomic perspective, we may identify holistic and parts-based feature extraction approaches, which extract specific face "signatures" by processing the entire face image or localized portions of it, respectively. In principle, parts-based representation may offer advantages in terms of stability to local deformations, lighting variations, and partial occlusion. A number of recent algorithms aim at obtaining face representations using a linear combination of space-localized images roughly associated with the components of typical faces such as eyes, nose, and mouth, as in Fig. 3.

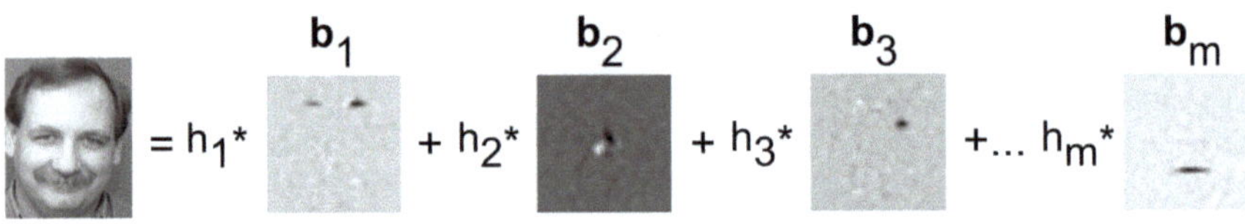

Fig. 3. Face representation using face localized basis images

The individual images form a (possibly non-orthogonal) basis, and the set of coefficients may be interpreted as the face "signature" related to the specific basis. In the following we present the main characteristics of four distinct solutions for obtaining such localized images. The general setting is as follows: the available N training images are organized as a matrix **X**, where a column consists of the raster-scanned p pixel values of a face. We denote by **B** the set of m basis vectors, and by **H** the matrix of projected coordinates of data matrix **X** onto basis **B**. If the number of basis vectors is smaller than the length of the image vectors forming **X** we get dimensionality reduction. On the contrary, if the number of basis images exceeds training data dimensionality we obtain overcomplete representations. As a consequence, we may write:

$$\mathbf{X} \simeq \mathbf{BH}\,. \tag{1}$$

where $\mathbf{X}\in\Re^{pxN}$, $\mathbf{B}\in\Re^{pxm}$, and $\mathbf{H}\in\Re^{mxN}$. Different projections techniques impose specific constraints on **B** and/or **H**, and some yield spatially localized basis images.

2.1 Non-negative Matrix Factorization (NMF)

Non-negative matrix factorization (NMF) [17] was recently introduced as a linear projection technique that imposes non-negativity constraints on both **B** and **H** matrices during learning. The method resembles matrix decompositions techniques such as positive matrix factorization, and has found many practical applications including chemometric or remote-sensing data analysis. The basic idea is that only *additive* combinations of the basis vectors are allowed, following the intuitive scheme of combining parts to form a whole. Referring to equation (1),

NMF imposes $\mathbf{B},\mathbf{H} \geq \mathbf{0}$. Based on the definition of specific objective functions measuring the quality of approximation in equation (1), multiplicative update procedures for **B** and **H** have been formulated as in equation (2). Matrices **B** and **H** are initialized with random strictly positive values, although superior initialization procedures have been proposed [18]. The above algorithm was implemented in MATLAB (code provided by Lee and Seung, available at *http:// journalclub.mit.edu*). Examples of basis vectors obtained by performing NMF on AR image database are presented in Fig. 4a. Unlike simulation results reported in [17], the images still maintain a holistic aspect, particularly in the case of poorly aligned images, as was previously noted by several authors.

$$\begin{aligned}
&H_{aj} \leftarrow H_{aj} \sum_i \left[\mathbf{B}^T\right]_{ai} \frac{X_{ij}}{[\mathbf{BH}]_{ij}}, \quad a=1\ldots m,\ j=1\ldots N \\
&B_{ia} \leftarrow B_{ia} \sum_j \frac{X_{ij}}{[\mathbf{BH}]_{ij}} \left[\mathbf{H}^T\right]_{ja}, \quad i=1\ldots p,\ a=1\ldots m\ . \\
&B_{ia} \leftarrow \frac{B_{ia}}{\sum_j B_{ja}}
\end{aligned} \tag{2}$$

2.2 Local Non-negative Matrix Factorization (LNMF)

In order to improve the localization of the basis provided by NMF, a local version of the algorithm was introduced in [19]. It imposes the following additional constraints: a) maximum sparsity of coefficients matrix **H**; b) maximum expressiveness of basis vectors **B**; c) maximum orthogonality of **B**. The following equations describe the updating procedure for **B** and **H**:

$$\begin{aligned}
&H_{aj} \leftarrow \sqrt{H_{aj} \sum_i \left[\mathbf{B}^T\right]_{ai} \frac{X_{ij}}{[\mathbf{BH}]_{ij}}}, \quad a=1\ldots m,\ j=1\ldots N \\
&B_{ia} \leftarrow B_{ia} \sum_j \frac{X_{ij}}{[\mathbf{BH}]_{ij}} \left[\mathbf{H}^T\right]_{ja}, \quad i=1\ldots p,\ a=1\ldots m\ . \\
&B_{ia} \leftarrow \frac{B_{ia}}{\sum_j B_{ja}}
\end{aligned} \tag{3}$$

Convergence speed is significantly lower compared to NMF, but the spatial localization of the basis vectors is clearly improved even for poorly aligned images, as suggested in Fig. 4b.

2.3 Independent Components Analysis (ICA)

Natural images are highly redundant. A number of authors argued that such redundancy provides knowledge [20], and that the role of the sensory system is to develop factorial representations in which the dependencies between pixels are separated into statistically independent components. While in PCA and LDA the basis vectors depend only on pairwise relationships among pixels, some authors consider that higher-order statistics are necessary for face recognition, and ICA is an example of a method sensible to such statistics. Basically, given a set of linear mixtures of several statistically independent components, ICA aims at estimating the mixing matrix based on the assumption of statistical independence of the components. There are 2 distinct possibilities to apply ICA to the problem of face recognition [21]:

a) We may organize the database into a large matrix and every image is a different column. In this case images are random variables and pixels are outcomes (independent trials). We are interested in independence of images or functions of images. Two i and j images are independent if when moving across pixels, it is not possible to predict the value taken by the pixel on image i based on the value taken by the same pixel on image j. This approach yields a set of spatially independent basis images, roughly associated with the components of faces such as eyes, nose, and mouth.
b) We may organize the images as the rows of the data matrix. In this case pixels are random variables and images represent outcomes. Pixel i and j are independent if when moving across the entire set of images it is not possible to predict the value taken by pixel i based on the corresponding value taken by pixel j on the same image. This approach produces a factorial face code, namely a representation in which the coefficients used to represent faces are statistically independent, which is one of the paradigms supposed to be used in biological nervous systems to encode complex objects that are characterized by high-order combinations of features [20].

These two distinct approaches are called Architecture I and II in [21]. In order to obtain spatially localized basis vectors, Architecture I should be used. The specific computational procedure includes two steps:

- Perform PCA to project original data into a lower dimensional subspace: this step both eliminates less significant information and simplifies further processing, since resulting data is decorrelated (and only higher-order dependencies are to be separated by ICA). Let $\mathbf{V}_{\mathrm{PCA}} \in \Re^{pxm}$ be the matrix whose columns represent the first m eigenvectors of the set of N training images, and $\mathbf{C} \in \Re^{mxN}$ the corresponding PCA coefficients matrix, we may write $\mathbf{X} = \mathbf{V}_{\mathrm{PCA}} * \mathbf{C}$.
- ICA is actually performed on matrix $\mathbf{V}_{\mathrm{PCA}}{}^{T}$, and the independent basis images are computed as $\mathbf{B} = \mathbf{W} * \mathbf{V}_{\mathrm{PCA}}{}^{T}$, where the *separating matrix* **W** is obtained with the InfoMax method [22] (since directly maximizing the independence condition is difficult, the general approach of most ICA methods aims at optimizing an appropriate objective function whose extrema occur when the unmixed components are independent; several distinct types of objective functions are commonly used, *e.g.* InfoMax algorithm maximizes the entropy of the components). The set of projected coordinates on ICA is computed as $\mathbf{H}^{\mathrm{T}} = \mathbf{C} * \mathbf{W}^{-1}$.

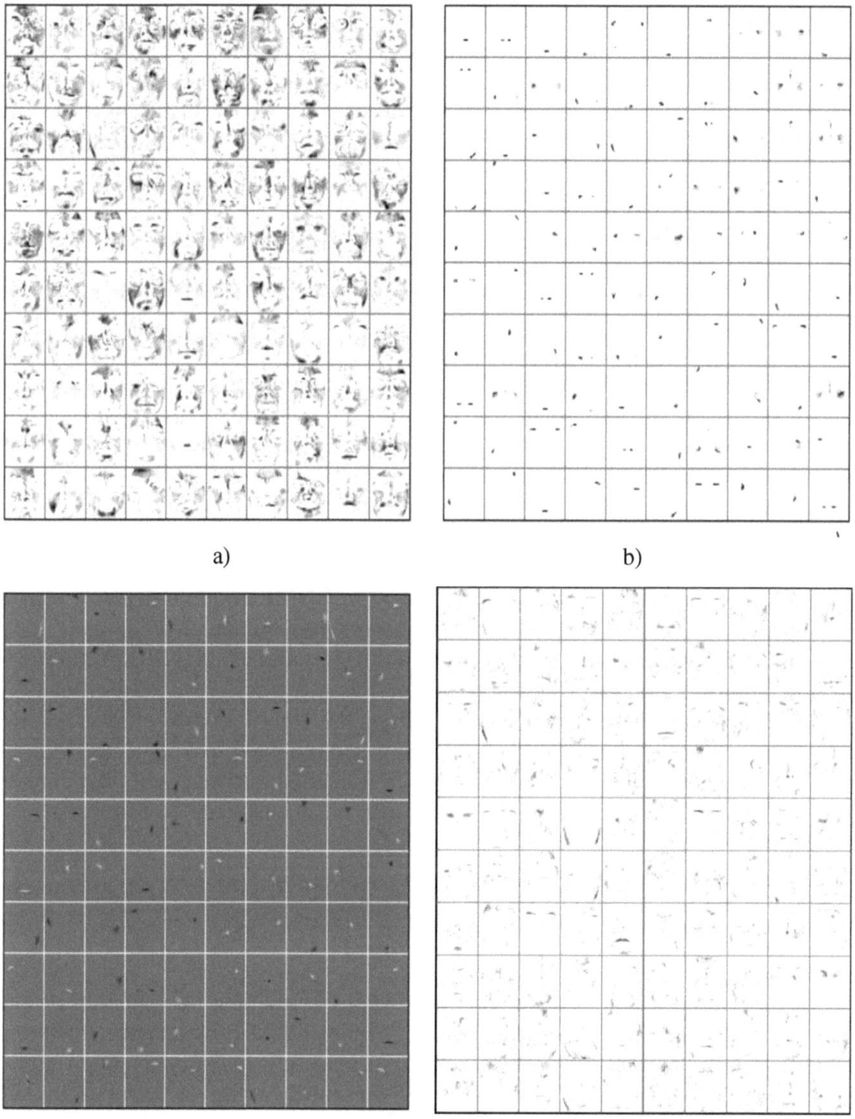

a) b) c) d)

Fig. 4. Examples of basis vectors for AR database: a) NMF; b) LNMF; c) ICA; d) NMFsc

Due to somehow contradictory comparative results between ICA and PCA presented in the literature, a systematic analysis has been reported in [23] in terms of algorithms and architectures used to implement ICA, the number of subspace dimensions, distance metric, and recognition task (facial identity vs. expression). Results indicate that specific ICA design strategies are significantly superior to standard PCA, although the task to be performed remains the most important factor. Nevertheless, the authors show that Architecture II yields better recognition results than (local features based) Architecture I. We have tested Architecture I with two different ICA algorithms, namely FastICA [24], and InfoMax (MATLAB code available at *www.cis.hut.fi/projects/ica/fastica* and *http://inc.ucsd.edu/~marni/code.html*, respectively). Examples of basis vectors obtained by ICA-InfoMax are presented in Fig. 4c.

2.4 NMF with Sparseness Constraints (NMFsc)

A random variable is called *sparse* if its probability density is highly peaked at zero and has heavy tails. Within the general setting expressed by equation (1), sparsity is an attribute of the activation vectors grouped in the lines of coefficients matrix **H**, the set of basis images arranged in the columns of **B**, or both. While standard NMF does yield a sparse representation of the data, there is no effective way to control the degree of sparseness. Augmenting standard NMF with the sparsity concept proved useful for dealing with overcomplete representations (that is, cases where the dimensionality of the space spanned by decomposition is larger than the effective dimensionality of the input space). While not present in standard NMF definition, sparsity is taken into account in LNMF and non-negative sparse coding [25]. In fact, the later enables the control over the (relative) sparsity level in **B** and **H** by defining an objective function that combines the goals of minimizing the reconstruction error and maximizing the sparseness level. Unfortunately, the optimal values of the parameters describing the algorithm are set by extensive trial-and-error experiments. This shortcoming is eliminated in a more recent contribution of the same author, which proposed a method termed NMF with sparseness constraints (NMFsc) [25]. Sparseness of a n-dimensional vector **x** is defined as follows:

$$sparseness(\mathbf{x}) = \frac{\sqrt{n} - \left(\sum |x_i|\right) / \sqrt{\sum x_i^2}}{\sqrt{n} - 1}. \tag{4}$$

The algorithm proceeds by iteratively performing a gradient descent step on the (Euclidean distance type) objective function, followed by projecting the resulting vectors onto the constraint space. Examples of basis images obtained after applying NMFsc on AR face database images are presented in Fig. 4d (MATLAB code available at *http://www.cs.helsinki.fi/patrik.hoyer/*).

Remark 1. A well-known result establishes that standard PCA may be efficiently implemented using a neural autoassociator architecture [26] (that is, a two-layer perceptron using identical input and target data and a lower dimensional hidden layer, as in Fig. 5a). On-line learning makes this approach more attractive than its algebraic counterpart, since the later requires repeatedly computing the covariance matrix and associated eigenvectors each time a new training vector becomes available. After convergence, the weights connecting the hidden and output layers are identical to the principal components of the data covariance matrix. The approach has been also considered for implementing the NMF technique [27],

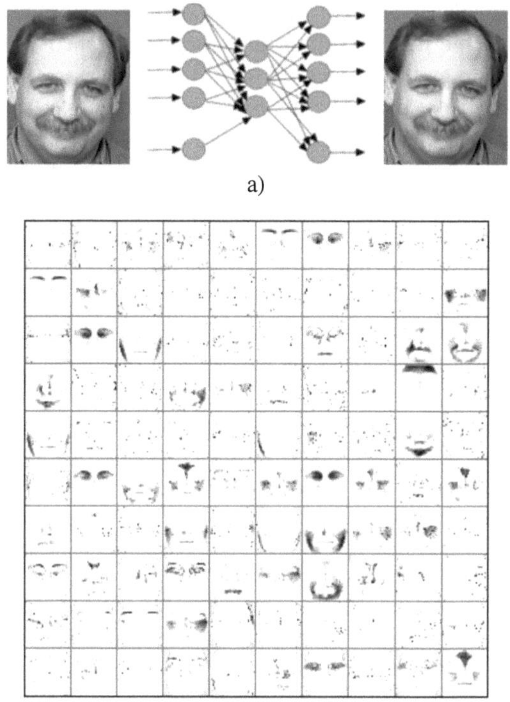

a)

b)

Fig. 5. a) Neural autoassociator; b) basis images obtained using non-negative autoassociator

imposing nonnegativity constraints on all activation and weights values. The backpropagation learning algorithm or one of its variants may be used for training the network, provided that the weights are set to zero whenever they tend to become negative. Examples of resulting basis vectors are given in Fig. 5b. Nevertheless, for high-dimensional data the approach is prone to overtraining, due to possible inappropriate relation between the training set dimension and the total number of parameters to be learned.

Several important aspects are worth discussing regarding the previous techniques:

- Sparsity/locality/parts-based representations define a set of extensively used interrelated notions whose distinct meanings should be emphasized. A random variable is called sparse if its probability density is highly peaked at zero and has heavy tails. As mentioned earlier, while not present in standard NMF definition, sparsity is taken into account in LNMF and non-negative sparse coding. Localization implies the concept of spatial neighborhood, and as such it does not appear explicitly in the formulation of any of the algorithms. Parts-based representation is intuitively related to sparseness in the basis functions, but it fundamentally depends on the statistics of the training data and the number of basis vectors (increasing the number of basis vectors leads to (better localized) smaller parts). There are still fundamental issues to be addressed such as the uniqueness of the decomposition and the significance of the resulting parts, which have been partially analyzed in [28].

- Selecting significant basis vectors is a more demanding task compared to classical linear projection approaches. While the magnitude of the ordered eigenvalues offers a simple selection criterion for standard PCA, in the case of non-negatively constrained algorithms we lack such simple selection principles. Typically, we may follow one of the following procedures: a) compute an LDA-like maximal discrimination capacity of the projection coefficients **H**, as suggested in [21] for the ICA approach; b) substitute the Euclidean distance metric with a more relevant one for parts-based representations, such as earth mover's distance (EMD) [29]. In case of non-orthogonal bases, a Riemannian metric-like distance could be adopted [30]; c) in order to reduce redundancy, the basis learned via one of the presented methods can be orthogonalized [31].

2.5 Locality Preserving Projections (LPP)

Linear subspace projection techniques such as PCA or LDA are unable to approximate accurately data lying on nonlinear submanifolds hidden in the face space. Although several nonlinear solutions to unveil the structure of such manifolds have been proposed (Isomap [32], LLE [33], Laplacian Eigenmaps [34]), these are defined only on the training set data points, and the possibility of extending them to cover new data remains largely unsolved (efforts towards tackling this issue are reported in [35]). An alternative solution is to use methods aiming at preserving the *local structure* of the manifold after subspace projection, which should be preferred when nearest neighbor classification is to be subsequently performed. One such method is Locality Preserving Projections (LPP) [36]. LPP represents a linear approximation of the nonlinear Laplacian Eigenmaps introduced in [34]. It aims at preserving the intrinsic geometry of the data by forcing neighboring points in the original data space to be mapped into closely projected data. The algorithm starts by defining a similarity matrix **S**, based on a (weighted) k nearest neighbors graph, whose entry $\mathbf{S}_{ij}$ represents the edge between training images (graph nodes) $\mathbf{x}_i$ and $\mathbf{x}_j$. Gaussian type weights of the form $\mathbf{S}_{ij} = e^{-\frac{\|\mathbf{x}_i - \mathbf{x}_j\|^2}{\sigma}}$ have been proposed in [36], although other choices (*e.g.*, cosine type) are also possible. Based on matrix **S**, a special objective function is constructed, enforcing the locality of the projected data points by penalizing those points that are mapped far apart. Basically, the approach reduces to finding a minimum eigenvalue solution to the following generalized eigenvalue problem:

$$\mathbf{XLX}^{\mathrm{T}}\mathbf{b} = \lambda\mathbf{XDX}^{\mathrm{T}}\mathbf{b}\,. \tag{5}$$

where $\mathbf{D} = \sum_i \mathbf{S}_{ij}$, and $\mathbf{L} = \mathbf{D} - \mathbf{S}$ (Laplacian matrix). The components of the subspace projection matrix **B** are the eigenvectors corresponding to the smallest eigenvalues of the problem above. Rigorous theoretical grounds are related to optimal

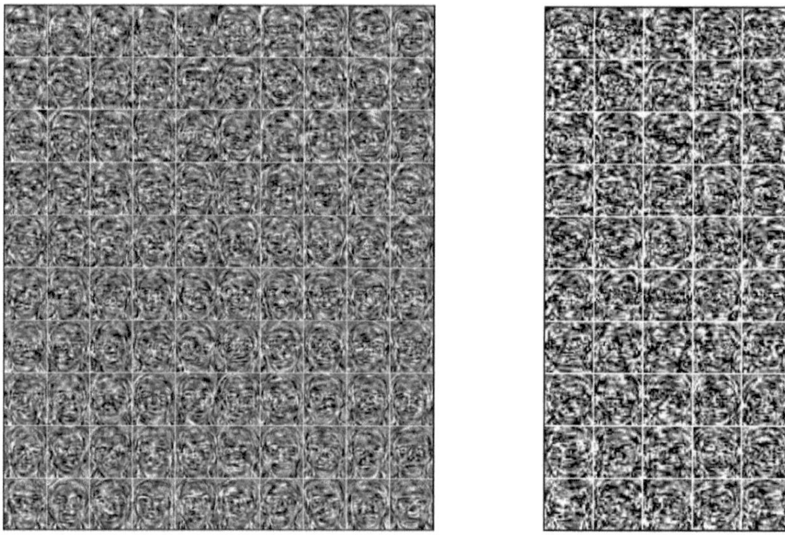

a) b)

Fig. 6. Examples of basis vectors for AR database: a) LPP; b) OPRA

linear approximations to the eigenfunctions of the Laplace-Bertrami operator on the manifold and are extensively presented in [36] (MATLAB code available at *http://people.cs.uchicago.edu/~xiaofei*). When applied to face image analysis the method yields so-called Laplacianfaces, examples of which are presented in Fig. 6a.

2.6 Orthogonal Projection Reduction by Affinity (OPRA)

Another interesting manifold learning algorithm has been recently proposed [37], which also starts by constructing a weighted graph that models the data space topology. This affinity graph is built in a manner similar to the one used in Local Linear Embedding (LLE) technique [33], and expresses each data point as a linear combination of (a limited number of) neighbors. The advantage of OPRA over LLE is that the mapping between the original data and the projected one is made explicit through a linear transformation, whereas in LLE this mapping is implicit, making it difficult to generalize to new test data. Compared to LPP, OPRA preserves not only the locality but also the geometry of local neighborhoods. Moreover, the basis vectors obtained by performing OPRA are orthogonal, whereas projection directions obtained by LPP are not. When class labels are available, as in our case, the algorithm is to be used in its supervised version, namely an edge is present between two nodes in the affinity graph only if the two corresponding data samples belong to the same class. Examples of basis images obtained using OPRA are given in Fig. 6b.

3 Experimental Setup and Results

Intensive computer simulations have been performed using three different face image databases, namely (a subset of the) AR database, Olivetti, and Yale. Specific details of the experimental setup are presented in the following, focusing on the description of the preprocessing techniques.

3.1 Description of the Face Image Databases

AR database[1] contains images of 116 individuals (63 males and 53 females). Original images are 768x576 pixels in size with 24-bit color resolution. The subjects were recorded twice at a 2- week interval, and during each session 13 conditions with varying facial expressions, illumination and occlusion were captured. In Fig. 7 we present examples from this database. As in [38], we used as training images 2 neutral poses of each person captured in different days (labeled AR011 and AR012 in Fig. 7), while the testing set consists of pairs of images for the remaining 12 conditions, AR02...AR13, respectively. More specifically, images AR02, 03, and 04 are used for testing the performances of the analyzed techniques to deal with expression variation (smile, anger, and scream), images AR05, 06, and 07 are used for illumination variability, and the rest of the images are related to occlusion (eyeglasses and scarf), with variable illumination conditions.

Olivetti database comprises 10 distinct images of 40 persons, and includes variations in pose, light conditions, scaling, and expression. Each image is represented

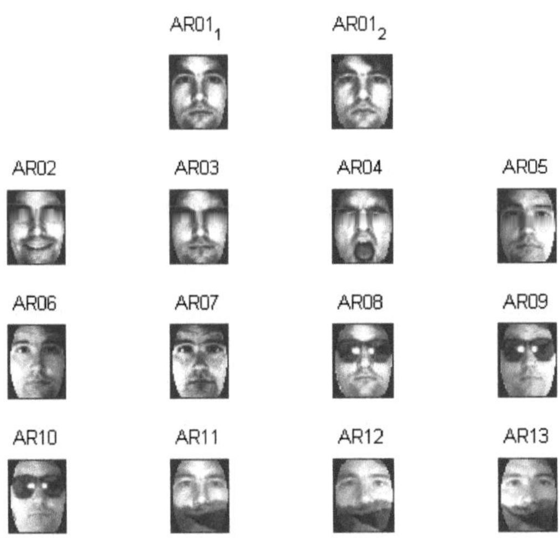

Fig. 7. Example of one individual from the AR face database. The conditions are: (1) neutral, (2) smile, (3) anger, (4) scream, (5) left light on, (6) right light on, (7) both lights on, (8) sunglasses, (9) sunglasses/left light, (10) sunglasses/right light, (11) scarf, (12) scarf/left light, (13) scarf/right light.

[1] The subset of the AR database was kindly provided by Dr. David Guillamet, as used in [38].

by 112x92 pixels, with 256 gray levels. All the images were taken against a dark homogeneous background with the subjects in an upright, frontal position, with tolerance for some tilting and rotation of up to about 20 degrees. There is some variation in scale of up to about 10%. In order to enable comparisons with previously reported results, we used 5 images per person for training, randomly selected from the available 10, and the rest for the testing phase. The training and test datasets were not overlapping, and 10 distinct trials were performed.

Yale face database contains 165 images with 11 different images for each of the 15 distinct subjects. The 11 images per subject are taken under different facial expression or configuration: center-light, with glasses, happy, left-light, without glasses, normal, right-light, sad, sleepy, surprised, and wink. Due to the small dimension of this database, most of the available results are reported for specific forms of υ-fold crossvalidation (usually, 1 to 10 images are selected for the testing set and all the rest are included in the training set to be used for computing a projection matrix). The procedure is repeatedly applied and a final recognition rate is obtained by averaging the individual scores. We have experimented by repeatedly selecting one common expression for each individual in the testing set and using the remaining 10 expressions for training.

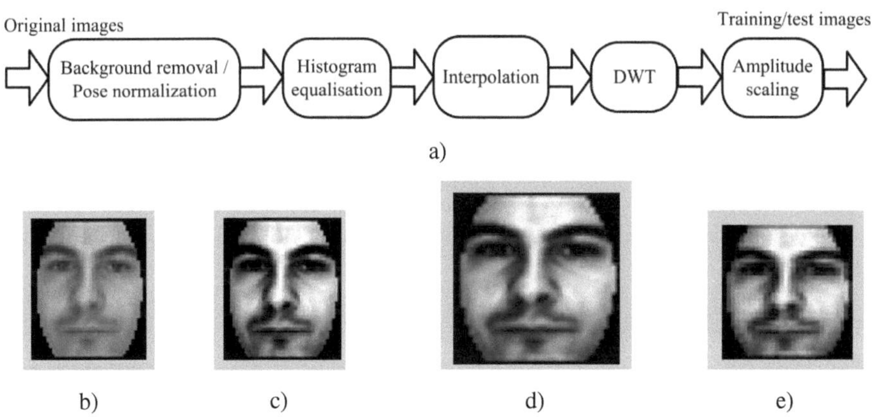

Fig. 8. Preprocessing of the original images: a) block diagram; b) background removal (48x40 pixels); c) histogram equalization (48x40 pixels); d) interpolation (64x64 pixels); e) DWT low-frequency coefficients (32x32 pixels)

3.2 Preprocessing Procedures

Prior to applying any of the techniques described in the preceding paragraph, the original images are preprocessed in order to facilitate improved recognition performances. The successive transformations are indicated in Fig. 8, and details are given in the following.

Pose normalization/Background removal

Both standard PCA and original NMF algorithm are known to be sensitive to poor alignment of the training data. Moreover, the external background may unfavorably affect the recognition performances. This is especially true for parts-based approaches, where insignificant details such as the tie or the hat may be systematically selected into the basis set. A theoretical justification for using only the pure face region has been presented in [39].

The subset of the AR database is the same as in [38], and was kindly provided by the author. First, a pose normalization has been applied in order to align all database faces, according to the (manually) localized eye positions. Next, only part of a face inside an elliptical region was selected, in order to avoid the influence of the background, as in Fig. 7. In [38], the size of each reduced image is 40x48 pixels, and when considering the elliptical region only, each image is represented using 1505 pixels. In order to cope with the "power-of-2" requirements of the Discrete Wavelet Transform (DWT) to be presented in the next section, we first interpolated the cropped images to yield 64x64 pixels resolution.

The images in the Yale database have 320x243 pixels resolution. After cropping and pose normalization, we got 128x128 images to be further processed by DWT.

In order to enable comparisons with extensive results reported in the literature, the images in the Olivetti database have not been pose normalized. Only rescaling through interpolation was performed, yielding 128x128 resolution.

Discrete Wavelet Transform (DWT)

Similar to other approaches, we perform a multiresolution decomposition of the original images based on the Discrete Wavelet Transform (DWT) and kept only the low-frequency components for further classification. Besides dimensionality reduction this procedure is also known to offer face expression invariance. The set of selected coefficients is termed *waveletface* [40]. We used Daubechies4 mother wavelet, and performed multilevel decompositions yielding 32x32 images for all databases.

3.3 Experimental Results

In this section we present simulation results for the three databases using the approaches described in Section 2. The performances are given in terms of recognition accuracy, and are compared to results previously reported in the literature for two leading computer vision techniques, namely Local Feature Analysis (implemented in a successful commercial product called FaceIt [38]), and Bayesian PCA [41]. There are several design items taken into account: a) the distance metric used: Euclidean (L2), Manhattan (L1), cos (cosine of the angle between the compared vectors, $\cos(\mathbf{x},\mathbf{y}) = \frac{\mathbf{x}\cdot\mathbf{y}}{\|\mathbf{x}\|\|\mathbf{y}\|}$); b) projection subspace dimension: the dimension of the feature space, equal to the number of basis vectors used, is set to 50, 100, 150, and 200 dimensions; c) basis images selection procedure: we have used the approach proposed in [21], computing the discriminability ratio for each coefficient in the decomposition (1), according to the formula $r = \frac{\sigma_{between}}{\sigma_{within}}$, where

$\sigma_{between} = \sum_j \left(\overline{x}_j - \overline{x}\right)^2$ is the variance of the class means, and $\sigma_{within} = \sum_j \sum_i \left(x_{ij} - \overline{x}_j\right)^2$ is the sum of the variances within each class. As stated previously, we may perform orthogonalization of the basis images obtained through one of the presented methods. Although the resulting basis violates the non-negativity constraint and is no longer spatially localized, this step reduces redundancy and may improve recognition accuracy.

Experimental results are separately reported in Tables 1-6 for distinct sources of variability, namely facial expression, changing illumination conditions, and occlusion. In order to yield overall comparative results of the algorithms used in the experiments, we focused on the AR database, due to its higher variability. We conducted a rank based analysis as follows: for each image/dimension combination, we ordered the performance rank of each algorithm/distance-measure combination (the highest recognition rate got rank 1, and so on), thus yielding a total of 12 rank numbers (3 images times 4 subspace dimensions) for each separate case. Then, we computed a sum of ranks for each algorithm over all the cases, and ordered the results (lowest sum indicates best overall performance). In Table 7 we give the order of the 10 top performers.

Facial expression recognition

The capacity of the methods to deal with expression variability was tested using the AR (images labeled AR02, 03, and 04), and Yale databases. Results are presented in Tables 1 and 2. The ICA-InfoMax approach outperforms the other competitors, and most notably it proves superior even to LDA technique, except for

Table 1. Recognition rates for AR database / expression variability (%)

	Dimension of projection subspace											
	m = 50			m = 100			m = 150			m = 200		
AR dataset:	**02**	**03**	**04**	**02**	**03**	**04**	**02**	**03**	**04**	**02**	**03**	**04**
NMF+L2	83.3	88.4	35	91	85.4	35.4	90.6	88	38.4	88	88	24.8
NMF+L1	83.7	91	33.7	91	90.6	36.7	91	92.3	34.2	91.4	93.6	27.3
NMF+cos	83.7	86.7	34.1	89.3	81.6	33.7	93.1	89.7	38.4	86.3	84.6	25.6
LNMF+L2	73.5	88.9	21.8	83.3	88.4	32.4	88.4	88.4	38	91	91.4	45.3
LNMF+L1	82.9	95.7	40.6	88.4	95.7	40.6	95.3	96.5	45.7	98.2	96.1	57.7
LNMF+cos	70.9	89.3	23	81.6	88	35.4	88	92.3	38	90.6	91.4	45.7
ICA+L2	99.1	97	59.8	99.1	96.5	64.5	99.1	98.3	63.6	94	97.8	65.3
ICA+L1	99.1	97	67	99.1	97.4	67	99.5	97.8	67	95.7	97.8	66.2
ICA+cos	98.3	96.5	67.5	99.1	96.5	69.6	98.7	97	70.5	94.4	96.1	70.9
NMFsc+L2	79	67.9	29.5	91	85.9	38.9	92.7	89.7	38.9	93.1	88.4	44.4
NMFsc+L1	88.9	86.7	41.8	95.7	91.8	44	96.1	92.7	46.5	93.6	90.6	46.5
NMFsc+cos	73.5	65.8	26.9	88	85.4	37.1	91.8	91	38.9	91.8	89.3	45.7
LPP	73.9	83.7	17	87.2	91.4	30.8	89.7	91.8	29.5	89.7	91.4	30.8
PCA	91	88	47.4	94.4	89.7	52.5	95.3	89.7	52.5	95.7	90.6	52.5
FaceIt [38]	96	93	78				not dependent on m					
Bayesian [41]	72	67	41				not dependent on m					

Table 2. Recognition rates for Yale database (%)

	Dimension of projection subspace		
	m = 50	m = 100	m = 150
NMF+L2	84.6	98	94
NMF+L1	86.6	98.6	93.3
NMF+cos	84	100	94.6
LNMF+L2	82.6	99.3	90.6
LNMF+L1	82.6	100	90.6
LNMF+cos	80	99.3	90.6
ICA+L2	98.6	99.3	98
ICA+L1	98	99.3	97.3
ICA+cos	100	100	100
PCA	88.6	90	86.6

AR04 expression (scream), which is the most difficult task to solve. Generally, greater basis dimensionality tends to be favored, except for the NA approach, where results degrade with increasing dimension due to overfitting. Orthogonalization of the basis images (results not shown here) marginally improves the performances, in combination with the L1 norm. L1 norm seems to yield best results, closely followed by cosine metric.

Changing illumination conditions

Changing illumination conditions are reflected in images AR05, 06, and 07, and results are given in Table 3. The ICA-InfoMax approach still works best, NMF is

Table 3. Recognition rates for AR database / illumination variability (%)

	Dimension of projection subspace											
	m = 50			m = 100			m = 150			m = 200		
AR dataset:	**05**	**06**	**07**	**05**	**06**	**07**	**05**	**06**	**07**	**05**	**06**	**07**
NMF+L2	67.9	7.7	58.5	71.8	23	77.3	73	30.7	76.9	79.5	35.9	88.4
NMF+L1	59.4	7.2	52.5	64.9	21.8	71.8	75.2	27.7	75.6	78.2	34.2	88
NMF+cos	68.3	8.5	56.4	72.2	22.6	75.6	78.2	31.2	72.6	81.6	44.8	91.4
LNMF+L2	40.1	20	40.1	64.1	20.5	64.5	72.6	19.2	68.8	74.3	31.6	72.2
LNMF+L1	32.9	16.2	38	48.3	17.5	62.8	60.2	16.6	65.3	67	23	67
LNMF+cos	45.7	13.6	40.1	63.6	20.5	65.8	67.9	18.8	70.9	76.5	29.9	71.3
ICA+L2	97.4	95.3	97.4	98.7	97.8	99.1	98.7	98.3	100	98.7	97.8	99.5
ICA+L1	97	97	97.4	98.3	97.4	98.3	98.7	98.3	99.5	98.3	98.3	100
ICA+cos	98.7	97.4	94	99.5	99.1	97.8	99.5	99.5	99.1	99.5	99.5	99.1
NMFsc+L2	44	9.8	11.1	56	22.6	15.3	76	22.2	23.5	71.3	34.2	27.3
NMFsc+L1	43.1	9.4	5.1	53	26.5	10.6	73	17.9	19.6	76	32	20.9
NMFsc+cos	55.1	11.9	22.6	61.9	27.7	24.3	77.3	25.6	34.6	73.9	36.7	37.6
LPP	79.5	72.6	56.8	91.5	93.2	87.2	94.4	93.1	91.4	95.3	92.7	89.7
PCA	72.6	16.2	59.8	75.2	19.6	66.6	76.9	20.5	67	77.7	20.5	68.8
FaceIt [38]	95	93	86	not dependent on m								
Bayesian [41]	77	74	72	not dependent on m								

Table 4. Recognition rates for AR database / occlusion (sunglasses) (%)

	Dimension of projection subspace											
	m = 50			m = 100			m = 150			m = 200		
AR dataset:	**08**	**09**	**10**	**08**	**09**	**10**	**08**	**09**	**10**	**08**	**09**	**10**
NMF+L2	7.7	7.2	5.5	21.8	9.4	9.4	23	9.4	7.7	26.5	10.2	6.8
NMF+L1	8.1	4.7	5.1	25.6	9.4	7.7	19.6	8.9	7.2	28.2	13.6	6.4
NMF+cos	9.8	7.7	2.1	19.6	11.5	7.2	26	11.1	7.7	19.2	8.9	7.2
LNMF+L2	11.1	3.8	3	13.6	6.4	3.8	13.2	8.9	7.2	19.6	10.2	11.1
LNMF+L1	17	5.5	3.8	18.3	13.2	2.1	17	11.9	6.8	24.3	11.9	8.9
LNMF+cos	9.8	5.1	2.5	14.1	6.4	2.5	13.2	9.8	7.2	18.3	11.5	9.8
ICA+L2	44.8	24.3	19.6	48.7	25.2	23.5	48.3	26.9	26.5	47.8	23.5	26.5
ICA+L1	42.7	23.9	17.5	51.7	26	20.5	50.8	26.9	26.9	51.7	27.3	29.5
ICA+cos	54.2	46.1	41	63.2	50.4	46.1	60.6	50	49.1	61.9	48.3	50.4
NMFsc+L2	10.2	5.5	6.4	9.8	8.1	6.8	11.5	7.7	8.1	17	11.5	8.5
NMFsc+L1	18.3	9.4	5.5	14.5	9.4	6.4	15.3	8.1	7.7	23.9	11.1	9.4
NMFsc+cos	9.8	4.2	5.1	9.4	7.2	6	9.8	7.2	6.4	18.3	11.1	7.7
LPP	8.9	4.7	4.7	15	6.8	8.9	17.9	11.5	8.9	18.3	12.4	9.4
PCA	13.6	11.1	7.2	15.8	11.9	7.7	16.6	12.4	8.1	17	12.4	8.1
FaceIt [38]	10	8	6	not dependent on m								
Bayesian [41]	34	35	28	not dependent on m								

Table 5. Recognition rates for AR database / occlusion (scarf) (%)

	Dimension of projection subspace											
	m = 50			m = 100			m = 150			m = 200		
AR dataset:	**11**	**12**	**13**	**11**	**12**	**13**	**11**	**12**	**13**	**11**	**12**	**13**
NMF+L2	10.2	8.1	3	4.2	6.4	3.4	8.5	8.5	5.9	10.6	7.7	4.2
NMF+L1	7.2	8.9	2.1	5.9	8.1	2.5	6.8	8.1	3.8	9.8	7.7	3
NMF+cos	10.2	9.4	3.4	5.5	5.9	4.2	8.9	9.4	5.9	11.1	8.5	6.8
LNMF+L2	2.1	2.5	3.8	2.5	3	3	3.4	3.8	3	3.4	4.7	3.4
LNMF+L1	1.7	1.7	2.5	2.5	3.4	3	3	3.4	3.8	3.8	6.4	3.4
LNMF+cos	2.5	2.1	2.5	3	3.8	3	3	3	2.5	3.4	3.8	3.4
ICA+L2	32.5	18.3	17.5	39.3	23.5	22.6	40.6	25.2	26.9	44	27.3	28.6
ICA+L1	37.1	22.2	17.9	47.4	26.5	28.2	51.7	34.6	33.7	57.2	35	31.6
ICA+cos	52.5	43.1	41.4	61.5	51.2	47.8	65.8	52.1	51.2	66.6	49.5	54.2
NMFsc+L2	7.2	9.8	5.1	11.1	11.9	3.4	15.3	12.8	5.5	17.9	14.5	6
NMFsc+L1	7.2	7.2	1.7	10.6	11.9	2.5	12.4	10.6	5.5	15.8	11.9	5.5
NMFsc+cos	6.8	8.5	4.7	8.5	11.1	2.5	9.4	14.1	6	12.4	13.2	5.1
LPP	30	12.4	8.9	38	21.8	18.8	41.4	25.2	23.5	41.8	8.1	21.3
PCA	4.2	4.7	2.1	4.2	4.2	1.7	3.8	4.2	1.7	3.8	4.2	2.5
FaceIt [38]	81	73	71	not dependent on m								
Bayesian [41]	46	43	40	not dependent on m								

superior to LNMF, and both are marginally better than standard PCA. The angle metric yield best results, closely followed by L1. Recognition rates improve with increasing basis dimension.

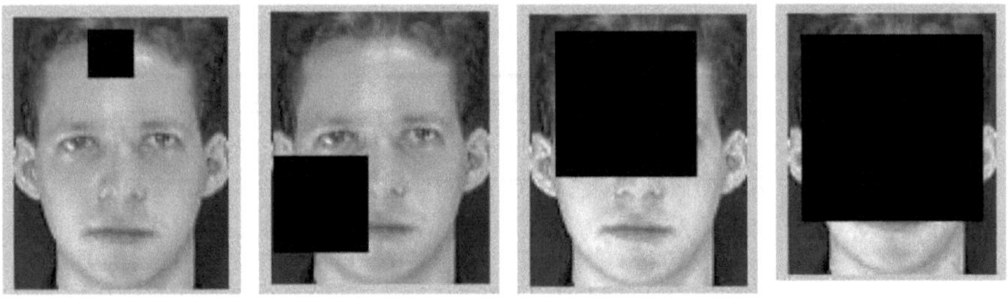

a)

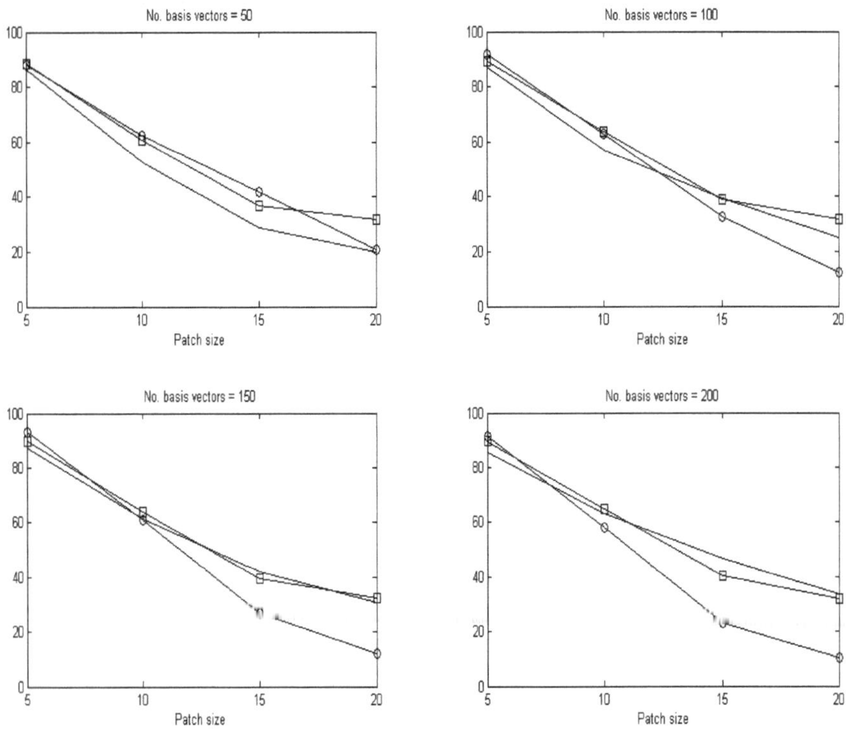

b)

Fig. 9. a) Occlusion simulated by rectangular patches with varying dimension; b) recognition rates for Olivetti database: NMF (solid line), LNMF (circle), ICA (square)

Occlusion

Occlusion is one of the situations that hopefully should be better tackled by parts-based techniques compared to holistic ones such as PCA. AR database provides 2 kinds of partially occluded images, using sunglasses (image AR08) and scarf (image AR11). Additional variable illumination conditions are provided as images AR09, 10, 12, and 13, respectively. Results presented in Tables 4 and 5 show significant

Table 6. Recognition rates for Olivetti database (%)

	Dimension of projection subspace			
	m = 50	m = 100	m = 150	m = 200
NMF+L2	90.8	91.2	89.6	87.3
NMF+L1	90.9	91.6	90.7	88.2
NMF+cos	91.6	92.2	91	89.9
LNMF+L2	90.4	93.4	93.2	92.8
LNMF+L1	92.3	95.1	94.4	94.3
LNMF+cos	89.1	92.9	91.7	91.1
ICA+L2	92	92.7	92.4	93
ICA+L1	92.3	93.3	92.8	93.7
ICA+cos	93.4	94.3	93.2	93.7
NMFsc+L2	89	91	89.9	90
NMFsc+L1	92	90.5	91.6	90.5
NMFsc+cos	91	92	90.8	92
LPP	91.1	90.7	89.9	90.7
OPRA	94.2	94.9	95	92.8
PCA	92.9	93.2	93.6	94.1

general decrease of the recognition performances, especially when the illumination conditions are changing. ICA-InfoMax proves again the best choice, although the Bayesian approach outperforms all other approaches. NMF and LNMF do indeed generally yield better results than PCA for specific subspace dimensions and metrics used. For the scarf images, LFA seems better than ICA-InfoMax, followed by the Bayesian approach, outperforming all the other methods by a long margin. Results for scarf images are better than for eyeglasses, indicating that eyes are important for recognition.

Occlusion was also simulated for Olivetti database, by using a rectangular patch of size sxs with $s \in \{5, 10, 15, 20\}$ placed at random locations, as in Fig. 9a. We used a training set of 5 images per person, randomly selected from the available 10, and the rest for the testing phase. Average recognition rates for 10 trials are given in Fig. 9b.

Pose variation

In Table 6 we give simulation results for images from Olivetti database, which present significant pose variation, while illumination conditions are better controlled. LNMF method yields best results when coupled with the cosine metric, while all other solutions have comparable performances, and show limited dependence on the subspace dimension.

4 Discussion and Conclusions

As mentioned in the previous paragraph, we conducted a rank based analysis on the results obtained for the AR database (due to its higher variability), and the ordered performance of the top 10 algorithm/distance measure combinations is given in Table 7. Some of the conclusions revealed by the results are as follows:

- Independent Components Analysis implemented by the InfoMax algorithm seems best suited for the recognition task, outperforming clearly the other solutions. It compares favorably with Local Feature Analysis and Bayesian techniques, two of the leading algorithms on the market. While explaining the exact reason for this remarkable performance needs further study, we may note that searching for most *informative features* (instead for most expressive ones, as in PCA, or most discriminant, as in LDA) has been previously proposed by some authors [42].
- Based on overall results, PCA is superior to most local-based representations, confirming the conclusions from [43]. If ignoring the scarf experiments results, PCA does better even than Bayesian and LFA approaches.
- Cosine and L1 metrics are almost always superior to L2. This agrees with extensive results reported in the literature for other face recognition approaches.
- The dependence of the recognition rates on the projection subspace dimension is not always clear, although larger dimensions tend to be generally favored. This could be partially a consequence of the limited significance of the basis selection procedure described in [21], especially when the number of available training images per class is small.

One interesting aspect is related to the degree of sparsity exhibited by each of the analyzed techniques. While this could be formally described by a formula as in [25], we prefer a more intuitive view by representing the histograms of the intensity values of the basis vectors. Diagrams in Fig. 10 indicate that all methods yield sparse basis (histograms are highly peaked at zero and have heavy tails), while LNMF shows steepest decrease in the amplitude values.

There are a number of important aspects still to be tackled if parts-based approaches are to become important tools in face (or general object) recognition applications. Reliable selection of significant basis vectors is still an open problem, if the number of training images per class is small. Basis vectors exhibiting

Table 7. Rank based analysis results

Algorithm/Distance	Expression rank	Illumination	Glasses	Scarf	Sum of ranks
ICA+cos	14	6	3	6	29
ICA+L1	10	8	9	3	30
ICA+L2	12	7	6	9	34
LPP	27	9	23	15	74
LNMF+L1	5	39	18	24	86
NMFsc+L1	13	31	22	31	95
PCA	15	25	24	39	103
NMFsc+cos	20	26	28	31	105
NMFsc+L2	22	29	30	27	108
LNMF+L2	17	40	37	32	126

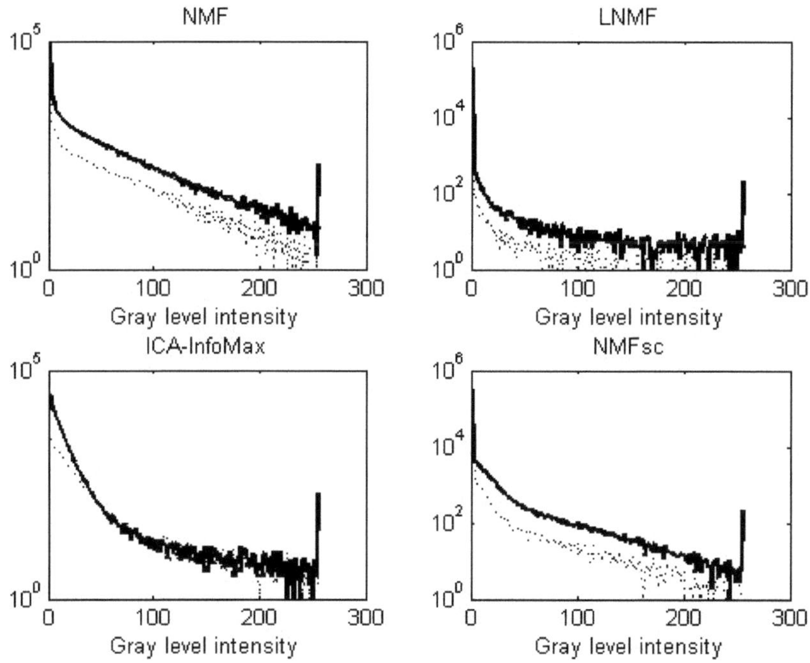

Fig. 10. Histogram of the basis vectors intensity values for different subspace dimensions (AR03 images): solid line: m = 200; dotted line: m = 50

invariance to common transformations such as translations and in-plane rotations would be desirable, as was partially addressed in [44]. Finally, identification of the conditions under which correct decompositions of faces into significant/generic parts emerge is a key problem to be further addressed.

References

1. IEEE Spectrum 41, p. 13 (2004)
2. Brunelli, R., Poggio, T.: Face Recognition: Features versus Templates. IEEE Trans. Pattern Anal. Machine Intell. 15, 1042–1052 (1993)
3. Turk, M., Pentland, A.P.: Eigenfaces for recognition. J. of Cognitive Neuroscience 3, 71–86 (1991)
4. Wiskott, L., Fellous, J.-M., Kruger, N., von der Malsburg, C.: Face Recognition by Elastic Bunch Graph Matching. IEEE Trans. Pattern Anal. Machine Intell. 17, 775–779 (1997)
5. Edwards, G.J., Taylor, C.J., Cootes, T.: Face recognition using the active appearance model. In: Burkhardt, H., Neumann, B. (eds.) ECCV 1998. LNCS, vol. 1407, pp. 581–595. Springer, Heidelberg (1998)
6. Penev, P., Atick, J.: Local feature analysis: A general statistical theory for object representation. Network: Computation in Neural Systems 7, 477–500 (1996)

7. Kong, S.G., Heo, J., Abidi, B.R., Paik, J., Abidi, M.A.: Recent advances in visual and infrared face recognition—a review. Computer Vision Image Understansding 97, 103–135 (2005)
8. Zhao, W., Chellappa, R., Rosenfeld, A., Phillips, P.J.: Face Recognition: A Literature Survey. ACM Computing Surveys, 399–458 (2003)
9. Lu, J., Plataniotis, K.N., Venetsanopoulos, A.N.: Face recognition using kernel direct discriminant analysis algorithms. IEEE Trans. Neural Networks 14, 117–126 (2003)
10. Yang, M.-H.: Kernel Eigenfaces vs. Kernel Fisherfaces: Face Recognition Using Kernel Methods. In: IEEE Int. Conf. Automatic Face and Gesture Recognition, pp. 215–220. IEEE Press, Washington (2002)
11. Yang, J., Frangi, A.F., Yang, J.-Y., Zhang, D., Jin, Z.: KPCA plus LDA: a complete kernel Fisher discriminant framework for feature extraction and recognition. IEEE Trans. Pattern Anal. Machine Intell. 27, 230–244 (2005)
12. Yang, J., Gao, X., Zhang, D., Yang, J.: Kernel ICA: An alternative formulation and its application to face recognition. Pattern Recognition 38, 1784–1787 (2005)
13. Heisele, B., Ho, P., Wu, J., Poggio, T.: Face recognition: component-based versus global approaches. Computer Vision and Image Understanding 91, 6–21 (2003)
14. Lucey, S., Chen, T.: A GMM parts based face representation for improved verification through relevance adaptation. In: Int. Conf. Computer Vision Pattern Recognition, pp. 855–861. IEEE Computer Society, Washington (2004)
15. He, X., Yan, S., Hu, Y., Niyogi, P., Zhang, H.J.: Face recognition using Laplacianfaces. IEEE Trans. Pattern Anal. Machine Intell. 27, 328–340 (2005)
16. Zhang, J., Li, S.Z., Wang, J.: Manifold Learning and Applications in Recognition. In: Tan, Y.P., Yap, K.H., Wang, L. (eds.) Intelligent Multimedia Processing with Soft Computing. Springer, Heidelberg (2004)
17. Lee, D.D., Seung, H.S.: Learning the parts of objects by non-negative matrix factorization. Nature 401, 788–791 (1999)
18. Wild, S.: Seeding Non-Negative Matrix Factorizations with the Spherical K-Means Clustering. Master Thesis, University of Colorado (2002)
19. Li, S.Z., Hou, X.W., Zhang, H.J.: Learning spatially localized, parts-based representation. In: IEEE Int. Conf. CVPR, pp. 1–6. IEEE Press, Washington (2001)
20. Barlow, H.B.: Unsupervised Learning. Neural Computation 1, 295–311 (1989)
21. Bartlett, M.S., Movellan, J.R., Sejnowski, T.J.: Face Recognition by Independent Component Analysis. IEEE Trans. Neural Networks 13, 1450–1464 (2002)
22. Bell, A.J., Sejnowski, T.J.: An information-maximization approach to blind separation and blind deconvolution. Neural Computation 7, 1129–1159 (1995)
23. Draper, B.A., Baek, K., Bartlett, M.S., Beveridge, J.R.: Recognizing faces with PCA and ICA. Computer Vision and Image Understanding 91, 115–137 (2003)
24. Hyvarinen, A.: Fast and robust fixed-point algorithms for independent component analysis. IEEE Trans. Neural Networks 10, 626–634 (1999)
25. Hoyer, P.O.: Non-negative matrix factorization with sparseness constraints. Journal of Machine Learning Research 5, 1457–1469 (2004)
26. Kramer, M.A.: Nonlinear principal components analysis using autoassociative neural networks. AIChE Journal 32, 233–243 (1991)
27. Ge, X., Iwata, S.: Learning the parts of objects by auto-association. Neural Networks 15, 285–295 (2002)
28. Donoho, D., Stodden, V.: When does non-negative matrix factorization give a correct decomposition into parts? In: NIPS, vol. 16. MIT Press, Cambridge (2004)

29. Guillamet, D., Vitria, J.: Evaluation of distance metrics for recognition based on non-negative matrix factorization. Pattern Recognition Lett. 4, 1599–1605 (2003)
30. Tipping, M.: Deriving cluster analytic distance functions, from Gaussian mixture models. In: 9th Int. Conf. on ANN, pp. 815–820 (1999)
31. Liu, W., Zheng, N.: Non-negative matrix factorization based methods for object recognition. Pattern Recognition Lett. 25, 893–897 (2004)
32. Tenenbaum, J.B., de Silva, V., Langford, J.C.: A global geometric framework for nonlinear dimensionality reduction. Science 290, 2319–2323 (2000)
33. Roweis, S., Saul, L.: Nonlinear dimensionality reduction by locally linear embedding. Science 290, 2323–2326 (2000)
34. Belkin, M., Niyogi, P.: Laplacian eigenmaps for dimensionality reduction and data representation. Neural Computation 15, 1373–1396 (2003)
35. Bengio, Y., Paiement, J.F., Vincent, P., Delalleau, O., Le Roux, N., Ouimet, M.: Out-of-Sample Extensions for LLE, Isomap, MDS, Eigenmaps, and Spectral Clustering. Neural Computation 16, 2197–2219 (2004)
36. He, X., Yan, S., Hu, Y., Niyogi, P.: Face recognition using Laplacianfaces. IEEE Trans. Pattern Anal. Machine Intell. 27, 328–340 (2005)
37. Kokiopoulou, E., Saad, Y.: Orthogonal neighborhood preserving projections: a projection-based dimensionality reduction technique. IEEE Trans. Pattern Anal. Machine Intell. 29, 2143–2156 (2007)
38. Guillamet, D., Vitrià, J.: Classifying Faces with Non-negative Matrix Factorization. In: 5th Catalan Conf. for Artificial Intell, pp. 24–31. IEEE Press, Washington (2002)
39. Chen, L.-F., Mark Liao, H.-Y., Lin, J.-C., Han, C.-C.: Why recognition in a statistics-based face recognition system should be based on the pure face portion: a probabilistic decision-based proof. Pattern Recognition 34, 1393–1403 (2001)
40. Chien, J.-T., Wu, C.-C.: Discriminant waveletfaces and nearest feature classifiers for face recognition. IEEE Trans. Pattern Anal. Machine Intell. 24, 1644–1649 (2002)
41. Moghaddam, B., Pentland, A.P.: Probabilistic visual learning for object representation. IEEE Trans. Pattern Anal. Machine Intell. 19, 696–710 (1997)
42. Rudra, A.: Informative Features in Vision and Learning. Ph.D. thesis, New York University (2002)
43. Phillips, P.J., Bowyer, K.W.: Empirical Evaluation Techniques in Computer Vision. Wiley-IEEE Press, Chichester (1998)
44. Eggert, J., Wersing, H., Koerner, E.: Transformation-invariant representation and NMF. In: IJCNN 2004, pp. 2535–2540. IEEE Press, Washington (2004)

Invariance Properties of Recurrent Neural Networks

Mihaela-Hanako Matcovschi and Octavian Pastravanu

Technical University "Gh. Asachi" of Iasi,
Department of Automatic Control and Applied Informatics
Bd. Mangeron no. 53A, Iasi, 700050, Romania
{mhanako,opastrav}@ac.tuiasi.ro

Abstract. The paper explores the flow (positively) invariant sets with respect to the trajectories of recurrent neural networks (RNNs). There are considered two types of sets, namely with arbitrary time-dependence and exponentially decreasing. The sets can have general shapes, defined by Hölder *p*-norms. The first part of the paper develops criteria for testing the existence of invariant sets. The second part analyzes the connections between the invariant sets and the stability of the RNN equilibrium points. Besides the novelty and the theoretical interest of the whole approach, the results corresponding to the usual p-norms ($p=1, 2, \infty$) yield numerically tractable procedures for testing the invariance properties.

1 Introduction

First, we present the notations used by this work.

For a vector $\boldsymbol{x} \in \mathbf{R}^n$, $\| \boldsymbol{x} \|_p$ is the Hölder vector *p*-norm defined for $1 \le p \le \infty$.

For a square matrix $\boldsymbol{M} = (m_{ij}) \in \mathbf{R}^{n \times n}$:

- $\| \boldsymbol{M} \|_p$ is the matrix norm induced by the vector norm $\| * \|_p$;
- $\mu_p(\boldsymbol{M}) = \lim_{\tau \downarrow 0} (\| \boldsymbol{I} + \tau \boldsymbol{M} \|_p - 1)/\tau$ is a matrix measure [1] associated with the matrix norm $\| * \|_p$;
- The matrix $\overline{\boldsymbol{M}} = (\overline{m}_{ij}) \in \mathbf{R}^{n \times n}$ is defined by $\overline{m}_{ii} = m_{ii}$, $i = 1, \ldots, n$; $\overline{m}_{ij} = | m_{ij} |$, $i \ne j$, $i, j = 1, \ldots, n$.
- If $\boldsymbol{M}$ is symmetric, $\boldsymbol{M} \prec= 0$ means $\boldsymbol{M}$ negative-semidefinite.

For both vectors and matrices:

- Θ^T denotes transposition;
- The inequalities $\Theta_1 \le \Theta_2$, $\Theta_1 < \Theta_2$ operate elementwise.

H.-N. Teodorescu, J. Watada, and L.C. Jain (Eds.): Intel. Sys. and Tech., SCI 217, pp. 105–119.
springerlink.com

Throughout the paper we use the following abbreviations whose meaning will be explained when the related notions are properly introduced:

- RNN – recurrent neural network;
- FI – flow invariant;
- w.r.t. – with respect to.

Next, we present the scenery for the problems studied in this paper.

Consider the continuous-time *recurrent neural networks* (RNNs) without delay described by

$$\dot{\boldsymbol{x}}(t) = \boldsymbol{B}\boldsymbol{x}(t) + \boldsymbol{W}\boldsymbol{f}\left(\boldsymbol{x}(t)\right) + \boldsymbol{u},\ t \geq 0,\ \ \boldsymbol{x}(t_0) = \boldsymbol{x}_0 \in \mathbf{R}^n, \tag{1}$$

where $\boldsymbol{x} = [x_1\ x_2\ \dots\ x_n]^T \in \mathbf{R}^n$ and $\boldsymbol{u} = [u_1\ u_2\ \dots\ u_n]^T \in \mathbf{R}^n$ are the state and input vectors, and $\boldsymbol{W} = \left[w_{ij}\right] \in \mathbf{R}^{n \times n}$, $\boldsymbol{B} = \text{diag}\{b_1, b_2, \dots, b_n\} \in \mathbf{R}^{n \times n}$, $b_i < 0$, $i = 1, \dots, n$. Each component f_i of the vector function $\boldsymbol{f}(\boldsymbol{x}) = [f_1(x_1)\ f_2(x_2)\ \dots\ f_n(x_n)]^T$, $\boldsymbol{f} : \mathbf{R}^n \to \mathbf{R}^n$, is differentiable and fulfils the *slope condition*:

$$\forall i \in \{1, \dots, n\},\ \ \exists L_i > 0,\ \ 0 \leq \frac{df_i(s)}{ds} \leq L_i,\ \ \forall s \in \mathbf{R}. \tag{2}$$

The considered neural networks are also known as *Hopfield neural networks*. The terminology "recurrent" suggests the closed-loop architecture.

The behavior of the RNN's state-space trajectories in the vicinity of the equilibrium points has been studied by many papers [1]-[5], placing emphasis on the stability properties under various hypotheses on the activation function $\boldsymbol{f}$. Within this context, the recent works [6]-[8] defined and explored a stronger type of stability (called componentwise) that, besides the fulfillment of the classical stability requirements, constrains the trajectories by *rectangular invariant sets*.

The investigation of invariant sets has been initiated by prominent mathematicians, such as Nagumo, Hukuhara, Brezis, Crandall, Martin, Yorke, working in the qualitative theory of differential equations; their contributions are presented by the first monograph on this field [9]. Further developments are reported by outstanding monographs such as [10]-[13].

Although stability, as a topic of RNN analysis, is close to flow invariance, the studies addressing the latter are rather scarce and focus on invariant sets with rectangular shape [6]-[8]. The present paper aims to expand the researches in [7], [8] and explores the existence of *invariant sets with arbitrary shape*.

Assume that RNN (1) has a finite number of equilibriums and let $\boldsymbol{x}_e$ be one of these. We consider two types of invariant sets described by arbitrary Hölder p-norms, $1 \leq p \leq \infty$.

- Sets with arbitrary time-dependence defined by

$$S^c_{p,\boldsymbol{H}(t)} = \left\{\boldsymbol{x} \in \mathbf{R}^n \mid \| \boldsymbol{H}^{-1}(t)(\boldsymbol{x} - \boldsymbol{x}_e) \|_p \leq c\right\},\ t \geq 0,\ c > 0, \tag{3}$$

where $\boldsymbol{H}(t)$ is a diagonal matrix with positive entries representing continuously differentiable functions:

$$\boldsymbol{H}(t) = \mathrm{diag}\{h_1(t),\cdots,h_n(t)\},\ h_i(t) > 0,\ \forall t \in \mathbf{R}_+,\ i = 1,\dots,n. \tag{4}$$

- Sets with exponential time-dependence defined by

$$S^c_{p,\boldsymbol{D}\mathrm{e}^{rt}} = \left\{\boldsymbol{x} \in \mathbf{R}^n \mid \| \boldsymbol{D}^{-1}(\boldsymbol{x} - \boldsymbol{x}_e) \|_p \le c\,\mathrm{e}^{rt}\right\},\ t \ge 0,\ c > 0, \tag{5}$$

where $\boldsymbol{D}$ is a diagonal matrix with constant positive entries:

$$\boldsymbol{D} = \mathrm{diag}\{d_1,\dots,d_n\},\ d_i > 0,\ i = \overline{1,n}, \tag{6}$$

and $r < 0$ is a negative constant.

In *geometrical terms*, the sets considered above are characterized by the following elements:

- The axes of coordinates translated to $\boldsymbol{x}_e$ play the role of *symmetry axes*, regardless of the Hölder p-norm, $1 \le p \le \infty$.
- The Hölder p-norm defines the *shape* of the set at any time $t \ge 0$. For the usual values $p \in \{1, 2, \infty\}$ the shape is a hyper-diamond, a hyper-ellipsis, or a hyper-rectangle, respectively.
- For a given constant $c > 0$, the *lengths* of the n semiaxes at any time are defined by $ch_i(t) > 0$ for $S^c_{p,\boldsymbol{H}(t)}$ (3) and $cd_i\,\mathrm{e}^{rt} > 0$ for $S^c_{p,\boldsymbol{D}\mathrm{e}^{rt}}$ (5).

The paper provides sufficient criteria for the invariance of the sets of form (3) or (5) with respect to the state-space trajectories of RNN (1). To simplify our exposition, we will express the state-space trajectories of RNN (1) with respect to the equilibrium $\boldsymbol{x}_e$, by considering the deviations

$$\boldsymbol{y} = \boldsymbol{x} - \boldsymbol{x}_e. \tag{7}$$

Thus, since $\boldsymbol{x}_e$ satisfies the equation $\boldsymbol{Bx}_e + \boldsymbol{Wf}(\boldsymbol{x}_e) + \boldsymbol{u} = \boldsymbol{0}$, with $\boldsymbol{u}$ a constant vector, the state-space equation (1) can be rewritten as

$$\boldsymbol{y}'(t) = \boldsymbol{By}(t) + \boldsymbol{Wg}(\boldsymbol{y}(t)),\quad t \ge 0, \tag{8}$$

where

$$\boldsymbol{g}(\boldsymbol{y}) = \boldsymbol{f}(\boldsymbol{y} + \boldsymbol{x}_e) - \boldsymbol{f}(\boldsymbol{x}_e),\quad \forall \boldsymbol{y} \in \mathbf{R}^n. \tag{9}$$

The hypothesis (2) on the function $\boldsymbol{f}$ implies that each component $g_i(y_i)$ of the vector function $\boldsymbol{g}$ defined by (9) is differentiable and satisfies the *sector condition*:

$$0 \le \frac{g_i(s)}{s} \le L_i,\quad \forall s \in \mathbf{R},\quad s \ne 0,\ i = 1,\dots,n. \tag{10}$$

As a consequence we will analyze the invariance properties of RNN (8) for which {0} is an equilibrium, and we are going to refer to the sets (3) and (5) adequately described, i.e.

- Sets with arbitrary time-dependence defined by

$$S^c_{p,\boldsymbol{H}(t)} = \left\{ \boldsymbol{y} \in \mathbf{R}^n \mid \| \boldsymbol{H}^{-1}(t)\boldsymbol{y} \|_p \le c \right\},\ t \ge 0,\ c > 0, \tag{11}$$

with $\boldsymbol{H}(t)$ given by (4);

- Sets with exponential time-dependence defined by

$$S^c_{p,\boldsymbol{D}e^{rt}} = \left\{ \boldsymbol{y} \in \mathbf{R}^n \mid \| \boldsymbol{D}^{-1}\boldsymbol{y} \|_p \le c\,e^{rt} \right\},\ t \ge 0,\ c > 0, \tag{12}$$

where $\boldsymbol{D}$ is given by (6) and $r < 0$ is a negative constant.

Definition 1. Let $1 \le p \le \infty$ and $c > 0$. Denote by $\boldsymbol{y}(t;t_0,\boldsymbol{y}_0)$, $t \ge t_0$, the state-space trajectory of RNN (8) initiated in $\boldsymbol{y}_0 \in \mathbf{R}^n$ at t_0. The set $S^c_{p,\boldsymbol{H}(t)}\,/\,S^c_{p,\boldsymbol{D}e^{rt}}$ defined by (11)/(12) is *Flow (positively) Invariant with respect to* (FI w.r.t.) RNN (8), if any trajectory initiated inside $S^c_{p,\boldsymbol{H}(t)}\,/\,S^c_{p,\boldsymbol{D}e^{rt}}$ remains inside $S^c_{p,\boldsymbol{H}(t)}\,/\,S^c_{p,\boldsymbol{D}e^{rt}}$ at any time, i.e.

(a) for the set $S^c_{p,\boldsymbol{H}(t)}$:

$$\begin{aligned} &\forall t_0 \in \mathbf{R}_+, \forall \boldsymbol{y}_0 \in \mathbf{R}^n,\ \| \boldsymbol{H}^{-1}(t_0)\boldsymbol{y}_0 \|_p \le c \Rightarrow \\ &\qquad \Rightarrow \forall t > t_0,\ \| \boldsymbol{H}^{-1}(t)\boldsymbol{y}(t;t_0,\boldsymbol{y}_0) \|_p \le c; \end{aligned} \tag{13}$$

(b) for the set $S^c_{p,\boldsymbol{D}e^{rt}}$:

$$\begin{aligned} &\forall t_0 \in \mathbf{R}_+, \forall \boldsymbol{y}_0 \in \mathbf{R}^n,\ \| \boldsymbol{D}^{-1}\boldsymbol{y}_0 \|_p \le c\,e^{rt_0} \Rightarrow \\ &\qquad \Rightarrow \forall t > t_0,\ \| \boldsymbol{D}^{-1}\boldsymbol{y}(t;t_0,\boldsymbol{y}_0) \|_p \le c\,e^{rt}. \end{aligned} \tag{14}$$

■

The remainder of the text is organized as follows. Section 2 explores the invariance of the sets $S^c_{p,\boldsymbol{H}(t)}$ of form (11) with respect to RNN (8). Section 3 explores the invariance of the sets $S^c_{p,\boldsymbol{D}e^{rt}}$ of form (12) with respect to RNN (8). Section 4 discusses the connection between the invariance properties and the stability of RNN (8). Section 5 illustrates the theoretical concepts by a numerical example. Section 6 formulates some conclusions on the importance of our work.

2 Invariant Sets with Arbitrary Time-Dependence

This section presents sufficient conditions for the invariance of the sets $S^c_{p,\boldsymbol{H}(t)}$ of form (11) w.r.t. the trajectories of RNN (8).

Theorem 1. Let $1 \le p \le \infty$. Assume there exist a set $\Omega_p \subseteq \mathbf{R}^n$ and a positive constant $\rho > 0$ such that $\forall t \in \mathbf{R}_+$, $S^{\rho}_{p,\boldsymbol{H}(t)} \subseteq \Omega_p$. By using the matrices $\boldsymbol{B} = \mathrm{diag}\{b_1, b_2, \ldots, b_n\}$ and $\boldsymbol{W} = \left[w_{ij} \right]$, $i, j = 1, \ldots, n$, in the definition of RNN (8), consider the matrix-valued function $\boldsymbol{A} : \mathbf{R}^n \to \mathbf{R}^{n \times n}$, $\boldsymbol{A}(\boldsymbol{y}) = \left[a_{ij}(y_j) \right]$, with the entries

$$
\begin{aligned}
a_{ii}(y_i) &= b_i + w_{ii}\varphi_i(y_i),\ i = 1, \ldots, n, \\
a_{ij}(y_j) &= w_{ij}\varphi_j(y_j),\ i \ne j,\ i, j = 1, \ldots, n,
\end{aligned}
\tag{15}
$$

where $\varphi_j(s) = \dfrac{g_j(s)}{s}$ if $s \ne 0$, and $\varphi_j(0) = \lim\limits_{s \to 0} \dfrac{g_j(s)}{s} = \left. \dfrac{dg_j(s)}{ds} \right|_{s=0}$. If

$$
\forall t \ge 0,\ \forall \boldsymbol{y} \in \Omega_p,\ \mu_p\left(\boldsymbol{H}^{-1}(t)\boldsymbol{A}(\boldsymbol{y})\boldsymbol{H}(t) - \boldsymbol{H}^{-1}(t)\dot{\boldsymbol{H}}(t) \right) \le 0 , \tag{16}
$$

then $\forall c \in (0, \rho]$, the sets $S^c_{p,\boldsymbol{H}(t)}$ of form (11) are FI w.r.t. the trajectories of RNN (8).

Proof: For an arbitrary, but fixed $c \in (0, \rho]$, we consider the function defined as:

$$
W^c(\boldsymbol{y}, t) : \Omega_p \times \mathbf{R}_+ \to \mathbf{R}_+,\ W^c(\boldsymbol{y}, t) = \| (c\boldsymbol{H})^{-1}(t)\boldsymbol{y} \|_p , \tag{17}
$$

and we show that

$$
D_t^+ W^c(\boldsymbol{y}(t), t) = \lim_{\tau \downarrow 0} \frac{W^c(\boldsymbol{y}(t+\tau), t+\tau) - W^c(\boldsymbol{y}(t), t)}{\tau} \le 0 \tag{18}
$$

along each trajectory of RNN (8). Next we prove that inequality (18) is a sufficient condition for the invariance of the set $S^c_{p,\boldsymbol{H}(t)}$ (11) w.r.t. the trajectories of RNN (8).

According to this plan, by using the notation $\boldsymbol{M}(\boldsymbol{y},t) = \boldsymbol{H}^{-1}(t)\boldsymbol{A}(\boldsymbol{y})\boldsymbol{H}(t) - \boldsymbol{H}^{-1}(t)\dot{\boldsymbol{H}}(t)$, we can write $c^{-1}\boldsymbol{H}^{-1}(t+\tau)\,\boldsymbol{y}(t+\tau) = c^{-1}\boldsymbol{H}^{-1}(t)\,\boldsymbol{y}(t) + \tau \dfrac{d}{dt}\left(c^{-1}\boldsymbol{H}^{-1}(t)\boldsymbol{y}(t) \right) + \tau \boldsymbol{O}(\tau)$, where $\dfrac{d}{dt}\left(c^{-1}\boldsymbol{H}^{-1}(t)\boldsymbol{y}(t) \right) = \boldsymbol{M}(\boldsymbol{y},t)\left(c^{-1}\boldsymbol{H}^{-1}(t)\boldsymbol{y}(t) \right)$ and

$\lim_{\tau\downarrow 0}\|\boldsymbol{O}(\tau)\|_p=0$. Hence, $W^c(\boldsymbol{y}(t+\tau),(t+\tau))=\|(\boldsymbol{I}+\tau\boldsymbol{M}(\boldsymbol{y},t))\left(c^{-1}\boldsymbol{H}^{-1}(t)\boldsymbol{y}(t)\right)+\tau\boldsymbol{O}(\tau)\|_p\leq$ $\|(\boldsymbol{I}+\tau\boldsymbol{M}(\boldsymbol{y},t))\left(c^{-1}\boldsymbol{H}^{-1}(t)\boldsymbol{y}(t)\right)+\tau\boldsymbol{O}(\tau)\|_p\leq\|\boldsymbol{I}+\tau\boldsymbol{M}(\boldsymbol{y},t)\|_p W^c(\boldsymbol{y}(t),t)+\tau\|\boldsymbol{O}(\tau)\|_p$ and $D_t^+W^c(\boldsymbol{y}(t),t)\leq \lim_{\tau\downarrow 0}\dfrac{\|I+\tau\boldsymbol{M}(\boldsymbol{y},t)\|_p-1}{\tau}W(\boldsymbol{y}(t),t)+\lim_{\tau\downarrow 0}\boldsymbol{O}(\tau)=\mu_{\|\,\|_p}(\boldsymbol{M}(\boldsymbol{y},t))W^c(\boldsymbol{y}(t),t)$.

This means inequality (16) implies $D_t^+W^c(\boldsymbol{y}(t),t)\leq 0$ (i.e. inequality (18)) along each trajectory of RNN (8) initialized inside Ω_p.

Assume that the set $S^c_{p,\boldsymbol{H}(t)}$ (11) is not FI w.r.t. the trajectories of RNN (8). This means there is a trajectory $\boldsymbol{y}^*(t)$ of RNN (8), which is initialized inside $S^c_{p,\boldsymbol{H}(t)}$, but leaves $S^c_{p,\boldsymbol{H}(t)}$. In other words, we can find a time instant t^* such that $\|(c\boldsymbol{H})^{-1}(t^*)\,\boldsymbol{y}^*(t^*)\|_p=1$ and $\|(c\boldsymbol{H})^{-1}(t)\,\boldsymbol{y}^*(t)\|_p>1$, for $t>t^*$. Along this trajectory, $W^c(\boldsymbol{y}^*(t),t)$ is strictly increasing in a vicinity of t^*, fact which contradicts the result obtained above, in the first part of the proof, which showed $D_t^+W^c(\boldsymbol{y}(t),t)\leq 0$. Hence we conclude that the set $S^c_{p,\boldsymbol{H}(t)}$ (11) is FI w.r.t. the trajectories of RNN (8).

Since the constant $c\in(0,\rho]$ was taken arbitrarily, the proof is completed. ■

The inequality (16) in Theorem 1 is very permissive in handling the nonlinearities of RNN (8) since the matrix-valued function $\boldsymbol{A}(\boldsymbol{y})$ defined by (15) incorporates all the information about the activation functions $g_i(y_i)$, $i=1,\ldots,n$. However the practical manipulation of $\boldsymbol{A}(\boldsymbol{y})$ is extremely difficult, reason for which we are looking for a condition ensuring the fulfillment of inequality (16), but relying on a constant matrix instead of $\boldsymbol{A}(\boldsymbol{y})$.

Corollary 1. Let $1\leq p\leq\infty$. Consider the constant matrix $\Theta=\left[\theta_{ij}\right]$, $i,j=1,\ldots,n$, defined by

$$\begin{aligned}&\theta_{ii}=\begin{cases}b_i & \text{, if } w_{ii}\leq 0,\\ b_i+w_{ii}L_i, & \text{if } w_{ii}>0,\end{cases}\quad i=1,\ldots,n,\\ &\theta_{ij}=|w_{ij}|\,L_j,\quad i\neq j, i,j=1,\ldots,n.\end{aligned}\tag{19}$$

If

$$\forall t\geq 0,\ \mu_p\left(\boldsymbol{H}^{-1}(t)\boldsymbol{\Theta}\,\boldsymbol{H}(t)-\boldsymbol{H}^{-1}(t)\,\dot{\boldsymbol{H}}(t)\right)\leq 0,\tag{20}$$

then $\forall c>0$, the sets $S^c_{p,H(t)}$ of form (11) are FI w.r.t. the trajectories of RNN (8).

Proof: First we show that

$$\forall t \geq 0,\ \forall \boldsymbol{y} \in \mathbf{R}^n,\ \mu_p\left(\boldsymbol{H}^{-1}(t)\boldsymbol{A}(\boldsymbol{y})H(t) - \boldsymbol{H}^{-1}(t)\dot{\boldsymbol{H}}(t)\right) \leq \\ \leq \mu_p\left(\boldsymbol{H}^{-1}(t)\boldsymbol{\Theta}\boldsymbol{H}(t) - \boldsymbol{H}^{-1}(t)\dot{\boldsymbol{H}}(t)\right) \leq 0, \tag{21}$$

where $\boldsymbol{A}(\boldsymbol{y})$ is the matrix-valued function defined by (15). Then, we apply Theorem 1 for arbitrary $c > 0$.

According to this plan and relying on inequalities (10), we can write, for $i = 1,\dots,n$, $a_{ii}(y_i) = b_i + w_{ii}\varphi_i(y_i) \leq \theta_{ii}$, $\forall y_i \in \mathbf{R}$, and for $i \neq j$, $i,j = 1,\dots,n$, $a_{ij}(y_j) = w_{ij}\varphi_j(y_j) \leq |a_{ij}(y_j)| \leq \theta_{ij}$, $\forall y_j \in \mathbf{R}$ Thus, by using the "bar" notation we get the componentwise matrix inequality $\boldsymbol{A}(\boldsymbol{y}) \leq \overline{\boldsymbol{A}(\boldsymbol{y})} \leq \Theta$, which yields $\boldsymbol{H}^{-1}(t)\boldsymbol{A}(\boldsymbol{y})\boldsymbol{H}(t) - \boldsymbol{H}^{-1}(t)\dot{\boldsymbol{H}}(t) \leq \overline{\boldsymbol{H}^{-1}(t)\boldsymbol{A}(\boldsymbol{y})\boldsymbol{H}(t) - \boldsymbol{H}^{-1}(t)\dot{\boldsymbol{H}}(t)} \leq$ $\boldsymbol{H}^{-1}(t)\boldsymbol{\Theta}\boldsymbol{H}(t) -$ $\boldsymbol{H}^{-1}(t)\boldsymbol{\Theta}\boldsymbol{H}(t) - \boldsymbol{H}^{-1}(t)\dot{\boldsymbol{H}}(t)$, for $\forall \boldsymbol{y} \in \mathbf{R}^n$ and $\forall t \geq 0$. If Lemmas 2 and 4 in [14] are applied to the preceding inequality, then we obtain: $\forall t \geq 0$, $\forall \boldsymbol{y} \in \mathbf{R}^n$, $\mu_p\left(\boldsymbol{H}^{-1}(t)\boldsymbol{A}(\boldsymbol{y})\boldsymbol{H}(t) - \boldsymbol{H}^{-1}(t)\dot{\boldsymbol{H}}(t)\right) \leq \mu_p\left(\overline{\boldsymbol{H}^{-1}(t)\boldsymbol{A}(\boldsymbol{y})\boldsymbol{H}(t) - \boldsymbol{H}^{-1}(t)\dot{\boldsymbol{H}}(t)}\right) \leq$ $\mu_p\left(\boldsymbol{H}^{-1}(t)\Theta\boldsymbol{H}(t) - \boldsymbol{H}^{-1}(t)\dot{\boldsymbol{H}}(t)\right)$, which completes the proof. ■

Remark 1. Despite the apparently awkward form of condition (20), for the usual p-norms one can derive simpler expressions:

For $p = \infty$, inequality (20) is equivalent to the following n differential inequalities

$$\sum_{j=1}^{n} \theta_{ij} h_j(t) \leq \dot{h}_i(t),\ i = 1,\dots,n. \tag{22}$$

For $p = 1$, the approach is *mutatis mutandis* similar to the one for $p = \infty$.

For $p = 2$, inequality (20) is equivalent to the matrix differential inequality

$$\Theta^T \boldsymbol{H}^{-2}(t) + \boldsymbol{H}^{-2}(t)\Theta + \frac{d}{dt}\left(\boldsymbol{H}^{-2}(t)\right) \prec= 0. \tag{23}$$

■

Remark 2. The equivalent forms of condition (20) formulated above for the usual p-norms can be exploited either for checking if the sets $S^c_{p,\boldsymbol{H}(t)}$ are invariant (when the functions $h_i(t),\ i = 1,\dots,n$ are pre-defined), or for finding invariant sets (by resolving the inequalities with respect to $h_i(t),\ i = 1,\dots,n$). ■

Remark 3. Inequality (22) was also obtained in our previous work [8] as a sufficient condition for the invariance of symmetrical rectangular sets. The construction procedure in [8] was different, relying on the subtangency condition.

This proves that Corollary 1 brings a substantial generalization to a result known only in a particular form. ■

3 Invariant Sets with Exponential Decrease

This section presents sufficient conditions for the invariance of the sets $S^c_{p,\boldsymbol{D}e^{rt}}$ defined by (12) w.r.t. the trajectories of RNN (8).

Theorem 2. Let $1 \le p \le \infty$. Assume there exist a set $\Omega_p \subseteq \mathbf{R}^n$ and a positive constant $\rho > 0$ such that $\forall t \in \mathbf{R}_+$, $S^{\rho}_{p,\boldsymbol{D}e^{rt}} \subseteq \Omega_p$. Consider the matrix-valued function $\boldsymbol{A}(\boldsymbol{y})$ defined by (15). If

$$\forall \boldsymbol{y} \in \Omega_p,\ \mu_p\left(\boldsymbol{D}^{-1}\boldsymbol{A}(\boldsymbol{y})\boldsymbol{D}\right) \le r, \tag{24}$$

then $\forall c \in (0, \rho]$, the sets $S^c_{p,\boldsymbol{D}e^{rt}}$ defined by (12) are FI w.r.t. the trajectories of RNN (8).

Proof: If the diagonal matrix $\boldsymbol{H}(t) = \boldsymbol{D}e^{rt}$ is used in inequality (16), we get $\forall \boldsymbol{y} \in \Omega_p$, $\mu_p\left(\boldsymbol{D}^{-1}\boldsymbol{A}(\boldsymbol{y})\boldsymbol{D} - r\boldsymbol{I}_n\right) \le 0$ which is equivalent to (24). The application of Theorem 1 guarantees the invariance of the sets $S^c_{p,\boldsymbol{D}e^{rt}}$. ■

The inequality (24) in Theorem 2 is very permissive in handling the nonlinearities of RNN (8) since the matrix-valued function $\boldsymbol{A}(\boldsymbol{y})$ defined by (15) incorporates all the information about the activation functions $g_i(y_i)$, $i = 1, \ldots, n$. However, the practical manipulation of $\boldsymbol{A}(\boldsymbol{y})$ is extremely difficult, reason for which we are looking for a condition ensuring the fulfillment of inequality (24), but relying on a constant matrix instead of $\boldsymbol{A}(\boldsymbol{y})$.

Corollary 2. Let $1 \le p \le \infty$. Consider the constant matrix Θ defined by (19). If

$$\mu_p\left(\boldsymbol{D}^{-1}\Theta\boldsymbol{D}\right) \le r, \tag{25}$$

then $\forall c > 0$, the sets $S^c_{p,\boldsymbol{D}e^{rt}}$ of form (12) are FI w.r.t. the trajectories of RNN (8).

Proof: Starting from the componentwise matrix inequality $\boldsymbol{A}(\boldsymbol{y}) \le \overline{\boldsymbol{A}(\boldsymbol{y})} \le \Theta$, $\forall \boldsymbol{y} \in \mathbf{R}^n$, proved in Corollary 1, we can write $\boldsymbol{D}^{-1}\boldsymbol{A}(\boldsymbol{y})\boldsymbol{D} \le \overline{\boldsymbol{D}^{-1}\boldsymbol{A}(\boldsymbol{y})\boldsymbol{D}} \le \boldsymbol{D}^{-1}\Theta\boldsymbol{D}$, $\forall \boldsymbol{y} \in \mathbf{R}^n$. If Lemmas 2 and 4 in [14] are applied to the above inequality, we get $\mu_p\left(\boldsymbol{D}^{-1}\boldsymbol{A}(\boldsymbol{y})\boldsymbol{D}\right) \le \mu_p\left(\overline{\boldsymbol{D}^{-1}\boldsymbol{A}(\boldsymbol{y})\boldsymbol{D}}\right) \le \mu_p\left(\boldsymbol{D}^{-1}\Theta\boldsymbol{D}\right)$, $\forall \boldsymbol{y} \in \mathbf{R}^n$, which, together with (25), show that condition (24) is fulfilled; hence, the proof is completed by Theorem 2. ■

Remark 4. For the usual p-norms, the sufficient condition (25) has numerically tractable forms.

For $p=\infty$, inequality (25) is equivalent to the following n algebraic inequalities

$$\sum_{j=1}^{n} \theta_{ij} d_j \leq r d_i, \ i=1,\ldots,n. \tag{26}$$

For $p=1$, the approach is *mutatis mutandis* similar to the one for $p=\infty$.

For $p=2$, inequality (25) is equivalent to the linear matrix inequality (LMI)

$$\Theta^T \boldsymbol{D}^{-2} + \boldsymbol{D}^{-2}\Theta - 2r\boldsymbol{D}^{-2} \preceq 0. \tag{27}$$

■

Remark 5. The equivalent forms of condition (25) formulated above for the usual p-norms can be exploited either for checking if the sets $S^c_{p,\boldsymbol{D}e^{rt}}$ are invariant (when the constants $d_i>0$, $i=1,\ldots,n$ and $r<0$ are pre-defined), or for finding invariant sets (by resolving the inequalities with respect to d_i, $i=1,\ldots,n$, and / or r). ■

Remark 6. Inequality (26) was also obtained in our previous work [7] as a sufficient condition for the invariance of symmetrical rectangular sets with exponential decrease. This proves Corollary 2 brings a substantial generalization to a result known only in a particular form. ■

4 Connection between Invariance Properties and Stability

This section shows that the invariance properties of RNN (8) in the sense of Definition 1 represent sufficient conditions for the stability of the equilibrium {0} of RNN (8). The time-dependence of the invariant sets (arbitrarily bounded, arbitrarily approaching 0, exponentially decreasing) implies different types of stability. The local or global character of stability is also studied.

Theorem 3. Let $1\leq p\leq\infty$ and let the functions $h_i(t)$, $i=1,\ldots,n$, in (4) be bounded.

(a) If there exists $\rho>0$ such that $\forall c\in(0,\rho]$ the sets $S^c_{p,\boldsymbol{H}(t)}$ are FI w.r.t. RNN (8), then the equilibrium {0} of RNN (8) is locally stable.

(b) If $\forall c>0$ the sets $S^c_{p,\boldsymbol{H}(t)}$ are FI w.r.t. RNN (8), then the equilibrium {0} of RNN (8) is globally stable.

Proof: Let $M>0$ be an upper bound of the positive functions $h_i(t)$, $i=1,\dots,n$, i.e. $h_i(t)<M$ for all $t\geq 0$.

(a) We show that $\forall\varepsilon>0$ and $\forall t_0\geq 0$ there exists $\delta(\varepsilon,t_0)= =\min\{\rho,\varepsilon/M\}\min_{i=1,\cdots,n}\{h_i(t_0)\}$ such that for $\forall \boldsymbol{y}_0\in\mathbf{R}^n$, $\|\boldsymbol{y}_0\|_p\leq\delta(\varepsilon,t_0)$ and $\forall t\geq t_0$ the inequality $\|\boldsymbol{y}(t;t_0,\boldsymbol{y}_0)\|_p\leq\varepsilon$ holds. Indeed, from $\forall\boldsymbol{y}_0\in\mathbf{R}^n$, $\|\boldsymbol{y}_0\|_p\leq\delta(\varepsilon,t_0)$ we get $\|\boldsymbol{H}^{-1}(t_0)\boldsymbol{y}_0\|_p\leq\|\boldsymbol{H}^{-1}(t_0)\|_p\|\boldsymbol{y}_0\|_p\leq\dfrac{1}{\min_{i=1,\cdots,n}\{h_i(t_0)\}}\delta(\varepsilon,t_0)=\min\{\rho,\varepsilon/M\}$ which, by the invariance property, yields, $\|\boldsymbol{H}^{-1}(t)\boldsymbol{y}(t,t_0,\boldsymbol{y}_0)\|_p\leq\min\{\rho,\varepsilon/M\}$, $\forall t\geq t_0$. Thus, $\|\boldsymbol{y}(t,t_0,\boldsymbol{y}_0)\|_p\leq\|\boldsymbol{H}(t)\|_p\|\boldsymbol{H}^{-1}(t)\boldsymbol{y}(t,t_0,\boldsymbol{y}_0)\|_p\leq M\min\{\rho,\varepsilon/M\}\leq\varepsilon$.

(b) The proof is similar to part **(a)**, with $\delta(\varepsilon,t_0)=\varepsilon/M\min_{i=1,\cdots,n}\{h_i(t_0)\}$, which ensures the global character of the stability.

If there exists a positive constant $m>0$ such that $\forall t\geq 0$, $m\leq h_i(t)$, $i=1,\dots,n$, then the equilibrium $\{0\}$ is uniformly stable [12] **(a)** in the local sense and **(b)** in the global sense. This is because for all $t_0\geq 0$ we can use a unique $\delta(\varepsilon)$, namely **(a)** $\delta(\varepsilon)=m\min\{\rho,\varepsilon/M\}$ and **(b)** $\delta(\varepsilon)=m\,\varepsilon/M$. ■

Theorem 4. Let $1\leq p\leq\infty$ and let the functions $h_i(t)$, $i=1,\dots,n$, in (4) meet the condition

$$\lim_{t\to\infty}h_i(t)=0,\quad i=1,\dots,n. \tag{28}$$

(a) If there exists $\rho>0$ such that $\forall c\in(0,\rho]$ the sets $S^c_{p,\boldsymbol{H}(t)}$ are FI w.r.t. RNN (8), then the equilibrium $\{0\}$ of RNN (8) is locally asymptotically stable.

(b) If $\forall c>0$ the sets $S^c_{p,\boldsymbol{H}(t)}$ are FI w.r.t. RNN (8), then the equilibrium $\{0\}$ of RNN (8) is globally asymptotically stable.

Proof: The stability of the equilibrium $\{0\}$ of RNN (8) results from Theorem 3, since condition (28) implies the functions $h_i(t)$, $i=1,\dots,n$, are bounded.

(a) We show that $\forall t_0\geq 0$ there exists $\gamma(t_0)=\rho\min_{i=1,\cdots,n}\{h_i(t_0)\}$ such that for $\forall\boldsymbol{y}_0\in\mathbf{R}^n$, $\|\boldsymbol{y}_0\|_p\leq\gamma(t_0)$ and $\forall t\geq t_0$, the equality $\lim_{t\to\infty}\|\boldsymbol{y}(t;t_0,\boldsymbol{y}_0)\|_p=0$ holds.

Indeed, from $\| y_0 \|_p \le \gamma(t_0)$ we get $\| \boldsymbol{H}^{-1}(t_0) \boldsymbol{y}_0 \|_p \le \| \boldsymbol{H}^{-1}(t_0) \|_p \| \boldsymbol{y}_0 \|_p \le \frac{1}{\min_{i=1,\cdots,n} \{h_i(t_0)\}} \gamma(t_0) = \rho$, which, by the invariance property, yields $\forall t \ge t_0$, $\| \boldsymbol{H}^{-1}(t) \boldsymbol{y}(t, \boldsymbol{y}_0, t_0) \|_p \le \rho$.

Thus, $\| \boldsymbol{y}(t, t_0, \boldsymbol{y}_0) \|_p \le \| \boldsymbol{H}(t) \|_p \| \boldsymbol{H}^{-1}(t) \boldsymbol{y}(t, t_0, \boldsymbol{y}_0) \|_p \le \| \boldsymbol{H}(t) \|_p \; \rho$. Hypothesis (28) implies that $\lim_{t \to \infty} \| y(t; t_0, y_0) \|_p = 0$. According to [12], the equilibrium $\{0\}$ of RNN (8) is locally asymptotically stable.

(b) The proof is similar to part **(a)**, for $\gamma(t_0) = k \min_{i=1,\cdots,n} \{h_i(t_0)\}$, with $k > 0$ arbitrarily taken, fact that ensures the global character of the asymptotic stability. ∎

Theorem 5. Let $1 \le p \le \infty$.

(a) If there exists $\rho > 0$ such that $\forall c \in (0, \rho]$ the sets $S^c_{p, \boldsymbol{D}e^{rt}}$ are FI w.r.t. RNN (8), then the equilibrium $\{0\}$ of RNN (8) is locally exponentially stable.

(b) If $\forall c > 0$ the sets $S^c_{p, \boldsymbol{D}e^{rt}}$ are FI w.r.t. RNN (8), then the equilibrium $\{0\}$ of RNN (8) is globally exponentially stable.

Proof: If $\boldsymbol{D}$ is a diagonal matrix defined according to (6), then $\| \boldsymbol{D}^{-1} \boldsymbol{y} \|_p$ represents a norm on $\mathbf{R}^n$, for which we use the notation $\| y \|_p^D$.

(a) We show that $\forall \varepsilon > 0$ small enough and $\forall t_0 \ge 0$ there exists $\delta(\varepsilon) = \varepsilon$ such that inequality $\| \boldsymbol{y}(t; t_0, \boldsymbol{y}_0) \|_p^{\boldsymbol{D}} \le \varepsilon e^{r(t-t_0)}$ holds for $\forall \boldsymbol{y}_0 \in \mathbf{R}^n$, $\| \boldsymbol{y}_0 \|_p^{\boldsymbol{D}} \le \delta(\varepsilon)$, and $\forall t \ge t_0$. Indeed, for $\forall \varepsilon > 0$ small enough and $\forall t_0 \ge 0$ we can find $c \in (0, \rho]$ such that $ce^{rt_0} = \varepsilon$. According to the invariance property, for $\forall t, t \ge t_0$ and $\forall \boldsymbol{y}_0 \in \mathbf{R}^n$, $\| \boldsymbol{D}^{-1} \boldsymbol{y}_0 \|_p = \| \boldsymbol{y}_0 \|_p^{\boldsymbol{D}} \le ce^{rt_0}$ we have $\| \boldsymbol{y}(t; t_0, \boldsymbol{y}_0) \|_p^{\boldsymbol{D}} = \| \boldsymbol{D}^{-1} \boldsymbol{y}(t; t_0, \boldsymbol{y}_0) \|_p \le ce^{rt} = \varepsilon e^{r(t-t_0)}$.

(b) The proof is similar to part **(a)**, with no restriction on $\varepsilon > 0$, which ensures the global character of the exponential stability. ∎

Remark 7. According to the above results, the existence of invariant sets guarantees the stability of the equilibrium $\{0\}$ of RNN (8), covering the cases discussed by Theorems 3 – 5. At the same time, by simple examples one can prove the converse parts of Theorems 3 – 5 are not true, meaning that the invariance properties are stronger than stability. For instance, let us consider a continuous-time RNN defined by (8) with

$$\boldsymbol{B}=\begin{bmatrix}-1 & 0\\ 0 & -5\end{bmatrix},\ \boldsymbol{W}=\begin{bmatrix}2 & -2\\ 2 & 2\end{bmatrix},\ \boldsymbol{y}=\begin{bmatrix}y_1\\ y_2\end{bmatrix},\ \boldsymbol{g}(\boldsymbol{y})=\begin{bmatrix}y_1\\ y_2\end{bmatrix}. \tag{29}$$

Due to the linearity of the activation functions, this RNN is equivalent with the linear differential system:

$$\begin{bmatrix}\dot{y}_1\\ \dot{y}_2\end{bmatrix}=\begin{bmatrix}1 & -2\\ 2 & -3\end{bmatrix}\begin{bmatrix}y_1\\ y_2\end{bmatrix} \tag{30}$$

for which the equilibrium $\{0\}$ is exponentially stable (since both eigenvalues are -1). On the other hand, according to Theorem 2 in [14], there exist sets $S^{c}_{p,\boldsymbol{D}e^{rt}}$ of form (14) invariant w.r.t. system (29) if and only if there exist $d_1>0,\ d_2>0,\ r<0$ such that:

$$\mu_p\left(\begin{bmatrix}d_1 & 0\\ 0 & d_2\end{bmatrix}^{-1}\begin{bmatrix}1 & -2\\ 2 & -3\end{bmatrix}\begin{bmatrix}d_1 & 0\\ 0 & d_2\end{bmatrix}\right)\le r. \tag{31}$$

For $p=\infty$ or $p=1$, condition (31) is equivalent with the linear inequalities (e.g. [1]):

$$\begin{cases}1+2(d_2/d_1)\le r\\ 2(d_1/d_2)-3\le r\end{cases}\quad \text{respectively} \quad \begin{cases}1+2(d_1/d_2)\le r\\ 2(d_2/d_1)-3\le r\end{cases} \tag{32}$$

that cannot have solutions of the form $d_1>0,\ d_2>0,\ r<0$.

For $p=2$, condition (31) is equivalent with the inequality (e.g. [1]):

$$\lambda_{\max}\left(\begin{bmatrix}1 & -2(d_2/d_1)\\ 2(d_1/d_2) & -3\end{bmatrix}+\begin{bmatrix}1 & 2(d_1/d_2)\\ -2(d_2/d_1) & -3\end{bmatrix}\right)\le r \tag{33}$$

that cannot have solutions of the form $d_1>0,\ d_2>0,\ r<0$.

Thus, although the equilibrium $\{0\}$ of the considered RNN is exponentially stable, there is no invariant set $S^{c}_{p,\boldsymbol{D}e^{rt}}$ for $p=1,2,\infty$, fact showing that the converse part of Theorem 5 is not true. ■

Remark 8. The particular case of Theorems 3-5 corresponding to $p=\infty$ (meaning invariant sets with hyper-rectangular form) was discussed in previous works [7], [8] under the name of componentwise, componentwise asymptotic, and, respectively, componentwise exponentially asymptotic stability of RNN (8). The current paper has the merit of developing a general framework which naturally accommodates these already known results. ■

5 Example

The following numerical example refers to the RNN described by (8) with $\boldsymbol{B}=\mathrm{diag}\{-5,-7\}$, $\boldsymbol{W}=\begin{bmatrix}-3 & -1\\ -1 & -3.5\end{bmatrix}$, and $g_1(y_1)=\mathrm{tansig}(y_1)$, $g_2(y_2)=\mathrm{tansig}(2y_2)$.

Thus, conditions (10) are satisfied with $L_1 = 1$ and $L_2 = 2$. We want to apply Corollary 2 for finding sets $S^c_{p,\boldsymbol{D}e^{rt}}$ of form (12) with $r = -2$, positively invariant w.r.t. the state-trajectories of this RNN. We first construct matrix Θ defined by (19) and obtain $\Theta = \begin{bmatrix} -5 & 2 \\ 1 & -7 \end{bmatrix}$, then we solve inequality (25) with respect to the diagonal matrix $\boldsymbol{D} = \text{diag}\{d_1, d_2\}$, $d_1, d_2 > 0$. We take into account only the particular cases when $p \in \{1, 2, \infty\}$ and bear Remark 4 in mind.

For $p = \infty$, the linear algebraic inequalities (26) lead to

$$\begin{aligned} -5d_1 + 2d_2 &\leq -2d_1, \\ d_1 - 7d_2 &\leq -2d_2. \end{aligned} \tag{34}$$

A solution to (34) is $d_1 = d_2 = 1$. Therefore, for arbitrary $c > 0$, the set $S^c_{\infty,\boldsymbol{D}e^{-2t}}$ with $\boldsymbol{D} = \text{diag}\{1, 1\}$ is positively invariant w.r.t. the trajectories of the considered RNN.

For $p = 2$, we have to solve the LMI (27) for computing the diagonal matrix $\boldsymbol{D} = \text{diag}\{d_1, d_2\} \succ= 0$. We use the Multi Parametric Toolbox for MATLAB [15] and obtain the solution $\boldsymbol{D} = \text{diag}\{1.92, 2.28\}$. Consequently, for $c > 0$, the set $S^c_{2,\boldsymbol{D}e^{-2t}}$ with $\boldsymbol{D} = \text{diag}\{1.92, 2.28\}$ is positively invariant w.r.t. the trajectories of the considered RNN.

The approach to $p = 1$ is similar to the one corresponding to $p = \infty$. We solve the linear algebraic inequalities

$$\begin{aligned} -5d_1 + d_2 &\leq -2d_1, \\ 2d_1 - 7d_2 &\leq -2d_2, \end{aligned} \tag{35}$$

and obtain $d_1 = d_2 = 1$. Therefore, for arbitrary $c > 0$, the set $S^c_{1,\boldsymbol{D}e^{-2t}}$ with $\boldsymbol{D} = \text{diag}\{1, 1\}$ is positively invariant w.r.t. the trajectories of the RNN.

6 Conclusions

The developed researches provide sufficient conditions for exploring the invariant sets with respect to the dynamics of RNNs. Criteria for testing the set invariance are formulated for two types of time-dependent sets, namely sets with arbitrary time-dependence and exponentially decreasing sets. The shapes of the sets are general, defined by Hölder p-norms. The basic results (Theorems 1 and 2) are not suitable for practical applications, since they operate with matrix-valued functions defined on subsets of the state space. Nevertheless, these basic results yield corollaries (Corollaries 1 and 2) with direct utility in practice, since their formulations rely on constant matrices (expressing majorants of the matrix-valued functions). For the usual p-norms ($p = 1, 2, \infty$) the numerical tractability of the tests is a straightforward task.

The paper also analyzes the connection between the invariance and the stability properties of RNNs, showing that the former are stronger than the latter. The existence of invariant sets with arbitrary time-dependence, bounded or approaching the equilibrium, ensures the uniform stability or, respectively, uniform asymptotic stability of the equilibrium. The existence of exponentially decreasing invariant sets ensures the exponential stability of the equilibrium.

The framework created by this work for studying the invariance properties of RNNs includes some previous results (obtained by different procedures) as a particular case corresponding to the infinity norm.

Acknowledgments. The Grant # 255 of the Executive Agency for Higher Education and Research Funding (CNCSIS-UEFISCSU) has supported part of the research presented in this paper.

References

1. Fang, Y., Kincaid, T.G.: Stability analysis of dynamical neural networks. IEEE Trans. Neural Networks 7, 996–1006 (1996)
2. Michel, A.N., Liu, D.: Qualitative Analysis and Synthesis of Recurrent Neural Networks. Marcel Dekker, Inc., New York (2002)
3. Cao, J., Wang, J.: Global asymptotic and robust stability of recurrent neural networks with time delays. IEEE Trans. Circuits Syst. I 52, 417–426 (2005)
4. Forti, M.: M-matrices and global convergence of discontinuous neural networks. Int. J. Circuit Theory and Appl. 35(2), 105–130 (2007)
5. Xu, J., Pi, D., Cao, Y.-Y., Zhong, S.: On Stability of Neural Networks by a Lyapunov Functional-Based Approach. IEEE Trans. Circuits and Systems I 54, 912–924 (2007)
6. Chu, T., Wang, Z., Wang, L.: Exponential convergence estimates for neural networks with multiple delays. IEEE Trans. Circuits Syst. I 49, 1829–1832 (2002)
7. Matcovschi, M.H., Pastravanu, O.: Flow-invariance and stability analysis for a class of nonlinear systems with slope conditions. Eur. J. Control 10, 352–364 (2004)
8. Pastravanu, O., Matcovschi, M.H.: Absolute componentwise stability of interval Hopfield neural networks. IEEE Trans. Syst. Man Cyb. Part B 35, 136–141 (2005)
9. Pavel, H.N.: Differential Equations: Flow Invariance and Applications; Research Notes in Mathematics, vol. 113. Pitman, Boston (1984)
10. Michel, A.N., Wang, K., Hu, B.: Qualitative Theory of Dynamical Systems. The Role of Stability Preserving Mappings. Marcel Dekker, Inc., New York (2001)
11. Motreanu, D., Pavel, N.H.: Tangency, Flow-Invariance for Differential Equations and Optimization Problems. Marcel Dekker, Inc., New York (1999)
12. Gruyitch, L.T., Richard, J.P., Borne, P., Gentina, J.C.: Stability Domains (Nonlinear Systems in Aviation, Aerospace, Aeronautics and Astronautics). Chapman & Hall/CRC, London (2004)

13. Carja, O., Vrabie, I.I.: Differential Equations on Closed Sets. In: Canada, A., Drabek, P., Fonda, A. (eds.) Handbook of Differential Equations: Ordinary Differential Equations, vol. 2, pp. 147–238. Elsevier BV/North Holland, Amsterdam (2005)
14. Pastravanu, O., Voicu, M.: Generalized matrix diagonal stability and linear dynamical systems. Linear Algebra and its Applications 419, 299–310 (2006)
15. Kvasnica, M.: Multi Parametric Toolbox (MPT) ETH Zurich (2007), `http://control.ee.ethz.ch/~mpt/`

Part III
Applications

7

Solving Bioinformatics Problems by Soft Computing Techniques: Protein Structure Comparison as Example

Juan R. González, David A. Pelta, and José L. Verdegay

Department of Computer Science and Artificial Intelligence (DECSAI)
University of Granada, E-18071, Granada, Spain
{jrgonzalez,dpelta,verdegay}@decsai.ugr.es

Abstract. Bioinformatics is a very interesting an active area that tackles difficult problems with lots of data that may have noise, missing values, uncertainties, etc. This chapter shows how the techniques that Soft Computing provides are appropriate to solve some Bioinformatics problems. This idea is then illustrated by showing several resolution techniques for one of the key problems of the Bioinformatics area: the Protein Structure Comparison problem.

1 Introduction

The development of several genome projects (including the human genome project) led to the generation of big amounts of data and to the apparition of new techniques and research lines. This is the context where Bioinformatics emerged as a strong research field, as it provides the tools to manage a wide and disperse set of data, suggest hypothesis and promote new experiments. Bioinformatics is devoted to the development and application of algorithms and methods to transform data on biological knowledge. The problems from Bioinformatics are often very difficult and they rely on big amounts of data coming from several sources. Moreover, the determination of the biological data through experimental techniques is not perfect and the obtained data often have some imprecisions. Therefore, there is a need to develop new techniques and methods that can cope with these complex problems and its uncertainty, so it becomes possible to model them appropriately and to solve the models with the highest possible throughput and quality. This will help to make possible and/or accelerate many biological researches.

On this scenario, Soft Computing is playing a crucial role as it provides techniques that are especially well suited to obtain results in an efficient way and with a good level of quality. Soft Computing can also be useful to model the imprecision and uncertainty that the Bioinformatics data and problems have. Moreover, the optimal solution to the models is not usually necessary as there can be irrelevant minimum or maximum, and what is really important is to obtain biologically relevant solutions and that can be done with suboptimal solutions like the ones provided by many of the Soft Computing techniques.

To develop these ideas, this chapter firstly introduces the Bioinformatics and Soft Computing areas. Then, to show how Soft Computing can be successfully applied to a Bioinformatics problem we will review some previous and current

H.-N. Teodorescu, J. Watada, and L.C. Jain (Eds.): Intel. Sys. and Tech., SCI 217, pp. 123–136.
springerlink.com

work done in the Protein Structure Comparison problem. The chapter ends with the conclusions and ideas for further research.

2 Bioinformatics

Bioinformatics is a multidisciplinary area that was born with the realization of several genome projects, including the human genome [13], [14]. The whole sequence of the human genome is publicly available, including the determination of the more than 20000 genes of our organism. But despite its importance, the determination of the genotype is only a first step to the comprehension of the organisms and their phenotypes. Therefore, the genome projects, far from meaning an end to the Genomic Science, provided an extra amount of information that led to more potential researches. Moreover, the high potential benefits of new discoveries in Medicine and Biotechnology (like the creation of new drugs) leads to very high investments to perform new researches and experiments. All these increasing research has motivated the apparition of experimental techniques that generate data with a high level of performance such as DNA sequencing, mass spectroscopy or microarray expression data analysis. And it all implies that there are now more possible research topics and an increasing need for new methods and techniques.

Bioinformatics is in the interface between Computer Science and Biology and can be considered as a tool to manage all the data and the execution of experiments. It is devoted to the development of algorithms and methods to transform data on biological knowledge. To do this Bioinformatics has to provide tools that can cross information coming from different sources such as sequences, structures, microarrays, textual information, etc. The problems where Bioinformatics is applied are diverse: database design, sequence and structure alignment and comparison, phylogenetic trees, protein structure prediction, fragment assembly, genome rearrangement, microarray expression data analysis, etc.

Structural Bioinformatics [12] is a subarea within Bioinformatics that focuses on two main goals: the creation of general purpose methods to manipulate information of biological macromolecules and the application of these methods to solve problems in Biology and to create new knowledge. These two goals are deeply interconnected because part of the validation of new methods consists on testing their performance when they are applied to real problems.

The current challenges on Biology and Bioinformatics require the development of new methods that can handle the high quantity of available data and the complexity of the scientific models and problems that the scientists has to construct to explain this data. It is also important to note that the data often comes from experimental determination or expert analysis and it is known that it can have errors or a certain level of uncertainty. To be successful in practice, the models and algorithms developed should take into account all these important characteristics.

Therefore, Bioinformatics area needs models for its problems that take into account the uncertainty of the data and methods that can provide relevant solutions to these problems in a reasonable amount of time. Soft Computing is a family of methods for problem resolution that can be successfully applied to this scenario as it provides both the techniques needed to model the uncertainty and the methods to get good quality solutions to these models in a fast pace that is crucial for the amount of data and experiments that are inherent to the area. In the next sections, we will describe the Soft computing techniques and how they are applied to solve a particular Bioinformatics problem. The Protein Structure Comparison problem will be presented as an example.

3 Soft Computing

The need to find the optimal solution to a problem or the best possible solution between the ones available justifies the construction and study of theories and methodologies that are well suited to the scientific area where the problem or question arises. One important type of problems are the optimization problems, which optimize the value that a function may reach on a previously specified set, and these and everything relating to them is covered by the area known as mathematical programming. When fuzzy elements are considered in mathematical programming, fuzzy optimization methods emerge, and these are perhaps one of the most fruitful areas of fuzzy-related knowledge, both from the theoretical and the applied points of view. Yet despite all its methods and models for solving the enormous variety of real practical solutions, as with conventional mathematical programming, it cannot solve every possible situation. While a problem may be expressed in fuzzy terms, it may not be possible to solve it using only fuzzy techniques. The ease of resolving ever larger real problems, the impossibility of discovering exact solutions to these problems in every case and the need to provide answers to the practical situations considered in a great many cases have led to the increasing use of heuristic-type algorithms which have proved to be valuable tools capable of providing solutions where exact algorithms are not able to. In recent years, a large catalogue of heuristic techniques has emerged inspired by the principle that satisfaction is better than optimization, or, in other words, rather than not being able to provide the optimal solution to a problem, it is better to give a solution which at least satisfies the user in some previously specified way, and these have proved to be extremely effective [15].

This is the scenario where Soft Computing appeared and taking as reference one of the most recent viewpoints [15] it can be seen as a series of techniques and methods so that real practical situations could be dealt with in the same way as humans deal with them, i.e. on the basis of intelligence, common sense, consideration of analogies, approaches, etc. In this sense, Soft Computing is a family of problem-resolution methods headed by approximate reasoning and functional and optimization approximation methods, including search methods. Soft Computing is therefore the theoretical basis for the area of intelligent systems and it is evident that the difference between the area of classic artificial intelligence and that of modern intelligent systems is that the first is based on hard computing and the second on Soft Computing. Following this viewpoint, on a second level, Soft Computing can be expanded into four main components, namely probabilistic reasoning, fuzzy logic and fuzzy sets, neural networks and metaheuristics, as shown in Fig. 1.

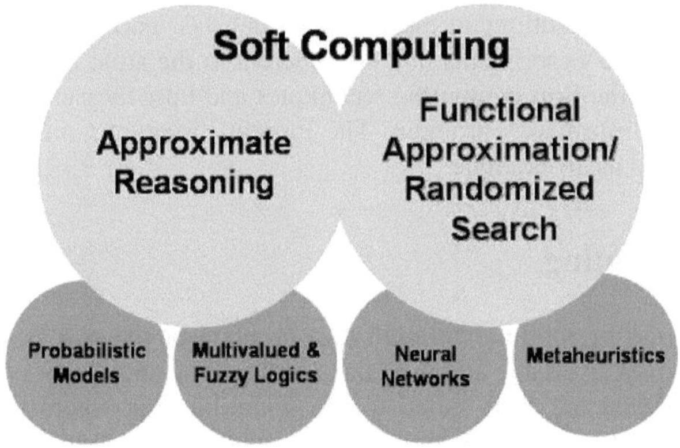

Fig. 1. The components of Soft Computing

The most relevant difference of this viewpoint over previous definitions is that previously just Genetic Algorithms (GA) where considered on the place now taken by metaheuristics in general. This can be explained by the popularity that GA have, but since it is just one of the possible metaheuristic techniques available, there is nothing that forbids considering all of them instead of just GA to get a broader range of options and flexibility, as well as to succeed on more problems where GA may not be the best option. It is possible to combine these components to build hybrid models and methods and due to the interdisciplinary, applications and results of Soft Computing immediately stood out over other methodologies such as the chaos theory, evidence theory, etc.

In conclusion, Soft Computing provides techniques to model and solve problems where exact algorithms are not able to provide solutions with a reasonable amount of resources (mostly time) and where there is imprecision or uncertainty. Since this is precisely the case with many of the Bioinformatics problems, we consider that the research on the Bioinformatics area can greatly benefit from Soft Computing and that the models and methods generated with this approach are being and are going to be very successful and help to accelerate the progress and the discovery of new knowledge. To illustrate it, the rest of this chapter will be devoted to the application of Soft Computing to one important Bioinformatics problem: The Protein Structure Comparison problem.

4 Application: The Protein Structure Comparison Problem

A protein is a complex molecule composed by a linear arrangement of amino acids. Each amino acid is a multi-atom compound. Usually, only the "residue" parts of these amino acids are considered when studying protein structures for comparison purposes. Thus a protein's primary *sequence* is usually thought-of as composed of "residues". Under specific physiological conditions, the linear arrangement of residues will *fold* and adopt a complex three dimensional shape. The shape thus adopted is called the *native state* (or tertiary structure) of the protein. In its *native state*, residues that are far away along the linear arrangement may come into proximity in three dimensional space in a fashion similar to what occurs with the extremes of a sheet of paper when used to produce complex origami shapes. The proximity relation between residues in a protein can be captured by a mathematical construct called a "contact map".

A contact map [8], [9] is a concise representation of a protein's 3D structure. Formally, a map is specified by a 0-1 matrix *S*, with entries indexed by pairs of protein residues, such that:

$$S_{i,j} = \begin{cases} 1 & if\ residues\ i\ and\ j\ are\ in\ contact \\ 0 & otherwise \end{cases} \tag{1}$$

Residues *i* and *j* are said to be in "contact" if their Euclidean distance is at most $\mathfrak{R}$ (a threshold measured in Angstroms) in the protein's native fold.

The comparison of proteins through their contact maps is equivalent to solving the maximum contact map overlap problem (MAX-CMO) [1], [3], a problem that belongs to the NP-Hard class of complexity.

Since the amount of protein data available is increasing at a fast rate (with around 40000 structures currently present on the Worldwide Protein Data Bank [10]), the problem of comparing a new protein against all the known proteins or even with a subset or a specific representative database is a very big task. So, despite there exist exact algorithms for the MAX-CMO model [11] they can not be applied to most instances because the amount of resources required would be prohibitive.

These considerations led us to a first way of using Soft Computing to contribute in the field by developing a simple and fast heuristic that could obtain good results for this model and provide biologically relevant solutions without the need to find the exact solutions to the model. This heuristic has been published recently on [5] where it is extensively tested showing how the proposed algorithm, that is based on the Variable Neighborhood Search metaheuristic can both obtain near-optimal results and solutions and similarity values that are biologically relevant for the purpose of classification. The heuristic is also shown to be competitive in classification performance with similarity measures coming from methods that compare proteins through different models like the ones based on distance matrixes [5], [18].

But although crisp contact maps are useful to compare proteins, it is also known that the errors on the determination of the 3D coordinates of the residues of a protein by X-Ray crystallography or NMR range from 0.01 to 1.27Å [16], what is close to the value for some covalent bonds. This kind of imprecision can not be modeled through the crisp contact maps but there exists an alternative formulation that uses fuzzy contact maps and a new model to compare such maps: the generalized fuzzy contact map overlap problem GMAX-FCMO [6], thus making another technique from Soft Computing (fuzzy sets) come into play. The use of fuzzy contact maps allows to soften the thresholds for the contacts to take into account the potential errors in the determination of coordinates and it also serves as a way to give different semantic to contacts that arise at different distance ranges. The comparison of proteins using fuzzy contact maps and the resolution of the GMAX-FCMO problem are particularly interesting for showing the benefits of Soft Computing on Bioinformatics and how to apply it, so we will focus on it on the rest of the chapter.

4.1 Fuzzy Contact Maps Model Description

Fuzzy contact maps were introduced in [6] with two aims: a) to take into account potential measurements errors in atom coordinates, and b) to allow highlighting features that occurs at different thresholds.

We define a *fuzzy contact* as that made by two residues that are *approximately*, rather than exactly, at a distance $\mathfrak{R}$. Formally, a fuzzy contact is defined by:

$$F_{i,j} = \mu(\overline{[\iota,\jmath]}, \mathfrak{R}) \quad (2)$$

where $\mu()$ is a particular definition of (fuzzy) contact, $\overline{[\iota,\jmath]}$ stands for the Euclidean distance between residues i and j, and $\mathfrak{R}$ is the threshold as for the crisp contacts. The standard, i.e. crisp, contact map is just a special case of the fuzzy contact map when a user-defined α-cut is specified.

Fig. 2 (a), (b) and (c) shows three alternative definitions for "contact". Each panel in the figure is a fuzzy contact map where a dot appears for each pair of residues having $F_{i,j} > 0$ (i.e. the support of the corresponding fuzzy set).

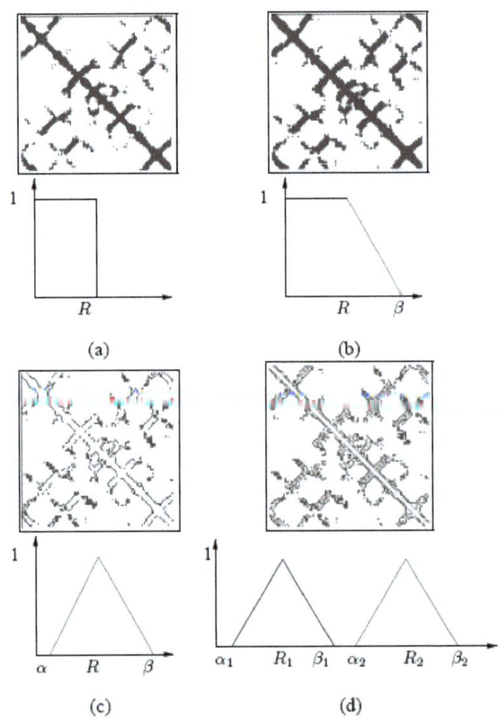

Fig. 2. Four examples of contact maps. In (a) the standard model; (b) the simplest fuzzy generalization; (c) another generalization; (d) a two threshold, two membership functions fuzzy contact maps.

Fuzzy contact maps are further generalized by removing the constraint (in the original model) of having only one threshold $\Re$ as a reference distance. In this way, besides having a membership value, a contact will have a "type". The formal definition of a General Fuzzy Contact is given by:

$$F_{i,j} = max\{\mu_1(\overline{[\iota,\jmath]},\Re_1),\mu_2(\overline{[\iota,\jmath]},\Re_2),\cdots,\mu_n(\overline{[\iota,\jmath]},\Re_n)\} \quad (3)$$

with the contact map C defined as:

$$C^{rxr} = (F_{i,j})\ with\ 0 \le i,j < r \quad (4)$$

i.e. up to n different thresholds and up to n different semantic interpretations of "contact" are used to define the $r\times r$ contact map being r the number of residues in the protein.

The benefits of using these fuzzy contact maps over the crisp ones have been probed before even with a very simple strategy for their comparison: the Universal Similarity Metric (USM). USM is a simple similarity metric that is able to approximate every other possible metric and that it is based on the concept of Kolmogorov complexity. It is non-computable but it was shown that it can be approximated by using the compressed sizes of the crisp contact maps of the proteins [17] with good similarity values for the purpose of classification. This initial work was then extended to apply USM over fuzzy contact maps [18] and it was shown that it also served to obtain similarity values that were good enough for classification. Moreover, the results obtained with the simplest fuzzy generalization of the contact maps were better than the results obtained using the crisp maps, what can be seen as an indication that the fuzzy model is more appropriate than the crisp one. These results were important per se and USM is a very fast method for comparison, but it has the problem that it only provides a similarity value between the proteins and not an alignment or correspondence between the residues of the two proteins. Therefore, to obtain full solutions and further improve the results it is needed to solve the optimization models like the GMAX-FCMO problem. Unfortunately, GMAX-FCMO has not been as extensively tested yet as the crisp MAX-CMO model, so now we will present an extended analysis of the fuzzy model optimization.

4.2 Protein Comparison through Fuzzy Contact Maps Overlap

A solution for the comparison of two contact maps under the crisp Maximum Contact Map Overlap model consists of an alignment or pairing between the nodes of the two contact maps. The value of an alignment is the number of cycles of length four that appear between the corresponding graphs after the pairings. This value is called the overlap (fitness) of the two contact maps and the goal is to maximize it.

For example, Fig. 3 shows a sample solution for the comparison of two contact maps of 5 and 7 residues respectively. In the crisp model, we can omit the colors of the arcs. Three residues are paired, as shown with a dotted line (first index corresponds to the bottom graph): 1↔1, 2↔4 and 3↔5, and the overlap value is two because there are two cycles of length four.

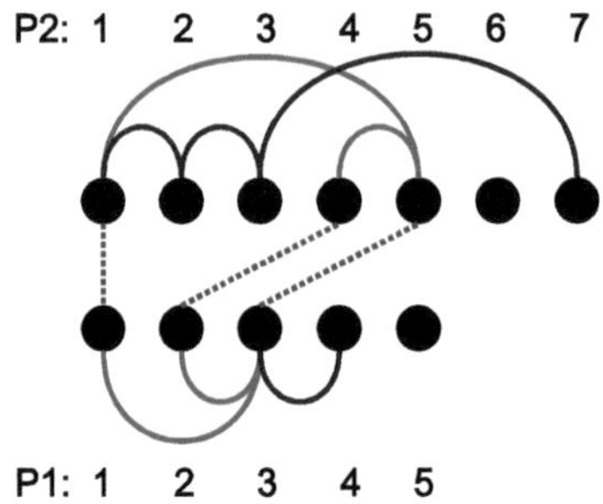

Fig. 3. Two levels of contacts in a fuzzy contacts graph

The first cycle is formed by the pairing 1↔1, the arc 1↔5∈P_2, the pairing 3↔5 and the arc 3↔1∈P_1. The second one begins with the pairing 2↔4, follows the arc 4↔5∈P_2, then the pairing 3↔5 and finally, the arc 3↔2∈P_1.

Solutions for the GMAX-FCMO model have exactly the same structure as in the crisp MAX-CMO, but the overlap value is computed differently. Now, the arcs have weights and types (colors), so the contribution of each cycle to the global fitness is calculated as a function of the membership values of the contacts involved and their types. In the original GMAX-FCMO model, both membership values are multiplied and then, if both contacts have the same type, then the contribution is added to the fitness. Otherwise, when contacts of different type are involved in a cycle, the contribution is subtracted from the fitness. In this way, alignments between contacts of different types are penalized.

So, as the contact maps in Fig. 3 are fuzzy, the contribution of one cycle to the fitness is the product of the membership value of the contact 1↔3 of the bottom protein and that of the contact 1↔5 of the upper protein, with positive sign because the two contacts are of the same type; and, this value is added to the product of the membership of contact 2↔3 and the membership of contact 4↔5.

It is important to note that GMAX-FCMO is also a problem of complexity class NP-hard because MAX-CMO is no more than a particular case for it. But it has not been as extensively tested as its "crisp" counterpart. In previous work we addressed the comparison of fuzzy contact maps against crisp contact maps [7] and we showed that if we first solved the problem trough GMAX-FCMO model, and then such solutions were measured as in MAX-CMO, the results obtained could be better than those obtained when MAX-CMO is solved directly.

Here we want to extend our research in the GMAX-FCMO in two aspects: a) proposing new alternatives to measure the cost of a solution, and b) to discuss the role of normalization, when the aim is to perform protein structure classification.

4.3 Experiments and Results

The aims of the experiments are: 1) to propose new alternatives for measuring the cost in the Generalized Maximum Fuzzy Contact Map Overlap problem, and 2) to analyze the role of normalization, when protein classification is performed.

4.3.1 Alternatives for Cost Calculations

The value of an overlap is the sum of the contribution of every cycle of length four. In the crisp model, every cycle contributes a unity to the overlap.

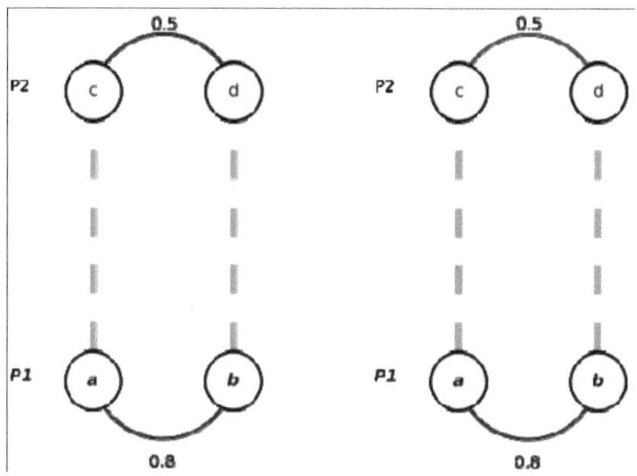

Fig. 4. A cycle of length four making contacts of the same (left) and different (right) types

Cycles in the fuzzy model has the appearance shown in Fig. 4. The contribution of a cycle is calculated as $C=\mu(a,b)\times\mu(c,d)\times F(t(a,b),t(c,d))$, where $t(a,b),t(c,d)$ stand for the type (color) of the contact between a,b and c,d respectively. The function F simply returns 1, if both contacts are of the same type, and -1 in other case. So, the costs of the cycles in the example are $0.8\times0.5\times1$ in Fig. 4 (left) and $0.8\times0.5\times-1$ in the right cycle.

In this experiment, we will use fuzzy contact maps generated from one membership function, thus avoiding the use of function F. Having this in mind, we propose the following set of alternatives to measure the contribution of an individual cycle:

1. Product: $C = \mu(a,b) \times \mu(c,d))$
2. Min: $C = min(\mu(a,b), \mu(c,d))$
3. Max: $C = max(\mu(a,b), \mu(c,d))$
4. Avg: $C = (\mu(a,b) + \mu(c,d))/2$

4.3.2 Normalization Alternatives

Overlap values per se, are not useful (at least in the crisp model) for classification purposes, as such values depend of the size of the proteins being compared. Once the GMAX-FCMO is solved, a normalization scheme should be applied and it is claimed that this scheme may play a crucial role in protein classification.

Following the ideas posed in [5], [7], we will use four alternatives in our experiments:

1. $Norm1(P_i, P_j) = overlap(P_i, P_j)/\min(contactsP_i, contactsP_j)$
2. $Norm2(P_i, P_j) = 2 * overlap(P_i, P_j)/(contactsP_i + contactsP_j)$
3. $Norm3(P_i, P_j) =$
$$\begin{cases} 0 & \text{if the contact difference is gretear than 75\%} \\ \text{Norm1}(P_i, P_j) & \text{otherwise} \end{cases}$$
4. $NormFuzzy(P_i, P_j) = overlap(P_i, P_j)/\max(selfSim(P_i), selfSim(P_j))$

where the self-similarity (*selfSim*) of a protein is the value of the overlap of a protein with itself.

4.3.3 Computational Experiments

The global idea of the experiment is as follows: we will conduct queries on a protein database, solving a GMAX-FCMO problem for comparing the query with each protein in the database. Then, we will have a list of overlap values that should be normalized and after that, we will analyze what are the most similar proteins in the database for every query performed (classes are known).

Our protein database consists on 150 selected protein structures from the Nh3D v3.0 test dataset [2]. This dataset has been compiled by selecting well resolved representatives from the Topology level of CATH database [19] and contains 806 topology representatives belonging to 40 architectures, which can be further classified in terms of "class". We selected as the query set the structures that have the nearest to average size for each of the 15 architectures with at least 10 topology representatives. Then, to build the test database we took all the proteins of these 15 architectures removing the query proteins and we picked randomly 10 proteins of each architecture. Each query was then compared against every structure in the test dataset.

For every protein, a fuzzy contact map is constructed using the membership function described in Fig. 5 which have a threshold of 8Å and where a decaying slope reduces the level of membership of a contact as the distance approaches from 6.5 to 8.

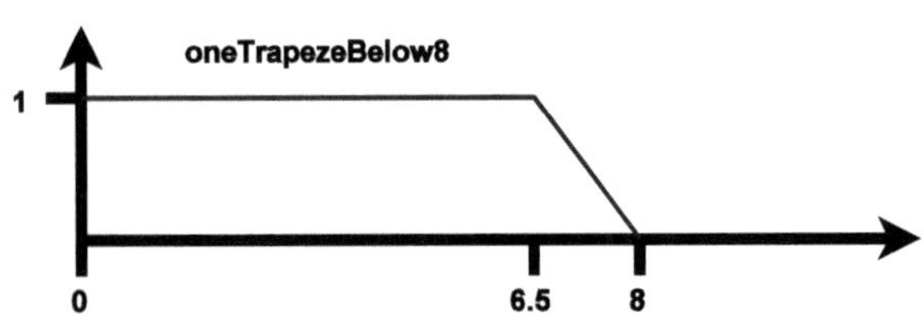

Fig. 5. Experimental fuzzy function

Then for solving GMAX-FCMO between the query and every protein, we used an adapted version of the previously mentioned Multi-Start Variable Neighborhood Search (VNS) metaheuristic developed for MAX-CMO [5], that is also publicly available online as one of the methods used in the ProCKSi server [4]. This algorithm follows a standard VNS algorithm structure with just a few changes: there is an extra Multi-Start loop to better explore the solution space and reduce the big influence of a single initial solution; and a "simplify" function that is used after every local search to remove pairings that do not contribute to the solution fitness (it helps to avoid the saturation of the solution with useless pairings). The algorithm also uses reduced solution evaluation to recompute the cost of a neighbor solution considering only the changes from the current one, thus significantly reducing the computational time needed.

For every alternative for measuring the cost, we run the VNS metaheuristic and we normalize the overlap values with every normalization alternative proposed. The results are analyzed using ROC curve analysis and the area under the curve (AUC) values, both in terms of classification at the level of architecture and at the level of class.

Table 1. AUCs for the classification at the level of arquitecture

	Fitness	**Norm1**	**Norm2**	**Norm3**	**Normfuzzy**
PRODUCT	0,565	0,468	0,625	0,542	0,622
MIN	0,571	0,479	0,637	0,552	0,631
MAX	0,565	0,470	0,628	0,546	0,623
AVG	0,569	0,476	0,636	0,551	0,629

Table 2. AUCs for the classification at the level of class

	Fitness	**Norm1**	**Norm2**	**Norm3**	**Normfuzzy**
PRODUCT	0,569	0,419	0,553	0,491	0,580
MIN	0,573	0,426	0,557	0,498	0,584
MAX	0,574	0,420	0,561	0,494	0,585
AVG	0,575	0,426	0,561	0,497	0,586

As we can see by looking at any column on Tables 1 and 2, the AUC values do not change significantly for any of the proposed cycle contributions. More precisely, the AUC values within each column have differences of at most 0.011, so the classification performance is mostly unaffected whatever the cycle contribution is. Considering this results it is clear that the actual value for the contribution of a cycle is not important as long as it adds to the final fitness value. This probably comes from the fact that the VNS algorithm will try to pair any residues that lead to more contributing cycles. Therefore, all the relevant pairings will get added as all of them improve the solution. In this manner, the solution obtained with the VNS will remain similar with all the different cycle contributions, and that is the reason why the classification performance (AUC values) is almost identical.

As could be inferred from the areas described in Tables 1 and 2 for different normalization kinds, the further analysis of the results using ROC curves shows that *Norm2* and *NormFuzzy* normalizations are the best options for post processing the overlap in order to classify the proteins set studied. This statement can be seen visually in Fig. 6 for proteins with the same architecture and in Fig. 7 for proteins

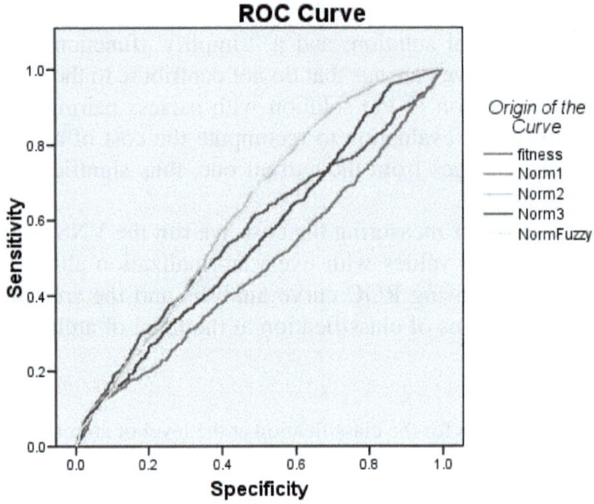

Fig. 6. ROC curves for the same architecture of proteins and all kinds of contributions

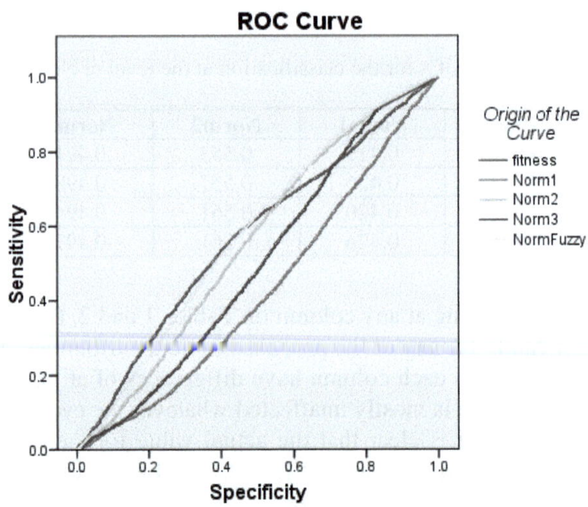

Fig. 7. ROC curves for the same class of proteins and all kinds of contributions

with the same class. These results are similar to the results obtained for the crisp model [5] in the sense that the normalization done is again an important factor on the fuzzy model. The differences are very significative among the normalizations with the aforementioned *Norm2* and *NormFuzzy* having reasonably good classification performance at the same time that *Norm1* and *Norm3* lead to worse than random performance (AUCs below 0.5).

5 Conclusions

This paper has analyzed the resolution of Bioinformatics problems by Soft Computing techniques. We have explained why Soft Computing is well suited to deal with the characteristics of the Bioinformatics problems and data. Therefore, many of the current research works done in Bioinformatics can be more productive and obtain better results if Soft Computing techniques are used.

To show how this can be done we have shown the application of this idea to the protein structure comparison problem. Firstly we described the application of a simple metaheuristic that has been useful to obtain biologically relevant results for the MAX-CMO model of protein structure comparison using limited resources. Then, we have presented the use of fuzzy contact maps for protein structure comparison. It has been shown that this generalization of the crisp contact maps was useful for classification and that the results obtained through fuzzy contact maps can outperform the results obtained through crisp contact maps. This demonstrates the suitability of Soft Computing at least for this Bioinformatics problem, but the GMAX-FCMO model had not been as extensively tested as MAX-CMO so we extended here the previous works by analyzing the influence of the computation of the contribution of each cycle and the normalization scheme when doing protein classification.

The results indicate that the strategy used to compute the contribution of each cycle to the solution is not relevant while the normalization is playing a key role. This emphasizes the importance of normalization as it has been proved to be very important both for the crisp and the fuzzy model.

As a future work we plan to conduct more analysis on the fuzzy model in two main lines: how to compute the value of a cycle when there is more than one type of contact; and which values for the distances should be used as the basis for the fuzzy functions that define each type of contact. The correct modelization of these two features will probably serve to obtain improved results and better performance thanks to the alignment of only the more meaningful residues.

Acknowledgements

This work is supported in part by Projects TIN2008-01948 from the Spanish Ministry of Science and Innovation and TIC-02970 from Consejería de Innovación, Ciencia y Empresa, Junta de Andalucia.

We also thank Lluvia Morales for her help with the computational experiments.

References

1. Caprara, A., Carr, R., Istrail, S., Lancia, G., Walenz, B.: 1001 optimal pdb structure alignments: integer programming methods for finding the maximum contact map overlap. J. Comput. Biol. 11(1), 27–52 (2004)
2. Thiruv, B., Quon, G., Saldanha, S.A., Steipe, B.: Nh3D: A Reference Dataset of Non-homologous protein structures. BMC Structural Biology 5(12) (2005)

3. Carr, B., Hart, W., Krasnogor, N., Burke, E., Hirst, J., Smith, J.: Alignment of protein structures with a memetic evolutionary algorithm. In: GECCO 2002: Proceedings of the Genetic and Evolutionary Computation Conference. Morgan Kaufman, San Francisco (2002)
4. Barthel, D., Hirst, J., Blazewicz, J., Burke, E., Krasnogor, N.: ProCKSI: a decision support system for Protein (Structure) Comparison, Knowledge, Similarity and Information. BMC Bioinformatics 8(416) (2007)
5. Pelta, D.A., González, J.R., Moreno-Vega, J.M.: A simple and fast heuristic for protein structure comparison. BMC Bioinformatics 9(161) (2008)
6. Pelta, D., Krasnogor, N., Bousono-Calzon, C., Verdegay, J.L., Hirst, J., Burke, E.: A fuzzy sets based generalization of contact maps for the overlap of protein structures. Journal of Fuzzy Sets and Systems 152(1), 103–123 (2005)
7. González, J.R., Pelta, D.A.: On Using Fuzzy contact Maps for Protein structure Comparison. In: IEEE International Fuzzy Systems Conference. FUZZ-IEEE (2007)
8. Mirny, L., Domany, E.: Protein fold recognition and dynamics in the space of contact maps. Proteins: Structure, Function, and Bioinformatics 26, 391–410 (1996)
9. Lifson, S., Sander, C.: Antiparallel and parallel beta-strands differ in amino acid residue preferences. Nature 282, 109–111 (1979)
10. Berman, H., Henrick, K., Nakamura, H., Markley, J.L.: The worldwide Protein Data Bank (wwPDB): ensuring a single, uniform archive of PDB data. Nucl. Acids Res. 35(suppl. 1), D301–D303 (2007)
11. Xie, W., Sahinidis, N.V.: A reduction-based exact algorithm for the contact map overlap problem. Journal of Computational Biology 14(5), 637–654 (2007)
12. Bourne, P., Weissig, H.: Structural Bioinformatics. Wiley-Liss, Inc., Chichester (2003)
13. Lander, E., Linton, L., Birren, B., Nusbaum, C., Zody, M., Baldwin, J., Devon, K., Dewar, K., Doyle, M., FitzHugh, W., et al.: Initial sequencing and analysis of the human genome. Nature 409(6822), 860–921 (2001)
14. Venter, J., Adams, M., Myers, E., Li, P., Mural, R., Sutton, G., Smith, H., Yandell, M., Evans, C., Holt, R., et al.: The Sequence of the Human Genome. Science 291(5507), 1304–1351 (2001)
15. Verdegay, J.L., Yager, R.R., Bonissone, P.P.: On heuristics as a fundamental constituent of soft computing. Fuzzy Sets and Systems 159(7), 846–855 (2008)
16. Laskowski, R.A.: Structural quality assurance. In: Bourne, P., Weissig, H. (eds.) Structural Bioinformatics. Wiley-Liss, Inc., Chichester (2003)
17. Krasnogor, N., Pelta, D.: Measuring the similarity of protein structures by means of the universal similarity metric. Journal of Bioinformatics 20(7), 1015–1021 (2005)
18. Holm, L., Park, J.: DaliLite workbench for protein structure comparison. Bioinformatics 16(6), 566–567 (2000)
19. Pearl, F., Bennett, C., Bray, J., Harrison, A., Martin, N., Shepherd, A., Sillitoe, I., Thornton, J., Orengo, C.: The CATH database: an extended protein family resource for structural and functional genomics. Nucleic Acids Research 31(1), 452–455 (2003)

Transforming an Interactive Expert Code into a Statefull Service and a Multicore-Enabled System

Dana Petcu and Adrian Baltat

Computer Science Department, Western University of Timisoara,
B-dul Vasile Parvan, 300223 Timisoara, Romania
petcu@info.uvt.ro
http://web.info.uvt.ro/~petcu

Abstract. Legacy codes are valuable assets that are difficult or even impossible to be rebuild each time when the underlying computing architecture is changed at conceptual or physical levels. New software engineering technologies, like the ones supporting the concept of service oriented architecture, promises to allow the easy use of the legacy codes. Despite this promise, the transition towards a service oriented architecture is not a straightforward task especially for the legacy codes with a rich user interface: the part mostly affected by the architectural change is the interface. This paper describes the transforming path that was followed in the case of a ten years old expert system for solving initial value problems for ordinary differential equations, emphasizing how the new interface should look like to preserve the code functionality despite the conceptual switch from a human user to a software code. The use of statefull Web services allows also preserving the large database of problems and methods of the expert code, as well as its special functionality that allows the extension of this database, in the benefit of any user of the service, human or another software code. This transformation path can be easily followed by other similar legacy codes, especially those designed for scientific computing. Moreover, the part of the expert system specially designed to deal with parallel computing techniques is extended in order to take advantage of the latest achievements in what concerns the hardware architectures, more precisely, to be able to exploit the advantages of multicore systems. In this context it is proved that parallelism across the numerical method that requires a small number of processors and can profit from the shared memory available to all cores can improve the response time of the expert code.

1 Introduction

The implementation of the latest software engineering concepts for software systems modernization, like service-oriented architectures, software-as-a-service, infrastructure-as-a-service, or utility computing, bring valuable options for extending the lifetime of legacy systems. Moreover, they allow the reducing of the costs of software maintenance by using software components running in remote computing centers. The availability of the technologies is pushing forward also the

H.-N. Teodorescu, J. Watada, and L.C. Jain (Eds.): Intel. Sys. and Tech., SCI 217, pp. 137–159.
springerlink.com

concept of pay-by-use that has is more difficult to adopt since it implies a fundamental change in the current ICT market. The technology demonstrators for various fields of applications can speed-up the adoption process. This is also the case of the demonstrators build for the scientific community if they are able to include in the new architectures the largely recognized and used legacy codes.

Service oriented architecture (SOA) is an already established architectural approach in which complex applications are composed from independent, distributed and collaborative components, namely the services. The big advantage is that the services can be written in different programming languages, can be deployed on various operating systems, hardware platforms or security domains, and still are able to cooperate. The key of the interoperability is the usage of standards in what concerns the service description, discovery and composition. The basic components are the service registers, the service providers and the service requestors, while the basic operations are the service publishing, the service search, the binding between the requestor and the service, and finally the service consume.

The SOA concept had in the last decade several implementations, like CORBA, Jini, JavaBeans etc. The recent most successful one is through the Web services, that use a communication model between the components based on the standard XML for specifying the data in a manner that is platform-, language-, hardware-, or software vendor- independent. Web Services Description Language (WSDL) is used to describe the service functionality and usage, Universal Description, Discovery and Integration (UDDI) for service registration and discovery, while other standards, like Simple Object Access Protocol (SOAP), for messages between services. Compared with the previous SOA concept implementations, the Web services are platform and language independent due to the usage of XML standard and allow the development of applications at Internet scale due to usage of HTTP for message exchanges. The disadvantages are the high overhead that hinder the development of application in real time, as well as the low versatility, e.g. a restricted form of the service invocation. Therefore, the Web services are adequate for weak coupled systems in which the client does not need to know anything about the Web service until it invokes it, as well as for applications at Internet scale. To deal with the low versatility of stateless Web services, several specifications have been added. We are in particular interested in this paper in the Web Service Resource Framework (WSRF) that allows to build statefull Web services. The state is maintained in a separate entity named resource, stored in server internal or external memory (it can be a variable, a file, a database, etc); a pair Web service and resource is named WS-resource, and can be addressed by a service requestor.

The most appropriate for the migration towards service-oriented architectures are the legacy systems conceived as black-boxes that are callable through a command line and having a fixed-format input and output. In the recent paper [23], we have discussed several cases that conform to these characteristics: as demonstrators, se-veral free-distributed legacy codes for image processing, symbolic computing, computational fluid dynamics, respectively evolutionary computing, were wrapped as WSRF-based services. The advantages of statefull services instead stateless Web services were proved also through the above mentioned usage examples based on client's asynchronous requests to the same WS-resource that is instantiated particularly for that client.

In this paper we are going beyond the simple case, analyzing the transformation path of a legacy code that has a rich user interface. We start with a short review of the techniques that were reported in the literature for porting the legacy codes into the new architectures. A more detailed analysis of this subject was presented recently in [6] by other authors. While [6] proposes a top-down approach, presenting a modular architecture of a generic wrapper and the steps of a migration process defining the flow of activities to be carried out for wrapping legacy system functionalities, our approach is a bottom-up one, from examples, through the general system. In particular, in this paper, we present a case study on wrapping as WSRF-based Web service an interactive legacy system, namely EpODE (acronym for Expert System for Ordinary Differential Equations), that provides numerical solutions for systems of ordinary differential equations, incorporates an expert system, and was designed ten years ago. From its new interface design, based on a wrapper meant for the interaction with software codes instead human user, we can draw some conclusions in what concerns the requirements of a general-purpose wrapper.

Section 3 presents the expert system, emphasizing its main characteristics which differentiate it from other tools and which, after ten years, are still unique, motivating its reuse as valuable component in actual architectures. The description of the statefull Web service wrapping EpODE's kernel follows in Section 4. An initial design architecture of the service was proposed in [25] and at that moment only the computational kernel of EpODE was available as service. The inclusion of the expert part as well as the rich database of problems and methods (as Resource in WSRF concept) has affected the Web service as well, and the changes are reflected in the Web service interface described in this paper compared with the one described in [25].

In order to explain the usefulness of the new service, details about the service interactions are provided through examples in Section 4. Potential usage scenarios and further developments are discussed in Section 5.

Section 6 is dedicated to a particular improvement of the legacy code in order to benefit from the availability of multicore architectures. The results presented in this paper extend the ones recently presented in the paper [24]. EpODE was initially designed to allow experimenting with different parallel computing techniques specific for the numerical methods proposed for solving large systems of ordinary differential equations. While the adaption to the new architectures is taking place for extend the lifetime of EpODE, it should take into consideration the current trends to increase the number of processors on a chip. It is well known that the extent to which software can be multithreaded to take advantage of the multicore chips is likely to be the main constraint on software performance in the future. Therefore we consider that the transformation of the expert system into a multicore-enabled one is a step further in its life extension. Moreover, this extension has a novelty character, since there is no report at this moment concerning the implementation of parallel techniques fro solving ordinary differential equations on multi-core architectures.

Numerical computations requiring computing power as well as large memory are well-suited candidates for deriving advantages from multicore architectures. In this context, it is necessary to design and implement new libraries and tools for parallel numeric computations, especially for parallel computing environments using multicore processors. One can notice that several parallel numeric computation packages were designed two decades ago assuming a shared-memory parallel computing and using multi-threading technique. The evolution towards distributed-memory parallel computers and clusters of workstations leaded to the impossibility to use the shared-memory parallel codes and to the need of designing and implementing new versions that are suited for distributed memory. In particular, for the case of computing the numerical solutions of ordinary differential equations this architectural change had a tremendous effect on the efficiency of parallel computing techniques that were used: the class of parallel methods considered to be well suited for parallel implementation has been changed from that applying parallelism across the method and across the problem (requiring a small number of processors sharing a common memory) towards that applying parallelism across the steps (requiring a large number of processors for almost independent iterative processes). The question that we pose in this paper is if we can reconsider again as efficient the parallelism across the method by exploiting multicore architectures. We prove in Section 6 that the answer is positive.

2 Wrappers for Legacy Codes

The transformation of a legacy system into a service is not a straightforward task. As stated also in [6], a main problem is that of establishing which part of the legacy system can be exposed as service. Another one is that of establishing how the transformation will be done technically.

The problem of integrating legacy software into modern distributed systems has two obvious solutions: the creation of wrapper components or the reengineering of the legacy software. Due to the complexity of most of the legacy systems, the first approach is more appropriate, as stated in [10, 30].

Consequently, the first class of techniques comprises the black-box reengineering techniques which integrate systems via adaptors that wrap legacy code as service. The legacy code is conceived as a box with specified input and output parameters and environmental requirements. This scenario is common especially when the source code is not available. Non-invasive approaches were recently discussed in papers like [1] and [12]. Recently reported technologies are OPAL [16] and O'SOAP [26] that also allows legacy command-line oriented applications to be deployed as stateless Web services without any modification. A solution for the particular case of interactive legacy systems is described in [5]. We have also proposed recently in [7] a technical solution for the migration of the well-known interactive software tools used in the particular field of symbolic computations.

The second class of techniques comprises white-box methods that require code analysis and modification to obtain the code components of the system to be presented as services. A technique from this class is based on invasive procedures on the legacy codes that usually improve the efficiency of legacy code. In such invasive approach, it is typically assumed that the application programmer has some programming background and would like to build service based applications using specific software libraries. The solutions that were proposed until now are based on the principles outlined in [18] and use Java wrapping in order to generate the interfaces automatically. Prominent examples are SWIG and MEDLI [15], JAVAW [13] and Soaplab [31]. Particularly for mathematical problems is JavaMath [29], a Java API for connecting to mathematical software systems. A request is made to a broker for a particular software system; the broker establishes a session to such system. OpenMath, an extensible XML-based standard for representing the semantics of mathematical objects, can be used. Despite the fact that a specific abstract interface for service access is given, there is little abstraction from service implementation.

Both above-described approaches are valid in different circumstances, depending on factors such as the granularity of the code, the assumed users and application area.

In this paper we make use of a technique from the third class mentioned in [6], of the grey-box techniques, that combines wrapping and white-box approaches for integrating the parts of the system that are more valuable: in our case these parts concern the computational kernel, the expert, and the problem and method database. Moreover, the computational kernel is extended to be able to exploit multicore architecture's benefits.

There are several research efforts aiming at automating the transformation of legacy code into a statefull Web service, in particular using Globus Toolkit implementation of WSRF concepts. These approaches are also either invasive (white-box) or non-invasive (black-box). The most remarkable non-invasive solution is represented by GEMLCA, the Grid Execution Management for Legacy Code [9,13]. The deployment process of a legacy code as GEMLCA service requires only a user-level understanding of the legacy application (e.g. parameters of the legacy code, kind of environment needed to run). GEMLCA provides the capability to convert legacy codes into Grid services by describing the legacy parameters and environment values in the XML-like file. A drawback is that it supposes that the legacy code is activated in a command-line style and does not exploit the possible successive interactions as described in [24] and in the case of the EpODE wrapper.

According to [28] software can be encapsulated at several levels: job level, transaction level, program level, module level and procedure level. The way that the wrapper interacts with the legacy software component depends on the native technology of the legacy component. The wrapper may use TCP/IP sockets, redirecting of the I/O streams for a process, input and output files or it may have to use JNI encapsulation. In our case we used input and output files as well as redirecting of the I/O streams.

3 EpODE's Components and Characteristics

EpODE, ExPert system for Ordinary Differential Equations, was designed ten years ago as a tool for solving numerically large systems of ordinary differential equations (ODEs). It has the following major components:

1. A user interface, the front end, allowing the description of initial value problems for ordinary differential equations, iterative methods, control of the solution computation process, and interpretation of the results of the computation; help facilities are provided in order to assist the user in its interaction with the tool.
2. A properties detection mechanism including the procedures for establishing the main properties of the systems of ordinary differential equations or the iterative numerical methods.
3. A mechanism for the selection of the solving procedure, implementing a decision tree for the selection of the class of iterative methods according to the properties of the initial value problem for ODEs and for the selection of one method from this class according to the solution accuracy requirements and time restrictions.
4. A sequential computing procedure, a generic solving procedure whose parameters are the current problem and the already selected numerical non-parallel method.
5. A parallel computing procedure, a generic solving procedure with message passing facilities for intercommunication of two or more computation process, implementing the techniques of parallelism across method, problem, or steps.

EpODE is considered to be an expert system due to its facilities to automatically identify the problem properties and to match them with the properties of the avai-lable solving methods. It can choose automatically not only the adequate method and computing parameters like the step size, but also it can switch to the usage of parallel computing techniques in the case when the estimated time for solving the problem is too high and there are more than one machine in the parallel virtual machine in PVM-based environment of the user.

After describing or loading a problem, the human user receives the following information about the problem's properties: it is linear or not, it is spare or not, it is separable or not, and in case of a positive answer, the set of subsystems, it is stiff or not, the stiff ratio and the estimated value of the biggest and the smallest real parts of the eigenvalues of the Jacobian matrix of the system evaluated at the lowest value of the integration interval. The separability property is a key issue in applying the parallelism across the problem.

After describing or loading a method, the human user receives the following information about the method's properties: it is implicit or not, onestep or multistep, onstage or multistage, onederivative or multiderivative, it is zero-stable or not, its error constant, its order, as well as a flow graph for applying the method that highlights if the method has a degree of parallelism higher than one (if some method equations can be solved simultaneously).

Given a problem and a method, the human user can proceed to obtain the numerical solution, giving a scheme for applying the method: the admissible error level, usage or not of a constant integration step, number of steps to be maintained constantly if a variable step is used, as well as the size of the integration step. The expert advises the user about the recommended scheme for computation according the problem's properties, the method's properties, and the admissible level of the error: the upper limitation of the step size, the usage of constant step size, and the estimated time for computation. In the case of unacceptable long time, the user can decide to apply parallelism across problem, method or steps, in the context that PVM is installed at user site and he or she manually adds available computers from user‘s machine neighborhood. After the solution computation, the numerical results can be visualized in a table form, in a two- or tree-dimensional graphical form, or saved in a textual form to be used by other software tools.

The human user can decide to allow the expert tool to select the method and the computing parameters. A simple decision tree is implemented to select the method to be loaded from the databases, according to the problem's properties and the admitted level of error.

EpODE can be used as a tool for describing, analyzing and testing new types of iterative methods for ODEs, mainly due to the method properties detector, as well as the immediate possibility to apply them on a large class of problems. In particular, it allows also the study of the methods that are proposed for parallel or distributed implementation using real or simulated parallel computing environments.

It is important to notice also that EpODE is freely distributed with a rich database of test problems and of solving methods, including the standard test problems presented in [19], the sequential or parallel methods described in classical books like [4], as well as methods proposed by EpODE's designer that were described in [20]. At least one hundred real and test problems can be loaded. Almost one hundred methods are also included in distribution: Runge-Kutta methods, multistep methods, multi-derivative multistep methods, block methods, hybrid methods, nonlinear multistep methods, general linear methods. These databases can be enriched by the human user with its own defined problems or methods, allowing an easy and comprehensive comparison with classical problems and methods that are already in the database.

At the time of EpODE's design the software tools that were frequently used were limited to a small number of equations and EpODE has eliminated this constraint by introducing a dynamic memory allocation scheme.

The main characteristics of EpODE distinguishing it from other ODE solving environments are the followings:

a) friendly interpreter for describing problems and solving methods;
b) the solvers are implemented in a uniform way: all solvers behave in a coherent way and have the same calling sequence and procedure;
c) the large extensible database of problems and methods.

Details about EpODE design are given in [21] and several experiments on clusters were reported in [22]. Tests were concern mainly with large systems of equations as well as stiff systems.

At the time of its design, EpODE was the unique tool that allowed the above mentioned facilities. Only a recent developed tool reported in [3] has similar facilities (without the ones for parallelism).

EpODE was written in C++ and two graphical user interfaces were provided, for Windows'95 and X Windows. One drawback currently affecting is usage is the fact that its interfaces are not portable. A transformation of its interface into a Web service wrapper does not only prolong its lifetime, but allows the extension of the user classes, from humans to software codes. The next section discusses this transformation. Most parts of the interface can be found in the operations set of the new service, and we consider that this is the right way to port the legacy code, with the exception of the case when the number of the operations is too high and can be a security threat to the remote system (these kind of problems have been raised in [7]).

Concerning the state-of-the-art in parallel techniques for solving ordinary differential equations, one should note that the rapid development of the hardware in the last ten years have affected the notion of the most adequate technique. In EpODE all three variants of parallelism were implemented and ten years ago was pointing that parallelism across steps is most efficient. A rerun of the experiments reported in [22] revealed that the current hardware improvements led to a response time of the computational procedures hundreds of times shorter. In these conditions the problem dimension for which the parallel computing techniques are efficient, in the sense that the computational time dominates the communication time, is increasing by at least ten times. In this context, Section 6 discusses the suitability of the well known parallel techniques on the new parallel computing architectures, the multicore-based one.

4 EpODE as a Statefull Web Service

A statefull Web service, WSEpode, was build as wrapper of EpODE's components. We have use the WSRF implementation included in Globus Toolkit 4. We applied the same technique that was reported in [23] for other legacy codes wrapped in the black-box style. The wrapper is written in Java. The Web service is registered in a service container according the WSRF specifications. Axis was used as implementation of the SOAP specification requested by WSRF. The file describing the service in WSDL was generated with the Java2WSDL tool of Axis.

The following EpODE's components were isolated to lead to a stand-alone code, named here CL-EpODE (command-line version):

1. The computational engine which provides the numerical solutions if the problem, the method, and the computation parameters are given;
2. The expert part that detects the problem and method properties and give advise about the computation parameters;
3. The loading and saving procedures for problems, methods, sessions descriptions and saving procedures for the computation results (text files).

The part that was isolated and not included in the code is the one related to the user interface, including also the result visualization module. The new partial code can work in a command-line mode requiring inputs from files and providing the results in another file. Its loops infinitely, waiting inputs from the terminal, expressing commands like load a problem, method or session file description, compute the solution, or save the current data.

The Web service was build in three phases. In the first phase the computational kernel was wrapped as service – the interface was described in [24] and allows a small set of operation to be invoked in order to describe the problem, the method and the computation parameters, as well as launch a computation and retrieve the results; the drawback was the fact that the problem and method were necessary to be written in Polish form of their expressions, the expression interpreter being part of the expert that was not ported at that moment. In the second phase the expert part was ported, allowing the description of the problem and method in a mathematical form, as well as the add of the new operations for problem and method properties detection. In the third phase the I/O procedures were ported, allowing the extent of the method and problem databases.

The current available operations of the Web service as well as their I/O data structures as appear in the WSDL file of the service are shortly presented in Table 1. The operations are allowing the following actions:

a) description of the problem (`setProblem`), the method (`setMethod`) and the computation parameters (`setSession`), with a well defined complex-type XML structure of the input data;
b) launch of the computation in expert or non-expert mode (`compute`);

Table 1. Instance service operations

Operation name	Action	Input name and data type		Output name and type	
setProblem	specify the problem	title indep dep depexpr t0 t1 iniexpr	xsd:string xsd:string, unbounded xsd:string, unbounded xsd:string, unbounded xsd:float xsd:float xsd:string, unbounded	setProblemReturn	xsd:boolean
setMethod	specify the method	sta fin plu staexpr finexpr pluexpr ecmpas ecimpl isolve save coef	xsd:string, unbounded xsd:string, unbounded xsd:string, unbounded xsd:string, unbounded xsd:string, unbounded xsd:string, unbounded xsd:string, unbounded xsd:string, unbounded xsd:boolean xsd:int xsd:float	setMethod Return	xsd:boolean
setSession	specify the computation parameters	cons nstepc erlevel h parallel	xsd:boolean xsd:int xsd:float xsd:float xsd:int	setSession Return	xsd:boolean
compute	computation request	expert	xsd:boolean	setCompute-Return	xsd:boolean
getProp-Problem	get the problem properties	-	-	getPropProb-lemReturn	xsd:string, unbounded
getPropMe-thod	get the method properties	-	-	getPropMe-thodReturn	xsd:string, unbounded
getStatus	get the computation status	-	-	getStatus-Return	xsd:boolean
getResults	get the computation results	-	-	getResultsRe-turn	xsd:string, unbounded

c) retrieve of the problem's (`getPropProblem`) and method's properties (`get-PropMethod`), as an array of strings of proprieties names or values;
d) retrieve of the computation status (`getStatus`) as a Boolean value, and the computation results (`getResults`) as an array of strings.

Details about the operation usage are given in the next section.

Before any specification of the problem to be solved, the service requestor must contact the factory service that will create a WS-resource: the instance of the service that allows the operations exposed in Table 1, and the resource that in this case refers to a database with methods and problems. The factory service has one operation, `createResource`.

The instance service will perform the following actions:

Action 1. As result of `setProblem`, `setMethod`, or `setSession` invocation, it writes in a specific text file the problem parameters and signals to the CL-EpODE, launched in a batch mode by the factory service, to load the description file (redirected I/O stream between the service instance and the legacy code);

Table 2. Resource operations

Operation name	Action	Input data type	Output data type
listProblem	get the list of problems registered at the resource	-	xsd:string, unbounded
listMethod	get the list of methods registered at the resource	-	xsd:string, unbounded
loadProblem	set a problem identified in the resource by its acronym	xsd:string	xsd:boolean
loadMethod	set a method identified in the resource by its acronym	xsd:string	xsd:boolean
loadSession	set a computation session identified in the resource by its acronym	xsd:string	xsd:boolean
saveProblem	store the currently set problem in the resource under the given acronym	xsd:string	xsd:boolean
saveMethod	store the currently set method in the resource under the given acronym	xsd:string	xsd:boolean
saveSession	store the currently set session in the resource under the given acronym	xsd:string	xsd:boolean

Action 2. As result of a `compute` invocation, the service signals to CL-EpODE the request. If the expert mode is used, the code will ignore the method and computation parameters that were set and will load the most appropriate method according to the internal implementation of the decision tree. If the non-expert mode is used and the problem, method or computation parameters where not set, the test problem (y'=-y, y(0)=1), or the explicit Euler method, or the implicit computational parameters (e.g. the step size is 0.01) will be used.

Action 3. During a computation (either of the problem or method properties, or a solution computation) the resource variable `Done` is set on `false`, its implicit value being `true`. As result of a `getStatus` invocation, the current setting of `Done` is returned to the service requestor.

Action 4. As result of `getResults` invocation the computation's results are read from the text file provided by CL-EpODE: each line is transformed in a string and the array of strings is send to the service requestor. If the status of `Done` is `false`, an empty array is returned.

One of the advantages of using EpODE was its facility to increase the database of problems and methods. In order to preserve this advantage, a I/O scheme was designed. The database is seen as a persistent resource lying at the instance service's side. Multiple instances (representing multiple service requestors) can invoke the same resource, in this way benefiting from the new knowledge that is accumulated. Moreover, different service requestors can extend the same database. Saving a problem or a method under the same name as an already existing one that correspond to a read-only description file is not allowed (to avoid the erase of the basic methods and problems). Table 2 describes the main operations that are allowed on the resource.

The invocation of `listProblem` or `listMethod` allows browsing the repertory of registered problems and methods, by their acronyms (identical with the name of the text files containing their description in textual form) that are the same as the ones used in the literature. The invocation of `loadProblem` or `loadMethod` lead to a signal to the CL-EpODE to load the description file specified by its name. The reverse operation, of saving, leads to a signal addressed to the legacy code to save a description file using as file name the string that is given at the operation invocation. A session refers to a combination of problem, method and computation parameters.

After the creation of the resource, the service requestor uses an endpoint reference to its personal WS-resource. The above described instance service operations are designed to be invoked in an asynchronous manner: the service requestor receives almost immediate a response to any operation invocation, and the order of the operation invocations is not predefined. Moreover, using the WSRF, it is possible to maintain the current settings of the legacy codes (its state).

We should underline that the above described wrapping process can be repeated for any legacy code if the following actions can be performed:

- one-to-one transformation of the actions associated with the user interface into service operations;
- isolation of the parts of the legacy code that are behind the actions associated with the user interface and build a new version of the code that accepts inputs from files or keyboard and store the results into specific files.

5 Examples of Input and Output Data

In what follows we describe several examples of input and output data that can be provided by and to a service requestor. To simplify the presentation, we assume that for all the following examples, the service requestor is a Java code.

We start with a set of problems. The first one is a small test problem for which the exact solution is known:

$$y_1'(t)=5y_2(t)-6y_1(t)+2\sin t,\ y_2'(t)=94y_1(t)-95y_2(t),\ t_0=0\le t\le 100=t_1,\ y_1(0)=0, y_2(0)=0$$

In order to invoke the `setProblem` operation of the instance service, the Java client code should set the problem parameters:

```
String title="Test problem", indep="t";
String[] dep={"y1","y2"},
         depexpr={"5*y2-6*y1-2*sin(t)","94*y1-95*y2"},
         iniexpr={"0","0"};
float t0=0, t1=100;
...
boolean b=WSEpode.setProblem(title,indep,dep,depexpr,
                              t0,t1,iniexpr);
```

These variables values are similar to the ones that one can find in the text files des-cribing problems stored in the initial EpODE's directories.

An invocation of the `getPropProblem` will return an array of strings with the values of the following variables:

```
boolean linear, separ, sparse;
double max_eigenvalue,min_eigenvalue,stiff_ratio;
int stiff_class;
String[] subsets;
```

For the above described simple problem, the array of strings will contain the values detected by the CL-EpODE's expert component:

```
linear: "true", separ: "false", sparse: "false",
min_eigenvalue: "1", max_eigenvalue: "100",
stiff_ratio: "100", stiff_class: "2", subsets: ""
```

Another test problem is the stiff problem B1 described in [11]:

$$y_1'=y_2-y_1,\ y_2'=-100y_1-y_2,\ y_3'=y_4-100y_3,\ y_4'=-100y_4-10000y_3,\ t_0=0\le t\le 20=t_1,$$
$$y_1(0)=1,\ y_2(0)=0,\ y_3(0)=1,\ y_4(0)=0$$

for which the Java code should mention:

```
String title="B1", indep="t";
String[] dep={"y1","y2","y3","y4"},  depexpr={
     "y2-y1","-100*y1-y2","y4-100*y3","-100*y4-1000*y3"},
     iniexpr={"1","0","1","0"};
float t0=0, t1=20;
```

In this case, the `getPropProblem` invocation will return with the array of strings:

```
Linear: "true", separ: "true", sparse: "true",
min_eigenvalue: "1", max_eigenvalue: "100",
stiff_ratio: "100", stiff_class:"2",
subsets: "y1,y2","y3,y4"
```

Note that the system is separable and the parallelism across the system can be applied.

A more complicated example that we use in the tests reported in the next section is obtained by the spatial discretization (using the method of lines) of the Medical Akzo Nobel problem [19] defined in the study of the penetration of radiolabeled antibodies into tumor tissue:

$$u_{1t}=[(x-1)^4 u_{1xx}+2(x-1)^3 u_{1x}]/16-100u_1u_2,\ u_{2t}=-100u_1u_2,\ u_1(x,0)=0,\ u_2(x,0)=1,$$
$$u_{1x}(1,t)=0,\ u_1(0,t)=2 \text{ for } 0\leq t\leq 5 \text{ and } 0 \text{ for } 5< t\leq 20,\ \ 0\leq x\leq 1,\ 0\leq t\leq 20,$$

In the case when the discretized system has 160 ODEs, the Java client will mention:

```
indep="t"; dep={"y001","y002",…,"y160"};
depeprx={ "80^3*(79*y003-160*y001+81*(1+sqrt((t-5)*(t-5))/
(5-t)))/(16*6561)-100*y001*y002",
"-100*y002*y001",
"79^3*(78*y005-158*y003+80*y001)/(16*6561)-100*y003*y004",
"-100*y004*y003",
"78^3*(77*y007-156*y005+79*y003)/(16*6561)-100*y005*y006",
…,
"2^3*(y159-4*y157+3*y155)/(16*6561)-100*y157*y158",
"-100*y158*y157",
"2*(y157-y159)/(16*6561)-100*y159*y160",
"-100*y160*y159"};
iniexpr={"0","1","0","1",…,"1"} ;
t0=0, t1=20
```

For this problem, the `getPropProblem` will return

```
linear: "false", separ: "false", sparse: "true",
min_eigenvalue: "0", max_eigenvalue: "900",
stiff_ratio: "inf", stiff_class: "5", subsets: ""
```

Another problem that we mention in the next section is the one obtained thought the same discretization procedure (the method of lines) in the case of the mathematical model of the movement of a rectangular plate under the load of a car passing across it (plate problem [14]):

$$u_{tt}+1000u_t+100\Delta\Delta u=f(x,y,t),\ u|_{\partial\Omega}=0,\ \Delta u|_{\partial\Omega}=0,$$
$$u(x,y,0)=0,\ u_t(x,y,0)=0,\ \ t_0=0\leq t\leq 7=t_1,\ \Omega=[0,2]\text{x}[0,4/3]$$

where f is the sum of two Gaussian curves with four wheels which move in the x-direction:

$$f(x,y,t)=200(\mathrm{e}^{-5(t-x-2)\,(t-x-2)}+\mathrm{e}^{-5(t-x-5)\,(t-x-5)})$$

Applying the method of lines, the partial derivatives in x and y directions are replaced with approximations evaluated at different spatial grid points. The procedure leads to an ODE system with independent variable t. The number of spatial grid points depends on the accuracy required in the PDE solution. As the accuracy requirements increases, the spatial grid needs to be refined and this leads to a larger

ODE system. In the case of the discretization on a very simple grid of 10 x 7 points in the domain Ω (more in x direction), 80 ODEs are generated for the interior points. The data that are prepared by a Java service requestor for WSEpode operation `setProblem` should be similar to:

```
indep="t"; dep={"y01","y02",…,"y80"};
depeprx={"y41","y42",…,"y80",
 "-1000*y01-25./4*6561*(20*y01-8*y02-8*y09+2*y10+y17+y03)",
 …
 "-1000*y09-25./4*6561*(20*y09-8*y10-8*y17-
  8*y01+2*y18+2*y02+y25+y11)+200*(exp(-5*(t-2./9-2)*
  (t-2./9-2))+exp(-5*(t-2./9-5)*(t-2./9-5)))",…};
iniexpr={"0","0", …,"0"};
t0=0, t1=7;
```

The properties returned by the `getPropProblem` invocation should be:

```
linear: "true", separ: "false", sparse: "true",
min_eigenvalue: "0", max_eigenvalue: "1400",
stiff_ratio: "inf", stiff_class: "5", subsets: ""
```

Concerning the method description, we should start with a very simple method, the implicit Euler rule:

$$Y^{(n+1)}=Y^{(n)}+hF(t_n,Y^{(n+1)}),\ t_n=t_0+nh,\ n=0\ldots N : t_N\leq t_1$$

The Java client code should be as the follows:

```
String title="Euler implicit method";
String[] sta={"x"},fin={"y"},plu={},
         staexpr={"y"}, finexpr={"x+h*fct(y)"}, pluexpr={},
         ecmpas={}, ecimpl={"x+h*fct(x)"};
boolean  isolve=true;
int save=0;
float coef=1;
…
boolean b=WSEpode.setMehod(title,sta,fin,plu,staexpr,
          finexpr,pluexpr,ecmpas,ecimpl,isolve,save,coef);
```

where `isolve` states that Newton iterations will be used for solving the implicit equations (instead simple iterations); `ecimpl` indicates the starting values for these iterations; `save` indicates which variable's values will be saved as results at each integration step (`fin`[0], i.e. `x` in this case); `coef` specifies the scale of the step size with which the independent variable will be incremented in order to proceed with the next integration step.

An invocation of `getPropMethod` returns a string array that contains the values of the following variables:

```
boolean implicit, multistep, multistage, multider,
zero_stab;
int order, par_degree;
double conster;
```

For Euler's implicit rule, it is expected that WSEpode's expert should provide the following values:

```
implicit: "true", multistep: "false", multistage: "false",
multider: "false", zero_stab: "true", order: "1",
par_degree: "1", conster: "0.5"
```

For the more complicated diagonally implicit Runge-Kutta method presented in [14] and denoted by DIRK4 in what follows:

$$Y^{(n+1)}=Y^{(n)}+h(11k_1+25k_2+11k_3+25k_4)/72,\ n=0,1,...$$
$$k_1=F(t_n,Y^{(n)}+hk_1)$$
$$k_2=F(t_n+3h,Y^{(n)}+3hk_2/5)$$
$$k_3=F(t_n+5h,Y^{(n)}+h(171k_1\text{-}225k_2+44k_3)/44)$$
$$k_4=F(t_n+2/5h,Y^{(n)}+h(39k_2\text{-}43k_1+12k_4)/20)$$

the Java client code should contain:

```
String title="DIRK4";
String[] sta={"x"},fin={"y"},plu={"k1","k2","k3","k4"},
         staexpr={"y"},
         finexpr={"x+h*(11*k1+25*k2+11*k3+25*k4)/72"},
         pluexpr={"fct(x+h*k1)","fct(x+3*h*k2/5",
         "fct(x+h*(171*k1-215*k2+44*k3)/44)",
         "fct(x+h*(39*k2-43*k1+12*k4)/20)"},
         ecmpas={},
         ecimpl={"fct(x+h*fct(x))", "fct(x+3*h*fct(x)/5)",
                 "fct(x)", "fct(x+2h*fct(x)/5)"};
boolean isolve=true; int save=0; float coef=1;
```

and the properties expected to be provided by WSEpode are:

```
implicit: "true", multistep: "false", multistage: "true",
multider: "false", zero_stab: "true", order: "3",
par_degree: "2", conster: "0.2194583"
```

Another method that is used in the next section is the block method, denoted here with BL2:

$$Y^{(n+5/3)}=Y^{(n+2/3)}+h(19F^{(n+2/3)}\text{-}24\ F^{(n+1/3)}+9\ F^{(n)})/4,\ n=0,1,...$$
$$Y^{(n+4/3)}=Y^{(n+1/3)}+h(19F^{(n+2/3)}\text{-}8\ F^{(n+1/3)}+3\ F^{(n)})/4$$
$$Y^{(n+1)}=Y^{(n)}+h(3F^{(n+2/3)}+F^{(n)})/4 \quad \text{with} \quad F^{(n+a)}=F(t_n+ah,Y^{(n+a)})$$

The Java client should mention:

```
String title="BL2";
String[] sta={"z,y,x"},fin={"u,v,w"},plu={},
  staexpr={"w","v","u"},
  finexpr={"x+h*(3*fct(z)+fct(x))/4,
          "y+h*(9*fct(z)-8*fct(y)+3*fct(x))/4",
          "z+h*(19*fct(z)-14*fct(y)+9*fct(x))/4"},
  pluexpr={},
  ecmpas={"x+h/6*(fct(x)+3*fct(x+4*h./9*fct(x+2*h/9*fct(x))))",
         "x+h/12*(fct(x)+3*fct(x+2*h/9*fct(x+h/9*fct(x))))"},
```

```
ecimpl={};
boolean isolve=false;
int save=0;
float coef=1;
```

and the returned properties will be:

```
implicit: "false", multistep: "true", multistage: "false",
multider: "false", zero_stab: "false", order: "3",
par_degree: "3", conster: "0.0046296"
```

For the following predictor-corrector method, denoted here with PC5:

$$Y_P^{(2n+2)}= Y_C^{(2n-2)}+4h\,F_P^{(2n)},\ Y_P^{(2n+1)}= Y_C^{(2n-2)}+3h\,(F_P^{(2n)} + F_P^{(2n-1)})/2,\ n=0,1,...$$
$$Y_C^{(2n)}= Y_C^{(2n-3)}-h(3F_P^{(2n)}-9F_P^{(2n-1)})/2,\ Y_C^{(2n-1)}= Y_C^{(2n-3)}-2hF_C^{(2n-2)},\ F^{(n+a)}=F(t_n+ah,Y^{(n+a)})$$

the Java service requestor code can contain:

```
String title="PC5";
String[]
  sta={"r","q","x","s"},fin={"y,z,u,v"},plu={},
  staexpr={"u","z","y"},
  finexpr={"s+2*h*fct(s),"s-h*(3*fct(q)-9*fct(r))/2",
           "x+4*h*fct(q)","x+3*h*(fct(q)+fct(r))/2"},
  pluexpr={},
  ecmpas={"s+2*h*fct(s)","s+3*h*fct(s)","s+h*fct(s)"},
  ecimpl={};
  boolean isolve=false;
  int save=0;
  float coef=3;
```

the expected returned properties being:

```
implicit: "false", multistep: "true", multistage: "false",
multider: "false", zero_stab: "false", order: "3",
par_degree: "4", conster: "0.4600795"
```

In the case of the simple test problem and the Euler's implicit rule, a `setSession` invocation can be done by a Java code as follows:

```
boolean cons=true;
int nstepc=1;
float erlevel=(float)0.00001;
float h=(float)0.01;
int parallel=0;
...
boolean b=WSEpode.setSession(cons,nstepc,erlevel,h,parallel);
```

which states that a constant step size will be used (in this case the `nstepc`, stating how many constant steps will be maintained before a step size change, is neglected). Moreover, the step size is 0.01, no parallelism technique should be applied, and the level of the admissible error is 10^{-5}.

After a `compute(false)` invocation and a check of the computation status with `getStatus`, `getResults` returns an array of strings that contains on each string the values of the dependent variables for one specific value of the independent variable. In the case of the above mentioned session, the results will look as follows:

```
"t=0.000000 y1=0.000000 y2=0.000000"
"t=0.010000 y1=0.000200 y2=0.000000"
"t=0.020000 y1=0.000586 y2=0.000186"
"t=0.030000 y1=0.001156 y2=0.000556"
...
"t=100.0000 y1=-1.195908 y2=-1.285780"
```

Comparing this solution with the exact one, one can conclude that the required error level was not attained. The reason consists in the fact that the setting h=0.01 has not take into account the accuracy restrictions. If `compute(true)` is used, the expert recommends h=0.0105 due to the stability restrictions, but an h=0.006 due to accuracy restriction, and consequently the solution accuracy will be improved.

The `parallel` variable is used to signal the option for parallel computing as follows: 0 – no parallelism, 1 – parallelism across the method, 2 – parallelism across the problem, 3 – parallelism across the steps, 4 – using multi-threading. For the sake of the initial testing, only the option for multi-threading applying parallelism across the method is active (case 4). Further developments will include into the computational kernel also facilities for multi-threading across the steps and multi-threading across the problem.

6 Comments on the Usage in Complex Scenarios

The main advantage of transforming the computational kernel into a Web service is the fact that the client of the Web service can be any software code that sends the input data in the requested format. We imagine the case when the Web service is called, for example, by another numerical software code that solves partial differential equations and during its solving procedure it transforms the problem into a large system of ordinary equations. Note that the largest ODE systems frequently used in testing ODE software tools are provided by such discretization processes [4].

Using the discovery and introspection tools currently available, as the one described in [7], WSEpode's functions can be discover and generic clients can be automatically generated and used by various software codes. For example, the symbolic computing tools that are not specialized in numeric computations can invoke WSEpode to perform a specialized task, like detecting the properties of an iterative method, or the stiffness of a problem. In this context we should point that the middleware recently developed and described in [7] allows the invocation from within several computer algebra systems of any operation of stateless or statefull Web services, providing only the address of the service registry. Further developments of the same middleware allow the simple composition of mathematical services based on workflows, assuming their registration to a specific type of registry for mathematical services [8].

A drawback of the usage of EpODE is the context of mathematical service composition is that it does not currently use OpenMath or MathML encoding for the mathematical objects that are transmitted. This issue will be treated in the near future by incorporating into the instance service parsers for OpenMath and MathML objects. Renouncing completely at current style of communicating the date is not completely possible since not all the current available mathematical services have adopted OpenMath or MathML as standards in the message exchange.

Mathematical Services Description Language (MSDL [2]) was introduced to describe mathematical Web services so that these services can be discovered by clients. It implements a service description model that uses a decomposition of descriptors into multiple inter-linked entities: problem, algorithm, implementation, and realization. The session concept of EpODE, including a problem, a method, and computation parameters, and its current implementation as complex data structure described in the service WSDL file is close to the specifications of the MSDL. This relationship will be analyzed deeper in the future.

Note that EpODE is currently wrapped as a black-box. The development of efficient tools that solves several issues with which EpODE deals in its particular way, not exactly the most efficient one, generates the question if it is no more adequate to create a new version of EpODE that is composed by several other services. For example the symbolic described Jacobian requested by EpODE computational procedures can be easily generated with a computer algebra system – an older software tool, ODEXPERT [17] uses Maple for this task, for example. An answer to this question should be provided in the near future. We can imagine a complex scenario in which several Web services are composed: one that is generating the ODE system, another for computing the Jacobian, both wrapped as Web services, are sending the necessary information for the EpODE's expert system (a separate Web service) that picks an appropriate method from its rich database and ask another Web service to perform the computation, which at his turn sends the numeric results to be interpreted by a visualization tool also wrapped as Web service.

7 Multi-threading Functionality for Multicore Architectures

The EpODE's part considered to be the most computational intensive consists in the generic numerical solving procedures for iterative methods applied to initial value problems for ordinary differential equations. The procedures for explicit and implicit numerical methods are generic in the sense that they do not depend on the specific problem or the particular method – the concrete problem and methods are given as parameters. Since there is no need of user interventions in the computational process, and simultaneously is a need for a fast response, this part of EpODE is well suited for transformation into a computational service lying on a remote high-performance server or cluster.

EpODE allows experiments with the well-known parallel techniques proposed to fasten the solving process of initial value problems for ordinary differential equations. As mentioned above and in [4] there are three classical techniques:

1. parallelism across the problem that depends on the degree on the sparsity of the system's Jacobian matrix;
2. parallelism across the method that depends on the number of the method equations that can be solved simultaneous;
3. parallelism across the steps that allows a higher degree of parallelism than the above techniques with the drawback of a heavy control of the convergency of the numerical solutions towards the exact one.

According to the technique of parallelism across the system various components of the system of ODEs are distributed amongst available processors. This technique is especially effective in applying explicit solving methods and when the system can be split into a number of independent systems (an uncommon case). EpODE detects the sparsity of the system and allows applying the technique of parallelism across the system. The efficiency results are not considerable affected by the hardware changes since the computations are almost independent.

According to the technique of parallelism across the method, each processor executes a different part of a method. This approach has the triple advantage of being application-independent (it does not require user intervention or special properties of the given systems of ODE), of avoiding load balancing problems, and of using a small number of processors. The main disadvantage is the limited speed-up. EpODE's expert detects the degree of parallelism across the method, the maximum number of method equations that can be solved simultaneous. The efficiency results are strongly affected by the kind of memory that is used in the parallel computing environment, as well as the ratio between the communication and computation times.

The parallelism across the steps is the only possibility for using large-scale parallelism on small problems. Contrary to the step-by-step idea, several steps are performed simultaneously, yielding numerical approximations in many points of the independent variable axis (the time). Some continuous time iteration methods are used to decouple the ODE system, and henceforth to discretize the resulting subsystems, by solving them concurrently. The number of discrete points handled simultaneously is defining the degree of parallelism. The main weakness of this approach is that the iteration process may suffer from slow convergence or even divergence. Despite the fact that EpODE implements also this technique, we have not perform yet efficiency tests to see how the new hardware architecture affects the efficiency results – this is a subject for further developments.

We focus our tests on the parallelism across the method that was ten years ago a viable solution in the case of large systems of ODEs. With the increase of the computational power faster than the communication speed, parallel computations based on parallelism across the method are now justified only in the case of systems with hundreds of equations. Indeed, we have re-run the experiments reported in [22] dealing with systems of almost one hundred equations on a new generation cluster and the results show that the parallel variant is no more efficient, as was ten years ago. The new cluster characteristics are: 7 HP ProLiant DL-385 with 2 x CPU AMD Opteron 2.4 GHz, dual core, 1 MB L2 cache per core, 4 GB DDRAM, 2 network cards 1 Gb/s, while the old cluster was build on a 100Mbs network of Sun Sparc-4 stations.

The question that is normally raised by the new trends in computer architecture is that if we can improve the response time of the computational procedure implementing parallelism across the method by using multithreading when running on multicore architectures. To be able to answer to this question, we have rewrite some parts of the EpODE's C++ code of the computational procedure of EpODE. The multithreading implementation is close to the one based on the PVM library used by EpODE initial code – instead PVM processes, threads are used, and instead message passing, threads are communicating through a matrix lying in the shared memory.

The answer to the above-mentioned question is positive: the time response of the computational procedure is clearly improved using the multithreaded version. Table 3 shows the time results in the case of two classical problems of 80, respectively 140 equations solved by representative methods from different classes of parallel methods, available through the rich database of methods provided by EpODE:

- DIRK4 is a 4-stage 4th order Diagonally Implicit Runge-Kutta method mention in Section 4, while FR2 is another implicit Runge-Kutta method;
- PC5 is the predictor-corrector scheme described as well in Section 4, while PC6 is another similar predictor-corrector scheme;
- BL2 is the one-stage block method described in Section 4.

Med160 is the discretization of the Medical Akzo Nobel problem using the method of lines, mentioned in Section 4, together with the Plate80 problem, obtained following the same procedure starting from a diffusion problem. The full description of these problems and methods can be found in [4, 14].

According to the tests results we recommend the followings:

- in the case of solving systems of hundred orders of equations it is recommended that multiple CPUs are used in the solving process – this option is take into account by the computational procedure if `parallel` is set on 1;
- when the system to be solve has a number of equations of ten orders, multithreading is recommended and this option is recognized by the computational procedure if `parallel` is set on 4.
- for smaller systems, the classical non-parallel version of the computational procedure should be used – `parallel` is set on 0.

Table 3. Response times of the computational procedure with or without threads

Problem	Method		No.	Time (s)	
	Acronym	Parallelism degree	steps	No threads	With threads
Plate80	DIRK4	2	70000	12.97	9.33
	FR2	2	700000	36.22	20.47
	PC5	4	70000	16.21	6.27
Med160	PC6	2	200000	50.15	39.13
	BL2	3	200000	66.68	31.27
	PC5	4	200000	41.87	17.58

8 Conclusions and Further Developments

In order to prolong the lifetime of a legacy code, an expert system for ordinary differential equations, we have used a grey-box technique for migrating it towards a service-oriented and multicore architectures. Its unique components, the computational procedure that allows the testing of new iterative methods and the expert module, were wrapped as Web service. This service can be accessed by any software client that respects the format of the input data. The migration opens new possibilities to exploit the facilities provided by the legacy code by combining it with other services to offer more complex computational scientific services.

The adaption to the new architectures of the legacy code is not completed. While the most part of the legacy code was successfully adapted to make efficient use of multicore architectures and successfully exposed as Web service, several other components (e.g. the method and problem database) are still remaining to be translated into their service version in the very short time. Complex usage scenarios, as the ones described in Section 6, should be the context of intensive tests to be completed before the free release of the service-oriented version of the legacy code.

Acknowledgments. The project no. 11064, PEGAF, of the Romanian PNII-Partnership Programme has supported partially the research for this paper.

References

1. Balis, B., Bubak, M., Wegiel, M.: A Solution for Adapting Legacy Code as Web Services. In: Getov, V., Kiellmann, T. (eds.) Component Models and Systems for Grid Applications, pp. 57–75. Springer, Heidelberg (2005)
2. Baraka, R., Caprotti, O., Schreiner, W.: A Web Registry for Publishing and Discovering Mathematical Services. In: Procs. EEE 2005, pp. 190–193 (2005)
3. Bunus, B.: A Simulation and Decision Framework for Selection of Numerical Solvers in Scientific Computing. In: Procs. Annual Simulation Symposium, vol. 39, pp. 178–187. IEEE Computer Press, Los Alamitos (2006)
4. Burrage, K.: Parallel and Sequential Methods for Ordinary Differential Equations, Numerical Mathematics and Scientific Computation. Oxford University Press, Oxford (1995)
5. Canfora, G., Fasolino, A.R., Frattolillo, G., Tramontana, P.: Migrating Interactive Legacy System to Web Services. In: Procs. 10th European Conference on Software Maintenance and Reengineering, pp. 23–32. IEEE Computer Press, Los Alamitos (2006)
6. Canfora, G., Fasolino, A.R., Frattolillo, G., Tramontana, P.: A Wrapping Approach for Migrating Legacy System Interactive Functionalities to Service Oriented Architectures. J. Syst. Software 8(4), 463–480 (2008)
7. Carstea, A., Frincu, M., Macariu, G., Petcu, D., Hammond, K.: Generic Access to Web and Grid-based Symbolic Computing Services. In: Procs. ISPDC 2007, pp. 143–150. IEEE Computer Press, Los Alamitos (2007)
8. Carstea, A., Macariu, G., Petcu, D., Konovalov, A.: Pattern Based Composition of Web Services for Symbolic Computations. In: Procs. ICCS 2008. LNCS. Springer, Heidelberg (2008) (in print)

9. Delaittre, T., Kiss, T., Goyeneche, A., Terstyanszky, G., Winter, S., Kacsuk, P.: GEMLCA: Running Legacy Code Applications as Grid Services. Journal of Grid Computing 3, 75–90 (2005)
10. Denemark, J., Kulshrestha, A., Allen, G.: Deploying Legacy Applications on Grids. In: Procs. 13th Annual Mardi Gras Conference, Frontiers of Grid Applications and Technologies, pp. 29–34 (2005)
11. Enright, W.H., Hull, T.E., Lindberg, B.: Comparing Numerical Methods for Stiff System of ODEs. BIT 15, 10–48 (1975)
12. Gannon, D., Krishnan, S., Slominski, A., Kandaswamy, G., Fang, L.: Building Applications from a Web Service Based Component Architecture. In: Getov, V., Kiellmann, T. (eds.) Component Models and Systems for Grid Applications, pp. 3–17. Springer, Heidelberg (2005)
13. Glatard, T., Emsellem, D., Montagnat, J.: Generic Web Service Wrapper for Efficient Embedding of Legacy Codes in Service-based Workflows. In: Procs. of the Grid-Enabling Legacy Applications and Supporting End Users Workshop, GELA, pp. 44–53 (2006)
14. Hairer, E., Wanner, G.: Solving Ordinary Differential Equations II. Stiff and Differential-Algebraic Systems. Springer, Heidelberg (1991)
15. Huang, Y., Taylor, I., Walker, D.W.: Wrapping Legacy Codes for Grid-based Applications. In: Procs. of the 17th International Parallel and Distributed Processing Symposium, workshop on Java for HPC, Nice (2003)
16. Krishnan, S., Stearn, B., Bhatia, K., Baldridge, K., Li, W., Arzberger, P.: Opal: Simple Web Services Wrappers for Scientific Applications. In: Procs. International Conference on Web Serves, ICWS, pp. 823–832 (2006)
17. Kamel, M.S., Ma, K.S., Enright, W.H.: ODEXPERT - An Expert System to Select Numerical Solvers for Initial Value ODE Systems. ACM Transactions on Mathematical Software 19(1), 44–62 (1993)
18. Kuebler, D., Eibach, W.: Adapting Legacy Applications as Web Services. IBM Developer Works (2002), `http://www-106.ibm.com/developerworks/webservices/library/ws-legacy/`
19. Lioen, W.M., de Sqart, J.J.B., van der Veen, W.A.: Test Set for IVP Solvers. Report NM-R9615, Centrum voor Wiskunde en Informatica, Amsterdam (1996)
20. Petcu, D.: Parallelism in Solving Ordinary Differential Equations. Mathematical Monographs, vol. 64. The Press of the West University of Timisoara (1998)
21. Petcu, D., Dragan, M.: Designing an ODE Solving Environment. In: Langtangen, H.P., Bruaset, A.M., Quak, E. (eds.) Modern Software Tools for Scientific Computing. Lectures Notes in Computational Science and Engineering, vol. 10, pp. 319–338. Springer, Heidelberg (2000)
22. Petcu, D.: Experiments with an ODE Solver on a Multiprocessor System. Computers & Mathematics with Appls. 42(8-9), 1189–1199 (2001)
23. Petcu, D., Eckstein, A., Giurgiu, C.: Using Statefull Web Services to Expose the Functionality of Legacy Software Codes. In: Procs. SACCS 2007, Iasi, pp. 257–263 (2007)
24. Petcu, D., Eckstein, A., Giurgiu, C.: Reengineering a Software System Implementing Pa-rallel Methods for Differential Equations. In: Procs. SEPADS, Cambridge, pp. 95–100 (2008)
25. Petcu, D.: Migrating an Expert System towards Service Oriented Architecture and MultiCore Systems. In: Teodorescu, H.N., Crauss, M. (eds.) Scientific and Educational Grid Applications, pp. 39–48. Politehnium (2008)

26. Pingali, K., Stodghill, P.: A Distributed System based on Web Services for Computational Science Simulations. In: Procs. 20th Intern. Conf. on Supercomputing, pp. 297–306 (2006)
27. Senger, M., Rice, P., Oinn, T.: Soaplab - A Unified Sesame Door to Analysis Tools. In: Cox, S.J. (ed.) Procs. UK e-Science, All Hands Meeting 2003, pp. 509–513 (2003)
28. Sneed, H.M.: Encapsulation of Legacy Software: A Technique for Reusing Legacy Software Components. In: Annals of Software Engineering, pp. 293–313. Springer, Heidelberg (2000)
29. Solomon, A., Struble, C.A.: JavaMath - an API for Internet Accessible Mathematical Services. In: Procs. Asian Symposium on Computer Mathematics (2001)
30. Solomon, A.: Distributed Computing for Conglomerate Mathematical Systems. In: Joswig, M., et al. (eds.) Integration of Algebra & Geometry Software Systems (2002)

9

Paradigmatic Morphology and Subjectivity Mark-Up in the RoWordNet Lexical Ontology

Dan Tufiş

Romanian Academy Research Institute for Artificial Intelligence
tufis@racai.ro

Abstract. Lexical ontologies are fundamental resources for any linguistic application with wide coverage. The reference lexical ontology is the ensemble made of Princeton WordNet, a huge semantic network, and SUMO&MILO ontology, the concepts of which are labelling each synonymic series of Princeton WordNet. This lexical ontology was developed for English language, but currently there are more than 50 similar projects for languages all over the world. RoWordNet is one of the largest lexical ontologies available today. It is sense-aligned to the Princeton WordNet 2.0 and the SUMO&MILO concept definitions have been translated into Romanian. The paper presents the current status of the RoWordNet and some recent enhancement of the knowledge encoded into it.

Keywords: lexical ontology, paradigmatic morphology, opinion mining, Romanian language, subjectivity priors.

1 Introduction

Most difficult problems in natural language processing stem from the inherent ambiguous nature of the human languages. Ambiguity is present at all levels of traditional structuring of a language system (phonology, morphology, lexicon, syntax, semantics) and not dealing with it at the proper level, exponentially increases the complexity of the problem solving. Most of the successful commercial applications in language processing (text and/or speech) use various shortcuts to syntactic analysis (pattern matching, chunking, partial parsing) and, to a large extent, dispense of explicit concern on semantics, with the usual motivations stemming from the computational high costs required by dealing with full syntax and semantics in case of large volumes of data. With recent advances in corpus linguistics and statistical-based methods in NLP, revealing useful semantic features of linguistic data is becoming cheaper and cheaper and the accuracy of this process is steadily improving. Lately, there seems to be a growing acceptance of the idea that multilingual lexical ontologies might be the key towards aligning different views on the semantic atomic units to be used in characterizing the general meaning of various and multilingual documents. Currently, the state of the art taggers (combining various models, strategies and

H.-N. Teodorescu, J. Watada, and L.C. Jain (Eds.): Intel. Sys. and Tech., SCI 217, pp. 161–179.
springerlink.com

processing tiers) ensure no less than 97-98% accuracy in the process of morpho-lexical full disambiguation. For such taggers a 2-best tagging[1] is practically 100% accurate. Dependency parsers are doing better and better and, for many significant classes of applications, even dependency linking (which is much cheaper than a full dependency parsing) seems to be sufficient. In a Fregean compositional semantics, the meaning of a complex expression is supposed to be derivable from the meanings of its parts, and the way in which those parts are combined. Therefore, one further step is the word-sense disambiguation (WSD) process. The WSD assigns to an ambiguous word ($\boldsymbol{w}$) in a text or discourse the sense ($\boldsymbol{s_k}$) which is distinguishable from other senses ($s_1, \ldots, s_{k-1}, s_{k+1}, \ldots, s_n$) potentially attributable to that word in a given context ($\boldsymbol{c_i}$). Sense inventories are specified by the semantic dictionaries, and they differ from dictionary to dictionary. For instance, in Merriam-Webster dictionary the verb ***be*** has listed 11 fine-grained senses and two coarse-grained senses. Longman Dictionary of Contemporary English glosses 15 fine-grained or 3 coarse-grained senses for the same verb. Cambridge Advanced Learner's Dictionary provides four fine-grained and two coarse-grained senses for the verb ***be***. Therefore, when speaking about word-sense discrimination one has to clearly indicate which sense inventory he/she is using. Word-sense disambiguation is generally considered as the most difficult part of the semantic processing required for deep natural language processing. In a limited domain of discourse this problem is alleviated by considering only coarse-grained sense distinctions, relevant for the given domain. Such a solution, although computationally motivated with respect to the universe of discourse considered, has the disadvantage of reduced portability and is fallible when the meanings of words are outside the boundaries of the prescribed universe of discourse.

Given the crucial role played by the dictionaries and lexical semantics in the overall description of a language system, it is not surprising the vast amount of work invested in these areas, during the time and all over the world, resulting in different schools, with different viewpoints and endless debates. Turning the traditional dictionaries into machine readable dictionaries proved to be a thorny enterprise, not only because of the technicalities and large amounts of efforts required, but mainly because of the conceptual problems raised by the intended computer use of knowledge and data initially created for human end-users only. All the implicit knowledge residing in a dictionary had to be made explicit, in a standardized representation, easy to maintain and facilitating interoperability and interchange. The access problem (how to find relevant stored information in a dictionary, with minimal search criteria) became central to computational lexicography. For psycho-linguists the cognitive motivations for lexical knowledge representations and their retrieval mechanisms were, at least, of equal relevance for building credible computational artefacts mimicking the mental lexicons. Multilinguality added a new complexity dimension to the set of issues related to dictionary structuring and sense inventories definition.

---

[1] In k-best tagging, instead of assigning each word exactly one tag (the most probable in the given context), it is allowed to have occasionally at most k-best tags attached to a word and if the correct tag is among the k-best tags, the annotation is considered to be correct.

2 Princeton WordNet

The computational lexicography has been tremendously influenced by the pioneering WordNet project, started in the early 80'ies at Princeton University by a group of psychologists and linguists led by George Miller [11].

WordNet is a special form of traditional semantic networks, very popular in the AI knowledge representation work of the 70'ies and 80'ies. George Miller and his research group developed the concept of a lexical semantic network, the nodes of which represented sets of actual words of English sharing (in certain contexts) a common meaning. These sets of words, called synsets (synonymy sets), constitute the building blocks for representing the lexical knowledge reflected in WordNet, the first implementation of lexical semantic networks. As in the semantic networks formalisms, the semantics of the lexical nodes (the synsets) is given by the properties of the nodes (implicitly, by the synonymy relation that holds between the literals of the synset and explicitly, by the gloss attached to the synset and, sometimes, by specific examples of usage) and the relations to the other nodes of the network. These relations are either of a semantic nature, similar to those to be found in the inheritance hierarchies of the semantic networks, and/or of a lexical nature, specific to lexical semantics representation domains. In more than 25 years of continuous development, Princeton WordNet [6] (henceforth PWN) reached an impressive coverage and it is the largest freely available semantic dictionary today.

The current version, PWN3.0[2], is a huge lexical semantic network in which almost 120,000 meanings/synsets (lexicalized by more than 155,000 literals) are related by semantic and lexical relations. The lexical stock covers the open class categories and is distributed among four semantic networks, each of them corresponding to a different word class: nouns, verbs, adjectives and adverbs. The notion of meaning in PWN is equivalent to the notion of concept and it is represented, according to a differential lexicographic theory, by a series of words which, in specific contexts, could be mutually substituted. This set of words is called a synset (synonymy set). A word occurring in several synsets is a polysemous one and each of its meanings is distinguished by a sense number. A pair made of a word and a sense number is generically called a word-sense. In the last version of PWN there are 206941 English word-senses. The basic structuring unit of PWN, the synset, is an equivalence relation over the set of word-senses. The major quantitative data about this unique lexical resource for English is given in the table 1 and table 2.

Table 1. POS distribution of the synsets and word s-senses in WPN 3.0

Noun literal/synset/ sense	**Verb literal/synset/ sense**	**Adjective literal/synset/ sense**	**Adverb literal/synset/ sense**	**Total literal/synset/ sense**
117798/82115/ 146312	11529/13767/ 25047	21479/18156/ 30002	4481/3621/ 5580	155287/117659/ 206941

[2] http://www.cogsci.princeton.edu/~wn/

Table 2. Polysemy in PWN 3.0

Noun literal/sense	Verb literal/sense	Adjective literal/sense	Adverb literal/sense	Total literal/sense
15935/44449	5252/18770	4976/14399	733/1832	26896/79450

Information in Table 1 shows that most of the literals, synsets and word-senses are given by the noun grammatical category: 117798 literals (75,85%), altogether with 146312 word-senses (70,70%) are clustered into 82115 synonymy equivalence classes (69,79% synsets).

The data in Table 2, show that only a small part of the lexical stock is polysemous, many nouns, verbs, adjectives and adverbs being monosemous. For instance, only 15935 nouns, that is 13.52%, occur in two or more synsets, all of them having 44449 word-senses, representing 30.37% of the total number of noun senses.

The relations among the synsets differ, being dependent on the grammar category of the literals in a synset. For each relation there is a reverse one. The major relations in PWN are: synonymy, hypernymy, meronymy (for nouns), troponymy, entailment (for verbs). Besides the semantic relations that hold between synsets, there are several other relations that relate word-senses, called lexical relations. Most relations for adjective and adverbs such as related nouns, verb participle, derivational are lexical relations. A very important lexical relation is antonymy which may hold between word-sense pairs of any of the four grammar categories. While one could speak about conceptual opposition between two synsets as in <rise:1 lift:3 arise:4...> and <descent:1 fall:2 go down:2 come down:2> or as in <rise:8, come up:10 uprise:2 ascend:3> and <set:10 go down:1 go under:8> the real antonymy relation in the examples above holds only between the pairs rise:1~fall:2 and rise:8~set:10.

An important feature, available in PWN from its early versions, is the word lemmatization which allows searching for lexical information using inflected word-forms.

The influence of the WordNet project in the domain of natural language processing was enormous and several other projects were initiated to complement information offered by PWN with various other types of information, useful for a large number of applications. Among the most important such initiatives was the alignment of PWN synsets to the concepts of SUMO&MILO upper and mid level ontology [13], which turned the ensemble PWN+SUMO&MILO into a proper lexical ontology. In spite of several other projects aimed at developing alternative lexical ontologies, no one can compete yet with the extended PWN. Another enhancement of PWN was the development of DOMAINS [1] hierarchical classification system, inspired by the DEWEY classification system, and attributing each synset of PWN a DOMAINS class. This additional labelling of the PWN synsets made possible to cluster the lexical stock into coarser grained semantic categories, operation which is extremely useful in word-sense disambiguation, document classification information retrieval etc.

Among the latest enhancements of the PWN was the development of the SentiWordNet [5], an explicit annotation of the all the synsets with subjectivity mark-up. Sentiment analysis has recently emerged as a very promising research area with multiple applications in processing arbitrary collections of text. Sentiment can be expressed about works of art and literature, about the state of financial markets, about liking and disliking individuals, organizations, ideologies, and consumer goods. In making everyday decisions or in expressing their opinions on what they are interested in, more and more people are interacting on the so-called social web, reading others' opinions or sharing their experiences or sentiments on a wide spectrum of topics. The review sites, forums, discussion groups and blogs became very popular and opinions expressed therein are getting significant influence on people's daily decisions (buying products or services, going to a movie/show, travelling somewhere, forming an opinion on political topics or on various events, etc.). Decision makers, at any level, cannot ignore the "word-of-mouth", as sometimes the social web is dubbed. Research in the area of opinion finding and sentiment analysis is motivated by the desire to provide tools and support for information analysts in government, commercial, and political domains, who want to automatically track attitudes and feelings in the news and on-line forums [32].

Irrespective of the methods and algorithms used in subjectivity analysis, they exploit the pre-classified words and phrases as opinion or sentiment bearing lexical units. Such lexical units (also called senti-words, polar-words) are manually

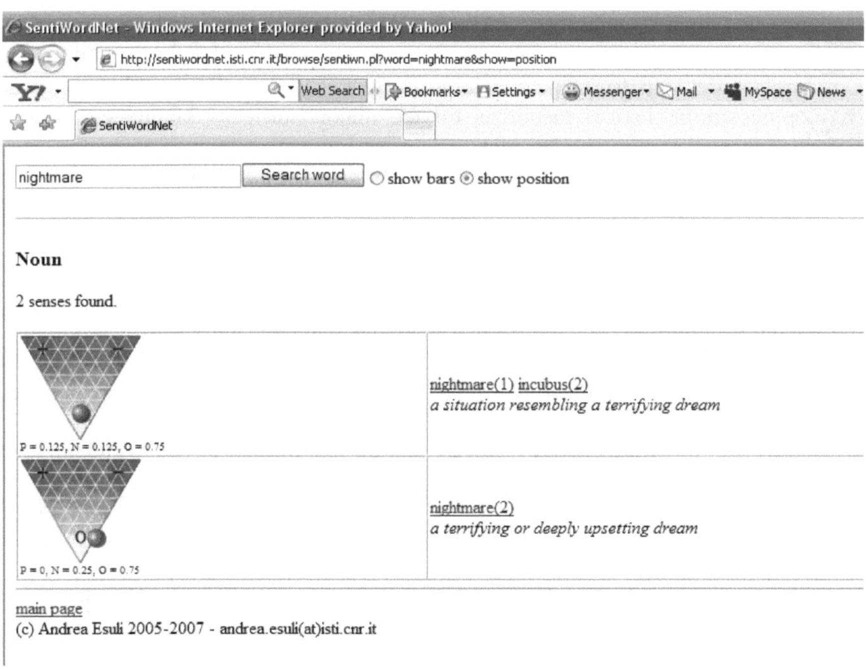

Fig. 1. SentiWordNet interface

specified, extracted from corpora or marked-up in specialized lexicons such as General Inquirer or SentiWordNet.

In SentiWordNet, each synset is associated with a triple <P:α N:β O:γ> where P denotes its Positive subjectivity, N represents the Negative subjectivity and O stands for Objectivity. The values α, β and γ are sub-unitary numbers summing up to 1 and representing the degrees of positive, negative and objective prior sentiment annotation of the synset in case. The SentiWordNet graphical interface, exemplified in Figure 1, is available at http://sentiwordnet.isti.cnr.it/browse/.

The Figure 1 shows that the subjectivity mark-up depends on the word-senses. The sense 2 of the word *nightmare* (which denotes a *cognition noun*, subsumed by the term *psychological feature*) has a higher degree of negative subjectivity than sense 1 (which denotes a *state noun*, subsumed by the meaning of the synset <condition:1, status:2> .

The convergence of the representational principles promoted both by the domain-oriented semantic networks and ontologies, and by PWN philosophy in representing general lexical knowledge, is nowadays a de-facto standard, motivated not by fashion, but by the significant improvements in performance and by the naturalness of interaction displayed by the systems that have adopted this integration.

3 Multilingual Wordnets: EuroWordNet and BalkaNet

As mentioned before, the impact of the PWN on the NLP systems for English language has been unanimously acclaimed by the researchers and developers of language processing systems and as a consequence, in 1996, the European Commission decided to finance a large project, EuroWordNet [29], aiming at developing similar lexical resources for the major European languages: Dutch, French, German, Italian and Spanish. The most innovative feature of this project, was the idea to have the synsets of the monolingual semantic dictionaries aligned via an Inter-Lingual Index (ILI), so that to allow cross-lingual navigation from the one language to the others [16]. Most of the ILI records corresponded to the indices of the PWN synsets, but there were also language specific synsets in each of the monolingual semantic networks. The ILI represented a conceptualization of the meanings linguistically realized in different language by specific synonymy sets. By exploiting the SUMO&MILO information attached to PWN's synsets, accessible via ILI index, the collection of monolingual semantic networks became the first multilingual lexical ontology.

To express the cross-lingual relations among the synsets in one language and the language-independent concepts of ILI, the EuroWordNet project (EWN henceforth) defined 20 distinct types of binary equivalence relations (EQ-SYN, EQ-HYPO, EQ-MERO etc.). While PWN was essentially focused on representing paradigmatic relations among the synsets, EWN considered the syntagmatic relations as well. As compared to PWN, the set of internal relations defined by EWN is much larger (90) including casual relations (Agent, Object, Patient, Instrument, etc.) and derivative lexical relations (XPOS-SYNONYMY: to adore/adoration).

After three successful years, the initial EWN project was extended for two more years with the task to include in the multilingual ontology four other languages: Basque, Catalan, Czech and Estonian.

A significant follow-up of EWN was the start-up in 2001 of the BalkaNet European project [17], meant as a continuation and development of the EuroWordNet methodology, bringing into the multilingual lexical ontology five new languages, specific to the Balkan area: Bulgarian, Greek, Romanian, Serbian and Turkish. The major objective of this project was to build core semantic networks (8,000 synsets) for the new languages and ensuring full cross-lingual compatibility with the other 9 semantic networks built by EWN. The philosophy of the BalkaNet architecture [23] was similar to EWN but it brought several innovations such as: more precise design methodologies, a common XML codification of the monolingual wordnets, the introduction of valence frames for verbs and deverbal nouns, the increased set of lexical relations (dealing with perfective/imperfective aspect and the rich inflectional morphology of the Balkan languages) allowing for non-lexicalized concepts, the definition of regional specific concepts etc. In BalkaNet (BKN henceforth) there were developed many public tools (language independent) for the development and validation of new wordnets such as: WordNet Management System (WMS), VISDIC, WSDTool, WNBuild, WNCorrect etc. (see for details [22]).

The concepts considered highly relevant for the Balkan languages (and not only) were identified and called BalkaNet Base Concepts. These are classified into three increasing size sets (BCS1, BCS2 and BCS3). Altogether BCS1, BCS2 and BCS3 contain 8516 concepts that were lexicalized in each of the BKN wordnets. The monolingual wordnets had to have their synsets aligned to the translation equivalent synsets of the PWN. The BCS1, BCS2 and BCS3 were adopted as core wordnets for several other wordnet projects such as Hungarian [10], Slovene [4], Arabic[2], [3], and many others.

The establishment of the Global WordNet Association[3] (2000) was another initiative that had a decisive role to the establishment of the concept of a wordnet as a, practically, standard way of representing lexical information. This association is an international forum of the wordnet developers and/or users and, biannually organizes the Global WordNet Conferences. Currently there are more than 50 projects aiming at developing wordnets for major languages of the world. Adopting the synsets of PWN as an interlingual sense inventory, it became possible to cross-lingually navigate among the semantic lexicons of language pairs, hardly to imagine a few years ago. One could say that the boost of the multilinguality research could be explained (at least partially) by the tremendous work carried out all over the world to develop wide coverage monolingual wordnets aligned to the PWN.

By the end of the BalkaNet project (August 2004) the Romanian wordnet, contained almost 18,000 synsets, conceptually aligned to Princeton WordNet 2.0 and through it to the synsets of all the BalkaNet wordnets. In [24], a detailed account on the status of the core RoWordNet is given as well as on the tools we used for its development.

[3] www.globalwordnet.org

After the BalkaNet project ended, as many other project partners did, we continued to update the Romanian wordnet and in the following we describe its latest developments.

4 Sense, Meaning, Concept: A Terminological Distinction

Let us define three terms relevant for the discussions to follow: "sense", "meaning" and "concept". Although closely related, and sometimes interchangeably used, these notions are slightly different distinguishing the perspective from which the encoded linguistic knowledge is considered.

The notion of *sense* is strictly referring to a word. The polysemy degree of a word is given by the number of senses the respective word has. A traditional explanatory dictionary provides definitions for each sense of a headword.

The notion of *meaning* generalizes the notion of *sense* and it could be regarded as a set-theoretic equivalence relation over the set of senses in a given language. In colloquial speech one says *this word has the same meaning with that word* while a more precise (but less natural) statement would be *the Mth sense of this word has the same meaning with the Nth sense of that word.* Synonymy, as this equivalence relation is called, is a lexical relation that represents the formal device for clustering the word-senses into groups of lexicalized meanings. The meaning is the building block in wordnet-like knowledge representations, and it corresponds in PWN, EWN, BKN and all their followers to the synset data type.

Each synset is associated with a gloss that covers all word-senses in the synonymy set. The meaning is thus a language specific realization of a *conceptualization* which might be very similar to conceptualizations in several other languages.

Similar conceptualizations are generalized in a language independent way, by what we call *interlingual concepts* or simply *concepts*. The meanings in two languages that correspond to the same concept are said to be equivalent. One could arguably say that the interlingual concepts cannot entirely reflect the meanings in different languages (be it only for the historical and cultural differences), however, concepts are very useful generalizations that enable communication between speakers of different natural languages. In multilingual semantic networks the interlingual level ensures the cross-lingual navigation from words in one language to words in the other languages. Both EWN and BalkaNet adopted as their interlingual concepts the meanings of PWN. This choice was obviously a matter of technological development and a working compromise: the PWN displayed the greatest lexical coverage and is still unparalleled by any other language. To remedy this Interlingua status of English, both EWN and BalkaNet considered the possibility of extending the ILI set by language specific concepts (or concepts specific to a group of languages).

5 The Ongoing RoWordNet Project and Its Current Status

The RoWordNet is a continuous effort going on for 8 years now and likely to continue several years from now on. However, due to the development methodology adopted in BalkaNet project, the intermediate wordnets could be used in various other projects (word-sense disambiguation, word alignment, bilingual lexical knowledge acquisition, multilingual collocation extraction, cross-lingual question answering, opinion mining, machine translation etc.).

Recently we started the development of an English-Romanian MT system for the legalese language of the type contained in JRC-Acquis multilingual parallel corpus [18], of a cross-lingual question answering system in open domains [14], [25] and an opinion mining system [25], [27]. For these projects, heavily relying on the aligned Ro-En wordnets, we extracted a series of high frequency Romanian nouns and verbs not present in the RoWordNet but occurring in JRC-Acquis corpus and in the Romanian pages of Wikipedia and proceeded at their incorporation into the RoWordNet. The methodology and tools were essentially the same as described in [24], except that the dictionaries embedded into the WNBuilder and WNCorrect were significantly enlarged.

The two basic development principles of the BalkaNet methodology [21], [23], that is Hierarchy Preservation Principle (HPP) and Conceptual Density Principle (CDP), were strictly observed. For the sake of self-containment, we restate them here.

Hierarchy Preservation Principle

If in the hierarchy of the language L1 the synset M_2 is a hyponym of synset M_1 (M_2 H^m M_1) and the translation equivalents in L2 for M_1 and M_2 are N_1 and N_2 respectively, then in the hierarchy of the language L2 N_2 should be a hyponym of synset N_1 (N_2 H^n N_1). Here H^m and H^n represent a chain of m and n hierarchical relations between the respective synsets (hypernymy relation composition).

Conceptual Density Principle (noun and verb synsets)

Once a nominal or verbal concept (i.e. an ILI concept that in PWN is realized as a synset of nouns or as a synset of verbs) was selected to be included in RoWordNet, all its direct and indirect ancestors (i.e. all ILI concepts corresponding to the PWN synsets, up to the top of the hierarchies) should be also included into RoWordNet.

By observing the HPP, the lexicographers were relieved of the task of establishing the semantic relations for the synsets of the RoWordNet. The hypernymy relations as well as the other semantic relations were imported automatically from the PWN. The CDP compliance ensures that no *orphan synsets* [23], i.e. lower-level synsets without direct ancestors, are created.

The tables below give a quantitative summary of the current Romanian wordnet (February, 2009). As these statistics are changing every month, the updated information should be checked at http://nlp.racai.ro/wnbrowser/Help.aspx. The RoWordNet is currently mapped on the various versions of Princeton WordNet: PWN1.7.1, PWN2.0 and PWN2.1. The mapping onto the last version PWN3.0 is also considered. However, all our current projects are based on the PWN2.0 mapping and in the following, if not stated otherwise, by PWN we will mean PWN2.0.

Table 3. POS distribution of the synsets in the Romanian wordnet

Noun literal/synset/ sense	Verb literal/synset/ sense	Adjective literal/synset/ sense	Adverb literal/synset/ sense	Total literal/synset/ sense
36686/40605/ 56457	6517/10587/ 16880	2840/3488/ 5754	769/841/ 1207	46943/55521/ 80298

Table 4. Internal relations used in the Romanian wordnet

hypernym	50395	category_domain	3257
near_antonym	3012	Also_see	1245
holo_part	5248	subevent	353
similar_to	2929	Holo_portion	392
verb_group	1484	causes	186
holo_member	1686	be_in_state	782

Table 5. PWN vs. RoWordNet ontological labeling (DOMAINS, SUMO, MILO)

LABELS	PWN2.0	RoWordNet
DOMAINS-3.1	168	166
SUMO	844	812
MILO	949	906
Domain ontologies	215	203

The BalkaNet development methodology, which we continue to observe, prescribed a top-down approach, beginning with the topmost synsets of the wordnet conceptual hierarchy (the most general concepts and, therefore, the most difficult to implement) downwards the leaf synsets (concepts or instances denoted by monosemous words with few or no synonyms). This basic strategy was complemented by corpus based detection of the most frequently used words and their inclusion into the lexical stock represented by the wordnet. In this way, in spite of containing 55,521 synsets (47,19% from the total number of synsets in PWN), the RoWordNet covers most of the DOMAINS-3.1, SUMO, MILO and domain ontologies concepts existing in PWN (Table 5). The large majority of the remaining synsets up to the PWN cardinality represent instances of the concepts already defined in our wordnet, or instances of a few concepts very specific to American culture.

The DOMAINS labelling (http://wndomains.itc.it/) classifies the PWN synsets into 168 distinct classes [1], [9]. The RoWordNet synsets cover 166 of the total number of these classes.

The SUMO&MILO upper and mid level ontology [13] is the largest freely available (http://www.ontologyportal.org/) ontology today. It is accompanied by more than 20 domain ontologies and altogether they contain about 20,000 concepts and 60,000 axioms. They are formally defined and do not depend on a particular application. Its attractiveness for the NLP community comes from the fact that SUMO, MILO and associated domain ontologies were manually mapped onto PWN. SUMO and MILO contain 1107 and respectively 1582 concepts. Out of these, 844 SUMO concepts and 949 MILO concepts were used to label almost all the synsets in PWN. Additionally, 215 concepts from some specific domain ontologies were used to label the rest of synsets (instances) in PWN. As one can see from Table 5, most of the SUMO, MILO and domain ontologies concepts occurring in PWN are lexicalized in the Romanian wordnet.

6 Recent Extensions of the Romanian WordNet

6.1 Paradigmatic Morphology Description of the Literals

The vast majority of dictionaries are based on normalized (lemma) form of the words. The wordnets are no exception to this lexicographic practice, the synsets being defined as lists of synonymic lemmas. However, for the effective use in NLP applications, especially for highly inflectional languages, lemmatization of words in an arbitrary text or generation of specific inflected forms from given lemmas have generally been recognized as very useful extensions to a wordnet system. As mentioned earlier, PWN had even from the first versions the facility to look-up an inflected word-form. Given the simple morphology of English the lemmatization was included in the search engine of PWN, that is outside linguistic specification of the lexical semantic network.

We preferred to declaratively encode the necessary information required to allow a linguistic processor to query the RoWordNet via an inflected word-form. Our solution relied on the paradigmatic morphology model [19] and the FAVR paradigmatic description of Romanian [20], an example of which is given in Figure 2. FAVR is a feature-based reversible description language that allows for the specification of the inflectional paradigms of a given language. The lemma headwords of a FAVR-based dictionary are associated with the corresponding inflectional paradigms. The FAVR descriptions are very compactly compiled into reversible data structures that could be used both for analysis and generation of inflected word-forms [20]. The original LISP implementation of the paradigmatic analyzer/generator was re-implemented [7] in a much faster C version.

This new paradigmatic morphological processor has been incorporated into RoWordNet service platform and the XML structure of a synset has been changed to accommodate the paradigmatic information for the literals occurring within the synset. In Figure 3 is exemplified a synset encoding which explicitly refers the paradigm in Figure 2.

To save space, in Figure 3 we exemplified a regularly inflecting word, that is one which does not change its root during declension or conjugation. If this is not the case, the PARADIGM field of a LITERAL in a synset will explicitly mention all the possible roots for the inflectional variants of the literal in case.

```
[PARADIGM: $nomneu3, INTENSIFY: none,
 [TYPE: {proper, common},
   [NUM: singular, GEN: masculine,
     [ENCL: no,[CASE: {nom, gen, dat, acc, voc }, TERM: ""]],
     [ENCL: yes,[CASE: {nom, acc}, TERM : "ul"],
                [CASE: {gen, dat}, TERM : "ului"],
                [CASE: voc,
                  [HUM: imperson, TERM : "ul"],
                  [HUM: person, TERM: "ule"]]]],
   [NUM: plural, GEN: feminine,
     [ENCL: no, [CASE: {nom, gen, dat, acc, voc }, TERM: "uri"]],
     [ENCL: yes, [CASE: {nom, acc}, TERM:" urile"],
                 [CASE: {gen, dat}, TERM: "urilor"],
                 [CASE: voc,
                   [HUM: imperson, TERM: "urile"],
                    [HUM: person, TERM: "urilor"]]]]]]
```

Fig. 2. A Romanian nominal paradigm specification

The encoding exemplified in Figure 3 is the most informative one from the linguistic point of view, since it allows both analysis and generation (in terms of attribute-value description) of a word-form. If only recognizing the word-forms is required (as is the case for the majority applications of a wordnet) one can dispense of the paradigmatic processor and the paradigmatic morphology, generating before-hand the entire paradigmatic family of a literal. For the example in Figure 3, this simpler version is shown in Figure 4.

```
<SYNSET>
   <ID>ENG20-07995813-n</ID>
   <POS>n</POS>
   <SYNONYM>
           <LITERAL>loc
             <PARADIGM>$nomneu3<ROOT>loc</ROOT></PARADIGM>
             <SENSE>4</SENSE>
           </LITERAL>
   </SYNONYM>
   <DEF>Spaţiu ocupat de cineva sau de ceva.</DEF>
   <BCS>1</BCS>
   <ILR>ENG20-08050136-n<TYPE>hypernym</TYPE></ILR>
   <DOMAIN>factotum</DOMAIN>
   <SUMO>GeographicArea<TYPE>+</TYPE></SUMO>
   <SENTIWN><P>0.0</P><N>0.0</N><O>1</O></SENTIWN>
</SYNSET>
```

Fig. 3. The structure of a Romanian synset containing paradigmatic abstract information

As one could observe, the reference to the paradigm and the root(s) disappeared and they were replaced with the paradigmatic family of the headword. Through a proper indexing mechanism any inflected form of the lemma loc could be used to select the XML representation of the synset exemplified in Figure 4 (and obviously all the other synsets containing in their PARADIGM field the inflected form used for interrogation).

```
<SYNSET>
   <ID>ENG20-07995813-n</ID>
   <POS>n</POS>
   <SYNONYM>
           <LITERAL>loc
             <PARADIGM>loc locul locului locule locuri locurile locurilor
             </PARADIGM>
             <SENSE>4</SENSE>
           </LITERAL>
   </SYNONYM>
   <DEF>Spaţiu ocupat de cineva sau de ceva.</DEF>
   <BCS>1</BCS>
   <ILR>ENG20-08050136-n<TYPE>hypernym</TYPE></ILR>
   <DOMAIN>factotum</DOMAIN>
   <SUMO>GeographicArea<TYPE>+</TYPE></SUMO>
   <SENTIWN><P>0.0</P><N>0.0</N><O>1</O></SENTIWN>
</SYNSET>
```

Fig. 4. The structure of a Romanian synset containing an inflectional paradigmatic family

6.2 Subjective Mark-Up and Opinion Mining

We mentioned in section 2 that the recent release of SentiWordNet [5] turned PWN into an ideal lexical resource for opinion mining. Most approaches to opinion mining rely on Bag-of-Words (BoW) models with pre-existing lists of lexical items classified into positive or negative words [30], with any word not classified either as positive or negative considered being neutral. The polar words (negative or positive) are used as seeds for further extending of the subjective lexica. However, recent experiments [29], [27] proved that polarity of the subjective words depends on their senses and that the syntax and punctuation (usually discarded in the BoW approaches) are very important for a reliable sentiment analysis of arbitrary texts. The SentiWordNet implementation answered the lexical requirement outlined by these recent experiments, namely it associated each synset (and thus each word-sense) in PWN with a prior subjectivity annotation [30].

To take full advantage of SentiWordNet annotations, one has to be able to contextually determine the senses of the subjective words and, to define a subjectivity calculus appropriate for computing the subjectivity score/polarity of the entire sentence, paragraph or text subject to sentiment analysis. Esuli and Sebastiani, the authors of SentiWordNet, did not make any reference to a specific way of using the lexical subjectivity priors, but a natural option is that this calculus must be compositional bringing the issue of deriving a sentential structure upon which the calculus might operate. The necessity for a structure-based compositional calculus can be easily supported by the effect of the so-called *valence shifters* or *intensifiers* [15], which may turn the polarity of an entire piece of text containing only positive or only negative sentiment words. Consider for instance the definition of *honest, honorable*: **not_disposed_to CHEAT**$^{-}$ or **DEFRAUD**$^{-}$, **not DECEPTIVE**$^{-}$ or **FRAUDULENT**$^{-}$; the polarity words are in upper case and the negative sign used as exponent indicate that their prior polarity is negative. Although the definition contains lots of negative words, the overall value of the text is definitely positive.

The way the SentiWordNet subjectivity priors were computed is largely described in [5], and it is based on pre-existing lists of senti-words and taking advantage of the PWN structure. In [27] we argued that some debatable synset annotations could be explained due to multiple reasons:

- Taxonomic generalization is not always working as in the examples below:

  Nightmare is bad & Nightmare is a dream, however, dream is not bad (per se)
  An emotion is something good (P:0,5) and so is love, but hate or envy are not!

- Glosses are full of valence shifters (BOW is not sufficient):

  honest, honorable: **not disposed to CHEAT**$^-$ or **DEFRAUD**$^-$, **not DECEPTIVE**$^-$ or **FRAUDULENT**$^-$
  intrepid: **invulnerable to FEAR**$^-$ or **INTIMIDATION**$^-$
  superfluous: serving **no USEFUL**$^+$ purpose; having **no EXCUSE**$^+$ for being
- Majority voting is democratic but not the best solution.

We also showed that a very good approach in bootstrapping initial seed lists of senti-words is combining the subjectivity information already existing in PWN via the SUMO/MILO and DOMAINS mappings. For instance, there are many SUMO&MILO concepts with a definite subjectivity load (*EmotionalState, Happiness, Psychological Process, SubjectiveAssesmentAttribute, StateOfMind, TraitAttribute, Unhappiness, War* etc.). Similarly, the DOMAINS categories such as *psychological_features, psychology, quality, military* etc., would bring evidence about the prior polarities of PWN synsets.

In spite of the drawbacks and inconsistencies mentioned in [27], the SentiWordNet remains one of the best resources for opinion mining. This is why, we decided to exploit the synset alignment between PWN2.0 and RoWordNet and import all the subjectivity mark-up. Our experiments brought evidence that SentiWordNet could be successfully used cross-lingually and that different synset labeling pertaining to subjectivity can be conciliated. Among the findings of our studies were the fact that the verb (and deverbal noun) argument structure is essential in finding out *who or what* is *good or bad* and that the prior polarity of some adjectives and adverbs are dependent on their head in context (*long*$^-$ *response time* vs. *long*$^+$ *life battery*; *high*$^-$ *pollution* vs. *high*$^+$ *standard*).

A new research avenue opened by the sense-based subjectivity priors is connotation analysis [28]. The CONAN system, relying on the subjectivity mark-up in PWN and in RoWordNet, has been developed to detect sentences which, when taken out from their original context and purposely put in a different context could be interpreted in a different way (sometimes funny, sometimes embarrassing). This system is language independent provided that a senti-wordnet (a wordnet with the subjective mark-up on the synsets) is available.

6.3 The Pre-processing and Correcting of the RoWordNet Definitions

As mentioned several times in this paper, a wordnet is a crucial language resource which can be used in many ways. The most frequent uses of wordnets exploit the relations among the synsets and synsets alignment to an external conceptual structure (a taxonomy such as IRST's DOMAINS or an ontology such as SUMO&MILO). Except a few remarkable works such those carried in Moldovan's group [12] or Princeton's new release PWN3.0, much less used are the synset glosses, in spite of their essential content. In order to develop a lexical chain algorithm similar to that of Moldovan and Novischi (2002) we needed to preprocess the glosses of the Romanian synsets. The pre-processing consisted in tokenizing, POS-tagging, lemmatizing and linking each gloss found in the Romanian WordNet.

POS-tagging and lemmatizing were performed using TTL [7] which outputs a list of tokens of the sentence, each with POS tag (morpho-syntactic descriptors) and lemma information. The RoWordNet glosses contain approx. 530K tokens out of which 60K are punctuation marks.

When performing POS tagging, the tagger identified more than 2500 unknown words. Most of them proved to be either spelling errors, or words written in disagreement with the Romanian Academy regulations (the improper use of the diacritical mark 'î' vs. 'â'). We automatically identified all spelling errors with an ad-hoc spell checker using Longest Common Sequences between errors and words in our 1 million word-form lexicon. After eliminating all spelling errors, we were left with 550 genuine unknown words which we added to our tagging lexicon along with their POS tags and lemmas. Although the planned lexical chain algorithm is still in progress, the preparatory work on the RoWordNet glosses allowed us to detect and correct a significant number of dormant errors.

Linking was achieved through the use of LexPar [7] which generates a planar, undirected and acyclic graph of the sentence (called a linkage) that mimics a syntactic dependency-like structure. We plan to use this structure to make a connection between words in the synset and words in its gloss. This way we will be able to outline several syntagmatic dependencies among the literals covered by RoWordNet.

6.4 The RoWordNet Web Interface

The RoWordNet can be browsed by a web interface implemented on our language web services platform (http://nlp.racai.ro/WnBrowser/). The browser uses graph hyperbolic representations (see Figure 5) and it visualizes in a friendly manner all the synsets in which one given literal appears together with its corresponding synonyms, the semantic relations for each of its senses, definition of each sense, the DOMAINS, SUMO, MILO and subjectivity mark-up.

Although currently only browsing is implemented, RoWordNet web service will, later on, include search facilities accessible via standard web service technologies (SOAP/WSDL/UDDI), such as distance between two word-senses, translation equivalents for one or more senses, semantically related word-senses, lexical chains etc.

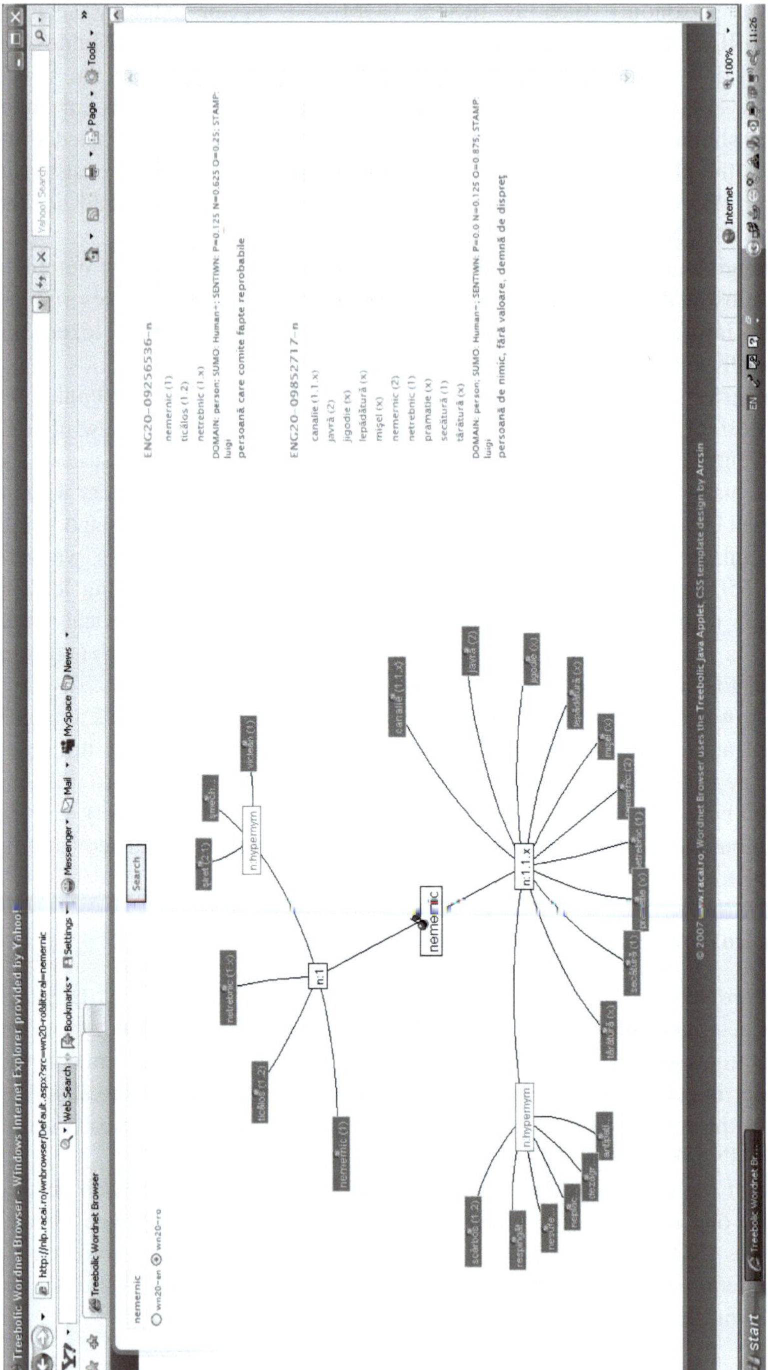

Fig. 5. Web interface to RoWordNet browser

7 Conclusions and Further Work

The development of RoWordNet is a continuous project, keeping up with the new updates of the Princeton WordNet. The increase in its coverage is steady (approximately 10,000 synsets per year for the last four years) with the choice for the new synsets imposed by the applications built on the basis of RoWordNet. Since PWN was aimed to cover general language, it is very likely that specific domain applications would require terms not covered by Princeton WordNet. In such cases, if available, several multilingual thesauri (IATE-http://iate.europa.eu/iatediff/about_IATE.html, EUROVOC-http://europa.eu/eurovoc etc.) can complement the use of wordnets. Besides further augmenting the RoWordNet, we plan the development of an environment where various multilingual aligned lexical resources (wordnets, framenets, thesauri, parallel corpora) could be used in a consistent but transparent way for a multitude of multilingual applications.

There are several applications we developed using RoWordNet as an underlying resource: word-sense disambiguation, word alignment, question-answering in open domains, connotation analysis etc. The state-of-art performances on these systems are undeniably rooted in the quality and the coverage of RoWordNet.

Currently we are engaged in the development of a statistical machine translation system, for which RoWordNet (and related enhancements such as the previously mentioned lexical chain algorithm) will be fundamental.

Acknowledgments. The work reported here was supported by the Romanian Academy program "Multilingual Acquisition and Use of Lexical Knowledge", the ROTEL project (CEEX No. 29-E136-2005) and by the SIR-RESDEC project (PNCDI2, 4th Programme, No. D1.1-0.0.7), the last two granted by the National Authority for Scientific Research. The SEE-ERA.net European project "Building Language Resources and Translation Models for Machine Translation focused on South Slavic and Balkan Languages" (ICT 10503 RP) and WISE - An Electronic Marketplace to Support Pairs of Less Widely Studied European Languages (BSEC 009 / 05.2007) were other supporting projects for the further development of the RoWordNet.

References

[1] Bentivogli, L., Forner, P., Magnini, B., Pianta, E.: Revising WordNet Domains Hierarchy: Semantics, Coverage, and Balancing. In: Proceedings of COLING 2004 Workshop on Multilingual Linguistic Resources, Geneva, Switzerland, August 28, pp. 101–108 (2004)

[2] Black, W., Elkateb, S., Rodriguez, H., Alkhalifa, M., Vossen, P., Pease, A., Fellbaum, C.: Introducing the Arabic WordNet Project. In: Sojka, P., Choi, K.-S., Fellbaum, C., Vossen, P. (eds.) Proceedings of the Third Global Wordnet Conference, Jeju Island, pp. 295–299 (2006)

[3] Elkateb, S., Black, W., Rodriguez, H., Alkhalifa, M., Vossen, P., Pease, A., Fellbaum, C.: Building a WordNet for Arabic. In: Proceedings of the Fifth International Conference on Langauge Resources and Evaluation, Genoa, Italy, pp. 29–34 (2006)

[4] Erjavec, T., Fišer, D.: Building Slovene WordNet. In: Proceedings of the 5th Intl. Conf. on Language Resources and Evaluations, LREC 2006, Genoa, Italy, May 22-28, pp. 1678–1683 (2006)
[5] Esuli, A., Sebastiani, F.: SentiWordNet: A publicly Available Lexical Resourced for Opinion Mining. In: LREC 2006, Genoa, Italy, May 22-28, pp. 417–422 (2006)
[6] Fellbaum, C. (ed.): WordNet: An Electronic Lexical Database. MIT Press, Cambridge (1998)
[7] Ion, R.: Word-sense Disambiguation Methods Applied to English and Romanian. PhD thesis (in Romanian). Romanian Academy, Bucharest (2007)
[8] Irimia, E.: ROG - A Paradigmatic Morphological Generator for Romanian. In: Vetulani, Z. (ed.) Proceedings of the 3rd Language & Technology Conference: Human Language Technologies as a Challenge for Computer Science and Linguistics, Poznań, Poland, October 5-7, pp. 408–412 (2007)
[9] Magnini, B., Cavaglià, G.: Integrating Subject Field Codes into WordNet. In: Gavrilidou, M., Crayannis, G., Markantonatu, S., Piperidis, S., Stainhaouer, G. (eds.) Proceedings of LREC-2000, Second International Conference on Language Resources and Evaluation, Athens, Greece, May 31-June 2, pp. 1413–1418 (2000)
[10] Miháltz, M., Prószéky, G.: Results and evaluation of Hungarian nominal wordnet v1.0. In: Sojka, P., et al. (eds.) Proceedings of the Second International Wordnet Conference (GWC 2004), pp. 175–180. Masaryk University, Brno (2004)
[11] Miller, G.A., Beckwith, R., Fellbaum, C., Gross, D., Miller, K.: Introduction to WordNet: An On-Line Lexical Database. International Journal of Lexicography 3(4), 235–244 (1990)
[12] Moldovan, D., Novischi, A.: Lexical chains for question answering. In: Proceedings of COLING 2002, pp. 674–680 (2002)
[13] Niles, I., Pease, A.: Towards a Standard Upper Ontology. In: Proceedings of the 2nd International Conference on Formal Ontology in Information Systems (FOIS 2001), Ogunquit, Maine, October 17-19, pp. 2–9 (2001)
[14] Puşcasu, G., Iftene, A., Pistol, I., Trandabăţ, D., Tufiş, D., Ceauşu, A., Ştefănescu, D., Ion, R., Orăşan, C., Dornescu, I., Moruz, A., Cristea D.: Developing a Question Answering System for the Romanian-English. In: Peters, C., Clough, P., Gey, F.C., Karlgren, J., Magnini, B., Oard, D.W., de Rijke, M., Stempfhuber, M. (eds.) CLEF 2006. LNCS, pp. 385–394. Springer, Heidelberg (2007)
[15] Polanyi, L., Zaenen, A.: In: Shanahan, J., Qu, Y., Wiebe, J. (eds.) Computing Attitude and Affect in Text: Theory and Applications. The Information Retrieval Series, vol. 20, pp. 1–9. Springer, Dordrecht (2004)
[16] Rodriguez, H., Climent, S., Vossen, P., Bloksma, L., Peters, W., Alonge, A., Bertagna, F., Roventini, A.: The Top-Down Strategy for Building EuroWordNet: Vocabulary Coverage, Base Concepts and Top Ontology. Computers and the Humanities 32(2-3), 117–152 (1998)
[17] Stamou, S., Oflazer, K., Pala, K., Christoudoulakis, D., Cristea, D., Tufiş, D., Koeva, S., Totkov, G., Dutoit, D., Grigoriadou, M.: BALKANET A Multilingual Semantic Network for the Balkan Languages. In: Proceedings of the International Wordnet Conference, Mysore, India, January 21-25, pp. 12–24 (2002)
[18] Steinberger, R., Pouliquen, B., Widiger, A., Ignat, C., Erjavec, T., Tufiş, D.: The JRC-Acquis: A multilingual aligned parallel corpus with 20+ languages. In: Proceedings of the 5th LREC Conference, Genoa, Italy, May 22-28, pp. 2142–2147 (2006)

[19] Tufiş, D.: It Would Be Much Easier If WENT Were GOED. In: Somers, H., Wood, M.M. (eds.) Proceedings of the 4th European Conference of the Association for Computational Linguistics, Manchester, pp. 145–152 (1989)
[20] Tufiş, D., Barbu, A.M.: A Reversible and Reusable Morpho-Lexical Description of Romanian. In: Tufiş, D., Andersen, P. (eds.) Recent Advances in Romanian Language Technology. Editura Academiei (1997)
[21] Tufiş, D., Cristea, D.: Methodological Issues in Building the Romanian Wordnet and Consistency Checks in Balkanet. In: Proceedings of LREC 2002 Workshop on Wordnet Structures and Standardisation, Las Palmas, Spain, May 2002, pp. 35–41 (2002)
[22] Tufiş, D. (ed.): Special Issue on BalkaNet. Romanian Journal on Information Science and Technology, Romanian Academy 7(2-3) (2004)
[23] Tufiş, D., Cristea, D., Stamou, S.: BalkaNet: Aims, Methods, Results and Perspectives: A General Overview. In: Tufiş, D. (ed.) Romanian Journal on Information Science and Technology. Special Issue on BalkaNet, Romanian Academy, vol. 7(2-3), pp. 9–43 (2004)
[24] Tufiş, D., Barbu, E., Mititelu, V., Ion, R., Bozianu, L.: The Romanian Wordnet. In: Tufiş, D. (ed.) Romanian Journal on Information Science and Technology. Special Issue on BalkaNet, Romanian Academy, vol. 7(2-3), pp. 107–124 (2004)
[25] Tufiş, D., Ştefănescu, D., Ion, R., Ceauşu, A.: RACAI's Question Answering System at QA@CLEF 2007. In: CLEF 2007 Workshop, Budapest, Hungary, 15 p. (2007), http://www.clef-campaign.org/2007/working_notes/CLEF2007WN-Contents.html
[26] Tufiş, D., Ion, R.: Cross lingual and cross cultural textual encoding of opinions and sentiments. Tutorial at Eurolan 2007: Semantics, Opinion and Sentiment in Text, Iaşi, July 23-August 3 (2007)
[27] Tufiş, D.: Subjectivity Priors in WordNet. In: Global WordNet Conference, Szeged, January 22-25 (2008) (invited panel talk)
[28] Tufiş, D.: Mind Your Words! You Might Convey What You Wouldn't Like To. Int. J. of Computers, Communications & Control III(suppl. issue: Proceedings of ICCCC 2008), 139–143 (2008), ISSN 1841-9836, E-ISSN 1841-9844
[29] Vossen, P. (ed.): A Multilingual Database with Lexical Semantic Networks. Kluwer Academic Publishers, Dordrecht (1998)
[30] Wiebe, J., Mihalcea, R.: Word-senses and Subjectivity. In: Proceedings of the 21st International Conference on Computational Linguistics and 44th Annual Meeting of the ACL, Sydney, July 2006, pp. 1065–1072 (2006)
[31] Wilson, T., Wiebe, J., Hoffman, P.: Recognizing Contextual Polarity in Phrase-Level Sentiment Analysis. In: Proceedings of Human Language Technology Conference and Conference on Empirical Methods in Natural Language Processing (HLT/EMNLP), Vancouver, October 2005, pp. 347–354 (2005)
[32] Wiebe, J., Wilson, T., Cardie, C.: Annotating Expressions and Emotions in Language. Language, Resources and Evaluation 39(2/3), 164–210 (2005)

10

Special Cases of Relative Object Qualification: Using the AMONG Operator

Cornelia Tudorie and Diana Ştefănescu

"Dunărea de Jos" University of Galaţi
2 Stiintei, 800146 Galati, Romania
{Cornelia.Tudorie,Diana.Stefanescu}@ugal.ro

Abstract. Fuzzy querying means selecting the database objects that more or less respect a non-Boolean condition. The relative object qualification is defined as a new kind of user's preference expressing; related with this, the AMONG operator is able to compute the fulfillment degree of the relative selection criteria. Two special cases of the relative object qualification are discussed in the paper; they are referring to qualifying database objects relatively to particular values of other attributes. Solutions to model and to evaluate them are proposed.

Keywords: Database, Flexible Query, Fuzzy Query, Relative Qualification, Linguistic Value, AMONG Operator.

1 Introduction

Database fuzzy querying is maybe the most attractive characteristic of an intelligent interface to databases. It is a way to express more flexibly the user's preferences and at the same time to rank the selected tuples by a degree of criteria satisfaction.

Some important advantages resulting from including vague criteria in a database query may be:

- Easy to express queries
- The possibility to classify database objects, by selecting them based on a linguistic qualification
- The possibility to refine the result, by assigning to each tuple the corresponding fulfillment degree (the degree of criteria satisfaction); in other words, to provide a ranked answer according to the user's preferences.

There are many scientific works regarding database fuzzy querying: general reference books (for example [1] and [3]), but also many articles in journals and communications at conferences (for example [4] and [9]).

In this context, the relative object qualification was proposed in our works, as a new kind of selection criteria, like in the query:

Retrieve the ***inexpensive*** *cars* **among** *the* ***high speed*** *ones*

H.-N. Teodorescu, J. Watada, and L.C. Jain (Eds.): Intel. Sys. and Tech., SCI 217, pp. 181–191.
springerlink.com

Related with this, we proposed also a new operator, AMONG, which is able to compute the fulfillment degree of the relative selection criteria.

The next section presents the general problem of the relative qualification of the database objects in a vague query, but also the model of the AMONG operator.

Separate two sections are dedicated to special cases expressing quite usually selection criteria in the real life. They are referring to qualifying database objects relatively to particular values of other attributes, in queries like:

Retrieve the clients in ***Galati*** *which get* ***large*** *quantities of our product* or
Retrieve the clients which get ***large*** *quantities of some product*

In order to evaluate such queries, dynamic modeling of the fuzzy terms is needed. In this idea, the following section presents solutions to evaluate queries like these. The AMONG operator is used to compute the fulfilment degree of the selection criterion in both new cases of relative qualification. Finally, some conclusions and future works are shown.

2 Relative Object Qualification

There are in the literature certain approaches of modeling user's preferences, which solve different situations, such as: accepting tolerance, accepting different weight of importance for the requirements in a selection criterion, accepting conditional requirements, etc (for example, [2]).

We has found a new class of problems: two gradual properties are combined in a complex selection criterion such that a second one is applied on a subset of database rows, already selected by the first one. We assume that the second gradual property is expressed by a linguistic value of a database attribute, which is a label from the attribute linguistic domain. In this case, modeling of the linguistic

Table 1. A Relational Database Table (CAR)

Name	...	Max Speed	Price	...
A A		236	46000	
A A 4		221	28450	
B Coupe		250	39000	
C 300M		230	32000	
L R D		130	28000	
M B S		240	69154	
M C		190	18200	
N V		132	15883	
O C S		120	26259	
O F		192	43615	
P 206		170	10466	
P 607		222	31268	
P 806		177	20633	
P 911 C		280	65000	

domain of the second attribute requires taking into account not the whole crisp attribute domain, but a limited subset, characteristic to the first criterion-selected database rows. The two simple selection criteria are not independent, but they are in a combination which expresses a user's preference. This is exactly the problem of the relative object qualification, introduced in [5] and [6].

Let us consider as an example, the following query, based on a complex fuzzy selection criterion, addressed to the CAR table (Table 1):

Retrieve the ***inexpensive*** *cars* **among** *the* ***high speed*** *ones.*

The query evaluation procedure observes the following steps:

Algorithm

1. The selection criterion ***high speed*** **cars** is evaluated, taking into account the definition in Fig. 1; an intermediate result is obtained, containing the rows where the condition $\boldsymbol{\mu}_{high}(t)>0$ is satisfied.
2. The underlying interval containing the price for the selected cars forms the **Price** sub-domain [28450, 69154].
3. The definitions of the linguistic values {***inexpensive, medium, expensive***} are scaled to fit the sub-domain [28450, 69154], instead of [10466, 69154] (see Fig. 2, where, in order to make the difference, the new linguistic values are labelled in capital letters).
4. The selection criterion ***INEXPENSIVE*** **cars** is evaluated taking into account the definition in Fig. 2. The fulfillment degree $\mu_{INEXPENSIVE}$ is computed for each row of the intermediate result from step 1.
5. The global fulfillment degree (μ) will result for each tuple and they are selected if μ(t)>0 (the shaded rows in Table 2).

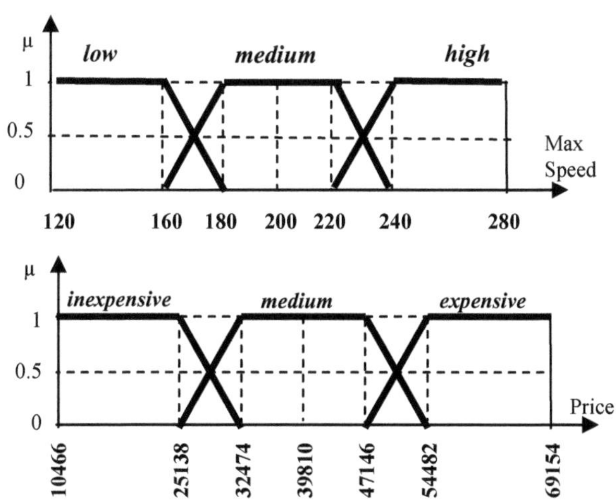

Fig. 1. Linguistic values defined on the Max Speed and Price attribute domains

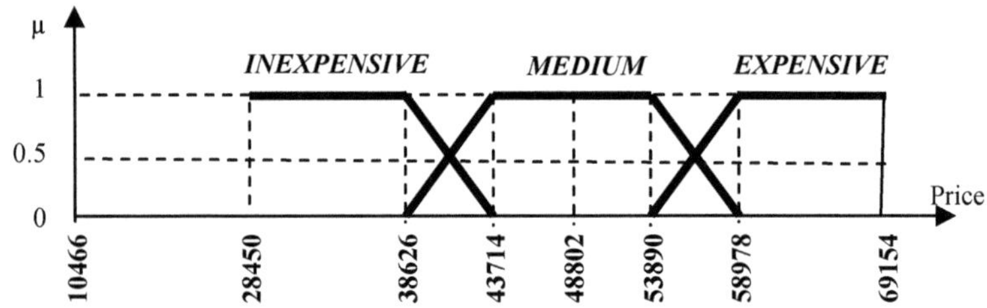

Fig. 2. Linguistic values defined on a sub-domain

Table 2. The "inexpensive cars among the high speed ones"

Name	Max Speed	Price	μ_{high}	$\mu_{INEXPENSIVE}$	μ
B Coupe	250	39000	1	0.92	0.92
M B S	240	69154	1	0.00	0.00
A A	236	46000	0.80	0.00	0.00
P 607	222	31268	0.10	1	0.10
A A 4	221	28450	0.05	1	0.05

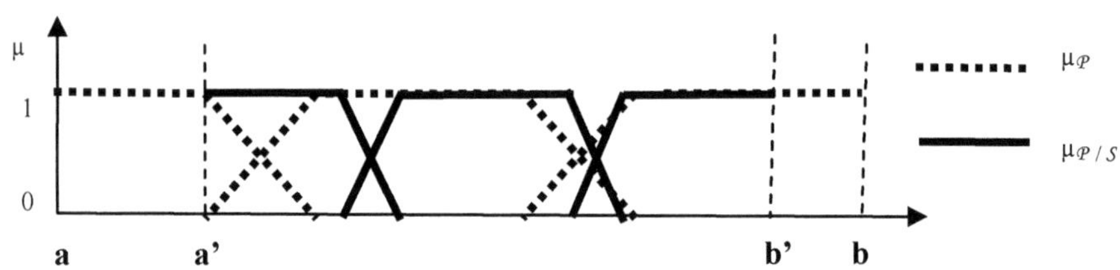

Fig. 3. Restriction of the attribute domain for a relative qualification

Definition. The **fuzzy model** of the **relative conjunction, AMONG**($\mathcal{P}$, $\mathcal{S}$), of two gradual properties, $\mathcal{P}$ and $\mathcal{S}$, associated with two attributes, A_1 and A_2, is defined by the mapping:

$$\mu_{\mathcal{P}\text{AMONG}\mathcal{S}} : D_1 \times D_2 \rightarrow [0,1] \text{ or}$$
$$\mu_{\mathcal{P}\text{AMONG}\mathcal{S}} : [a_1,b_1] \times [a_2,b_2] \rightarrow [0,1], \qquad (v_1,v_2) \mapsto min\,(\mu_{\mathcal{P}/\mathcal{S}}(v_1), \mu_{\mathcal{S}}(v_2))$$

The same fulfillment degree, defined on a database table R is:

$$\mu_{\mathcal{P}\text{AMONG}\mathcal{S}} : R \rightarrow [0,1],\ t \mapsto min\,(\mu_{\mathcal{P}/\mathcal{S}}(t), \mu_{\mathcal{S}}(t)) = min\,(\mu_{\mathcal{P}/\mathcal{S}}(t.A_1), \mu_{\mathcal{S}}(t.A_2))$$

where t is a tuple ; D_1 and D_2 are the the definition domains for the the attributes A_1 and A_2 ; $t.A_1$ and $t.A_2$ are the values of the attributes A_1 and A_2 for the tuple t; $\mu_{\mathcal{S}}$ is the membership function defining the $\mathcal{S}$ gradual property and $\mu_{\mathcal{P}/\mathcal{S}}$ is the fulfillment degree of the first criterion ($\mathcal{P}$) relative to the second one ($\mathcal{S}$).

After the first selection, based on the property $\mathcal{S}$, associated with the attribute A_2, the initial domain [a,b] of the attribute A_1 becomes more limited, i.e. the interval [a', b'] (Fig. 3).

Thus, if

$$\mu_{\mathcal{P}} : [a,b] \rightarrow [0,1] \quad , \quad v \mapsto \mu_{\mathcal{P}}(v)$$

then

$$\mu_{\mathcal{P}/\mathcal{S}} : [a',b'] \rightarrow [0,1] \quad , \quad v \mapsto \mu_{\mathcal{P}/\mathcal{S}}(v)$$

so that

$$\mu_{\mathcal{P}/\mathcal{S}} = \mathbf{f} \circ \mu_{\mathcal{P}}$$

where **f** is the transformation

$$f : [a',b'] \rightarrow [a,b]$$

$$f(x) = a + \frac{b-a}{b'-a'}(x - a') \ . \tag{1}$$

Therefore

$$\mu_{\mathcal{P}/\mathcal{S}}(v) = \mu_{\mathcal{P}}\left(a + \frac{b-a}{b'-a'}(v - a') \right) . \tag{2}$$

Definition. The **algebraic model** of the **AMONG operator** is:

$$\mu_{\mathcal{P}\,\mathrm{AMONG}\,\mathcal{S}} : R \rightarrow [0,1]$$

$$\mu_{\mathcal{P}\,\mathrm{AMONG}\,\mathcal{S}}(t) = \mathit{min}\left(\mu_{\mathcal{P}}\left(a_1 + \frac{b_1 - a_1}{b_1' - a_1'}(t.A_1 - a_1') \right), \mu_{\mathcal{S}}(t.A_2) \right) \tag{3}$$

where $[a_1', b_1'] \subseteq [a_1,b_1]$ is the sub-interval of the A_1 corresponding to the table $Q_{\mathcal{S}}(R)$ (obtained by the first selection, on the attribute A_2, using property $\mathcal{S}$).

Properties, remarks, and comments on the AMONG operator can be found in [8].

The answer to the query in our example is Table 2, where $\mu_{INEXPENSIVE}$ stands for $\mu_{\mathcal{P}/\mathcal{S}}$, i.e. $\mu_{\text{inexpensive/high}}$, and μ stands for the global selection criterion, computed as μ_{AMONG}. The membership function $\mu_{\mathcal{P}/\mathcal{S}}$ stands for a transformation of the initial membership function $\mu_{\mathcal{P}}$ and is obtained by translation and compression, as in (2).

The above procedure is easy to implement, if we consider it as a sequence of several operations. An efficient method to evaluate this kind of query is proposed by the "Unified Model of the Context", where the knowledge base (fuzzy definitions of the linguistic vague terms) is incorporated in the database (see [5]).

3 Relative Qualification to Other Crisp Attribute

At least one special situation is relatively frequent, when the linguistic values must be dynamically defined for an attribute sub-domain, obtained after a crisp selection. It is about a complex selection criterion that includes a gradual property referring to the database rows already selected by a crisp value. It is a special case of the relative qualification.

Let us imagine a table (Table 3), containing all the sales transactions of a national company. The following query must take into account the principle that generally, in different cities (from the biggest to the little ones) the amount of the sales is different.

Retrieve the clients in ***Galati*** *which get* ***large*** *quantities of our product*

Table 3. The transactions in sales of a national company

Client	...	Quantity	City	...
A		70	Galati	
B		30	Galati	
C		230	Bucharest	
I		145	Bucharest	
L		130	Galati	
M		8	Tecuci	
N		17	Tecuci	
O		120	Galati	
P		166	Galati	
R		222	Bucharest	
S		28	Tecuci	

The selection criterion "***large*** *quantity*" has a different meaning for different cities. In the same way, the query evaluation procedure follows the same steps as in the previous section:

Algorithm

1. The crisp selection criterion **city='Galati'** is classically evaluated and an intermediate result is obtained (Table 4).
2. The interval containing the **quantity** for the selected sales forms the quantity sub-domain [30, 166], instead of [8, 230].
3. The linguistic values {***small, medium, large***} will be defined on the new sub-domain (Fig. 4).
4. The fuzzy selection criterion ***large*** **quantity** is evaluated according to the new definitions and the fulfillment degree will result for each tuple. (Table 5).

Table 4. The Transactions in Galati City

Client	...	Quantity	City	...
A		70	Galati	
B		30	Galati	
L		130	Galati	
O		120	Galati	
P		166	Galati	

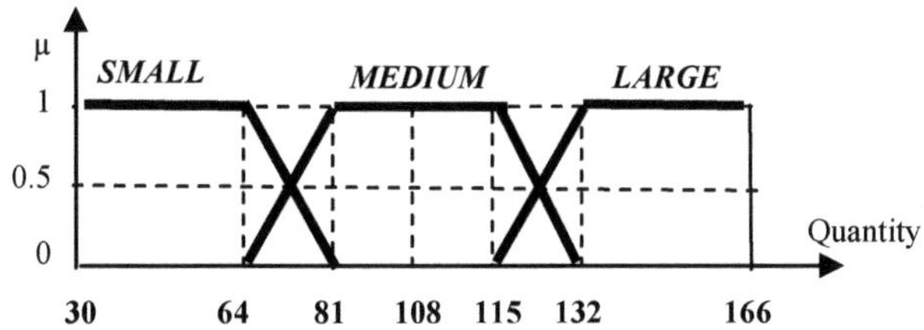

Fig. 4. Linguistic values defined on a sub-domain

Table 5. Transactions of Large Quantities in Galati City

Client	...	Quantity	City	...	μ
P		166	Galati		1
L		130	Galati		0.88
O		120	Galati		0.29

One can remark that a *large* quantity at Galati (for example 130) is less than the minimum at Bucharest (i.e. 145). This is the reason for which the definitions of the linguistic values must be adapted to the context; that means that the **qualification** (*large quantity*) is **relative to the other crisp attribute** (city).

In this case, **the simplified algebraic model of the AMONG operator** becomes:

$$\mu_{\mathcal{P}\text{AMONG}\,\mathcal{S}} : R \to [0,1] \quad , \qquad t \mapsto \mu_{\mathcal{P}/\mathcal{S}}(t.A_1)$$

$$\mu_{\mathcal{P}\text{AMONG}\,\mathcal{S}}(t) = \mu_{\mathcal{P}}\left(a_1 + \frac{b_1 - a_1}{b_1' - a_1'}(t.A_1 - a_1') \right) \tag{4}$$

where $[a_1', b_1'] \subseteq [a_1, b_1]$ is the sub-interval of the attribute A_1 values, in the table $Q_{\mathcal{S}}(R)$ (obtained by the first selection, on the attribute A_2, using property $\mathcal{S}$).

One can observe that, this time, the property $\mathcal{S}$ brings its contribution in expressing the criteria satisfaction degree, only by the limitation of the domain for the property $\mathcal{P}$, $[a_1, b_1]$ becomes $[a_1', b_1']$.

Important remark

If the above query is interpreted as a classical conjunction and not as a relative qualification, the answer will be absolutely different. Therefore, a more suggestive formulation of the query should be:

Retrieve the clients which get ***large*** *quantities of our product*
among the clients in **Galati**

4 Relative Qualification to Group on Other Attribute

The queries in the previous paragraph consider that all sales refer to the same product ("our product"), i.e. the quantities can be compared. But, let us consider now that all sales, for different products, are stored in database (Table 6).

In this case, the query:

*Retrieve the clients which get **large** quantities of **soap***

expresses a qualification relative to a crisp attribute and can be evaluated like above.

But if the query is:

*Retrieve the clients which get **large** quantities*

then the values of the *quantity* attribute, for different products, can't be compared. It is impossible. This example suggests the evaluation of the ***large** quantity* criterion, by taking into account one product at a time, i.e.

*Retrieve the clients which get **large** quantities of some product*

Table 6. Transactions of Various Product Sales

Client	...	Quantity	Product	...
A		70	soap	
B		6	soap	
C		226	envelope	
M		4	vacuum cleaner	
N		162	envelope	
O		14	vacuum cleaner	
P		2	soap	
R		102	soap	
S		1	vacuum cleaner	
T		2	vacuum cleaner	
U		18	envelope	

Table 7. Transactions of Large Quantities Sales

Client	...	Quantity	Product	...	μ
C		226	envelope		1
O		14	vacuum cleaner		1
R		102	soap		1
N		162	envelope		0.53
A		70	soap		0.46

For each product, the linguistic value is defined on the interval of the quantity values, existing in the database, but only for that product. According to the definitions in Fig. 5, the answer will be in the Table 7. One can remark the "higher weight" of the 14 vacuum cleaners than the 162 envelopes.

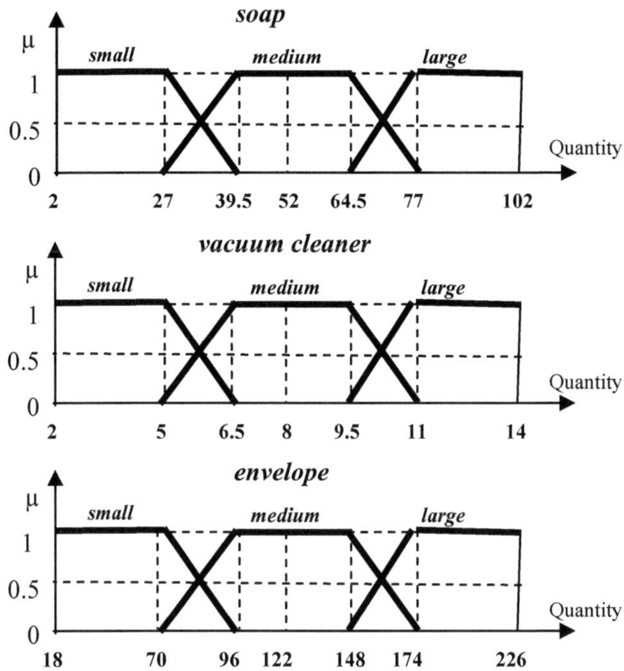

Fig. 5. Linguistic values defined on sub-domains of the quantity attribute, for each product

5 Dynamic Modeling of the Linguistic Values

Previous sections have presented certain types of queries that require dynamically defining of the linguistic values, by partitioning an attribute sub-domain, already obtained by a previous selection.

Actually, the main problem of the relative qualification is how to dynamically define the linguistic values on the subdomains (step 3 of the above algorithm), depending on an instant context? Some procedures for automatic discovering of the linguistic values definitions can be implemented, having a great advantage: details regarding effective attribute domain limits, or distributions of the values, can be easily obtained thanks to directly connecting to the database. Methods for automatic extraction of the linguistic values definitions from the actual database attribute values and solutions for uniformly modeling the context (database and knowledge base) were proposed in [5].

One example of algorithm is presented in the following. We assume that there are usually three linguistic values, modeled as trapezoidal membership functions. Any generalization is possible.

Obtaining the definition for the three linguistic values l_1, l_2, and l_3 on a database attribute starts from the predefined values α and β, and the attribute crisp domain limits, I and S; these ones are coming from the database content.

For example:

$$\alpha = \frac{1}{8}(S - I) \qquad \text{and} \qquad \beta = 2\alpha = \frac{1}{4}(S - I) \tag{5}$$

The membership functions for l_1, l_2, and l_3 are:

$$\mu_{l_1}(v) = \begin{cases} 1 \ , & I \le v \le I+\beta \\ 1 - \dfrac{v-(I+\beta)}{\alpha} \ , & I+\beta \le v \le I+\beta+\alpha \\ 0 \ , & v \ge I+\beta+\alpha \end{cases}$$

$$\mu_{l_2}(v) = \begin{cases} 0 \ , & I \le v \le I+\beta \\ \dfrac{v-(I+\beta)}{\alpha} \ , & I+\beta \le v \le I+\beta+\alpha \\ 1 \ , & I+\beta+\alpha \le v \le I+2\beta+\alpha \\ 1 - \dfrac{v-(I+2\beta+\alpha)}{\alpha} \ , & I+2\beta+\alpha \le v \le I+2\beta+2\alpha \\ 0 \ , & v \ge I+2\beta+2\alpha \end{cases} \tag{6}$$

$$\mu_{l_3}(v) = \begin{cases} 0 \ , & I \le v \le I+2\beta+\alpha \\ \dfrac{v-(I+2\beta+\alpha)}{\alpha} \ , & I+2\beta+\alpha \le v \le I+2\beta+2\alpha \\ 1 \ , & v \ge I+2\beta+2\alpha \end{cases}$$

where v=t.A is a value in the domain D = [I,S]) of an attribute A of a table R.

The fuzzy query evaluation is possible by building an equivalent crisp query. The knowledge (fuzzy model of the linguistic terms) is used first for the SQL query building and after that for the computing of the fulfillment degree for each tuple. The ***context*** is defined in this case as the pair database and the knowledge base corresponding to it.

One of the most important points of an interface to databases is the performance, more specifically, the response time in query evaluation. In order to have good performance, an efficient solution needs to model the context in a uniform approach, as a single database, incorporating the fuzzy model of the linguistic terms, or their description in the target database. So, a unified model of the context was proposed in [5]. That is an extended database, containing both the target data, but also the knowledge corresponding to it.

A ***Dynamic Context*** means including in the database only of the data necessary to dynamically define the linguistic terms at the moment of (or during) the querying process.

According to the proposed model of the dynamic context, the vague query evaluation consists in building only one crisp SQL query, so as to provide the searched database objects and at the same time the degree of criteria satisfaction, for each of them.

6 Conclusions

The relative qualification consists in two vague selection conditions, in a special relationship: the first gradual property, expressed by a linguistic qualifier, is interpreted and evaluated relatively to the second one; accordingly, the fulfillment degree is computed using the particular operator, AMONG. The main idea of the evaluation procedure is to dynamically define sets of linguistic values on limited attribute domains, determined by previous fuzzy selections. This is why it is not useful to create the knowledge base with the fuzzy definitions before, but to define the vague terms included in queries each time they are needed. Some implementations, that validate all these ideas, are developed and are running in the laboratory of our Department ([7]).

Two special cases of the relative qualification are presented in this paper. It is about complex selection criteria which include gradual property referring to already selected database rows or group by a crisp value. Solutions to evaluate such vague queries using the AMONG operator were proposed. They are inspired by real situations where humans need to express their particular preferences.

Future work will explore the implications of the new proposed kind of query in real fields, like Business Intelligence, OLAP, or Data Mining; but also other application fields of the new connective AMONG.

References

1. Bosc, P., Lietard, L., Pivert, O., Rocacher, D.: Gradualité et imprécision dans les bases de données. Ellipses, Paris (2004)
2. Dubois, D., Prade, H.: Using fuzzy sets in flexible querying: Why and how? In: Christiansen, H., Larsen, H.L., Andreasen, T. (eds.) Workshop on Flexible Query-Answering Systems, pp. 89–103. Roskilde, Denmark (1996)
3. Galindo, J., Urrutia, A., Piattini, M.: Fuzzy Databases: Modeling, Design and implementation. Idea Group Publishing, Hershey (2006)
4. Kacprzyk, J., Zadrozny, S.: Computing with words in intelligent database querying: standalone and Internet-based applications. Information Sciences, vol. 134, pp. 71–109. Elsevier, Amsterdam (2001)
5. Tudorie, C.: Contributions to interfaces for database flexible querying. PhD Thesis. University "Dunărea de Jos", Galaţi, Romania (2006)
6. Tudorie, C., Bumbaru, S., Segal, C.: New Kind of Preference in Database Fuzzy Querying. In: Proceedings of the 11th International Conference on Information Processing and Management of Uncertainty in Knowledge-based Systems, IPMU 2006, Paris, pp. 1389–1395 (2006)
7. Tudorie, C.: Laboratory software tools for database flexible querying. In: Proceedings of the 11th International Conference on Information Processing and Management of Uncertainty in Knowledge-based Systems, IPMU 2006, Paris, pp. 112–115 (2006)
8. Tudorie, C.: Qualifying Objects in Classical Relational Database Querying. In: Galindo, J. (ed.) Handbook of Research on Fuzzy Information Processing in Databases. Idea Group Publishing, Information Science Reference, Hershey, pp. 218–245 (2008)
9. Zadeh, L.A.: From Computing with Numbers to Computing with Words. In: Annals of the New York Academy of Sciences, vol. 929, pp. 221–252 (2001)

Effective Speaker Tracking Strategies for Multi-party Human-Computer Dialogue

Vladimir Popescu[1,2], Corneliu Burileanu[2], and Jean Caelen[1]

[1] Grenoble Institute of Technology, France
{vladimir.popescu, jean.caelen}@imag.fr
[2] "Politehnica" University of Bucharest, Romania
cburileanu@messnet.pub.ro

1 Introduction

Human-computer dialogue is already a rather mature research field [10] that already boiled down to several commercial applications, either service or task-oriented [11]. Nevertheless, several issues remain to be tackled, when unrestricted, spontaneous dialogue is concerned: barge-in (when users interrupt the system or interrupt each other) must be properly handled, hence Voice Activity Detection is a crucial point [13]. Moreover, when multi-party interactions are allowed (i.e., the machine engages simultaneously in dialogue with several users), supplementary robustness constraints occur: the speakers have to be properly tracked, so that each utterance is mapped to a certain speaker that had produced it. This is needed in order to perform a reliable analysis of input utterances [2].

Spoken human-computer dialogue systems can be seen as advanced applications of spoken language technology. A dialogue system represents a voiced and relatively natural interface between the user and a software application. Thus, spoken dialogue systems subsume most of the fields in spoken language technology, including speech recognition and synthesis, natural language processing, and dialogue management (planning).

A dialogue system involves the integration of several components, which generally provide the following functions [3]:

- **speech recognition:** conversion of an utterance (represented as a sequence of acoustic parameters), into a word sequence;
- **language understanding:** analysis of a word sequence in order to obtain a meaning representation for this sequence, in the dialogue context;
- **dialogue management:** system-human interaction control, as well as the coordination of the other components of the dialogue system;
- **task management:** interfacing of the dialogue management and language understanding modules, with the application domain for the tasks performed by the system;
- **answer generation:** computation of the sequence of words constituting the answer generated by the system, situating it in the discourse context represented by the dialogue history, and in the pragmatic context, represented by the relationship between user and machine, as well as by their social roles;

H.-N. Teodorescu, J. Watada, and L.C. Jain (Eds.): Intel. Sys. and Tech., SCI 217, pp. 193–218.
springerlink.com

- **speech synthesis:** conversion of the text representing the system's answers, into an acoustic waveform.

Among these components, some (dialogue and task management, partially language understanding and answer generation) are language independent and (in part) dependent on the application domain, whereas others (speech recognition and synthesis) depend on the language, being (in principle) independent of the application domain. Thus, for the components in the first category, reuse in new languages is, to a great extent, possible, if the application domains are kept, whereas the components in the second category have to be developed for each new language, in a manner that is independent of the application.

The task of the speech recognition component in a spoken dialogue system consists in converting the utterance (in acoustic form) came from the user, into a sequence of discrete units, such as phonemes (sound units) or words. A major obstacle in accomplishing a reliable recognition resides in speech signal variability, which results from the following factors:

- *linguistic variability*: consists in the effects of several linguistic phenomena that the speech signal undergoes, such as phonetic co-articulation (i.e., the fact that the same phoneme can have different acoustic realizations in different contexts, determined by the phonemes neighboring the sound concerned);
- *speaker variability*: consists in the effects of inter- and intra-speaker acoustic differences; inter-speaker differences are determined by physical factors, such as the particular shape of the vocal tract, the age, sex or origin of the human subjects (the fact that a speaker may not be native in the language being used for communication); intra-speaker differences are determined by the fact that the same word can be uttered in several ways by the same speaker, according to her or his emotional or physical state, or to the pragmatic (and situational) context of the utterance - a word can be uttered more emphatically in order to stress a certain idea;
- *channel variability*: consists in the effects of the environmental noise (which can be either constant or transient) and of the transmission channel noise (e.g., microphones, telephone lines, or data channels – "Voice over IP").

The speech recognition component in a typical dialogue application has to take into account several additional issues:

- *speaker independence*: since the application is normally used by a wide variety of individuals, the recognition module may not be trained for one single speaker (or for a few speakers) supposed to use the system, as it is the case, for instance, in voice dictation applications. Thus, speech has to be collected from an acoustically representative set of speakers, and the system will use these data in order to recognize utterance came from (potential) users, whose voices were not used during training. This is why the performances of the speaker independent recognition process are generally poorer than for speaker dependent recognition.
- *size of the vocabulary*: the number of words that are "intelligible" to the the dialogue system depends on the application considered, as well as on the dialogue

(management) complexity [10]. Thus, a strictly controlled and rather inflexible dialogue may constrain the user to a small vocabulary, limited to a few words expressing the options available in the system; yet, in more natural and flexible dialogues, the vocabulary accepted by the system can count for several thousands of words (for instance, the PVE – "Portail Vocal pour l'Entreprise" system, developed in France as a voice portal for enterprises, uses an about 6000 words recognition module [3]).

- *continuous speech*: the users are expected to be able to establish a conversation with the spoken dialogue system, using unconstrained speech and not, for instance, commands uttered in isolation. The issue of establishing the limits of the words is extremely difficult for continuous speech, since in the acoustic signal there is no physical border between them. Hence, linguistic or semantic information can be used in order to separate the words in users' utterances.
- *spontaneous speech*: since users' utterances are normally spontaneous and non-planned, there are generally characterized by disfluencies, such as hesitations or interjections (e.g., "humm"), false starts, in which the speaker begins an utterance, stops in the middle and re-starts, or extralinguistic phenomena, such as cough. The speech recognition module must be able to extract, out of the speech signal, a word sequence allowing the semantic analyzer to deduce the meaning of the user's utterance.

Dialogues between a computer and only *one* human partner are studied in a rather mature research field [10] and several commercial applications or systems exist in this respect, the situations where the computer is supposed to get involved in a dialogue with *several* humans at the same time, are still too little studied in a systematic manner. Several possibilities exist, towards multi-party human-computer dialogue:

- **multi-session** human-computer dialogue, where the machine gets involved in parallel dialogues with several humans; these dialogues are independent in that the speakers do not interact with each other and do not have access to the dialogues between the machine and the other speakers. This type of interaction is particularly interesting for situations involving concurrent access to a limited set of resources (e.g. meeting room reservation in a company); therefore, in this case there are several *classical* dialogues, on which the computer should maintain a coherent representation. Even if there is not a real multi-party dialogue, there is rather little work worldwide in this respect. For instance, current state of the art is represented by the PVE system [3]. In this system, multiple sessions are handled, at the dialogue control level, through a game theoretic approach, where machine contribution sequences are evaluated via gains that are dependent, at the same time, on the task context (amount of resources, speakers' roles, etc.) and on the speech acts performed by the speakers.
- **multi-party** human-computer dialogue, where the machine gets involved in *simultaneous* dialogues with several speakers; as in multi-session dialogue, the machine has to keep a coherent view on the dialogues; yet, there is a major difference in regards to the latter situation: in multi-party interaction, the dialogues are simultaneous, all the speakers being at the same place and having

access to *all* speakers' utterances. This is why modeling (and *formalizing*) this type of interaction is particularly difficult. However, since around 2000 there is more and more (substantial) research work in this respect, trying either to study the portability of models designed for traditional dialogues, to multi-party dialogue [5], or to analyze multi-party dialogue corpora in order to determine the differences between traditional and multi-party dialogues [16], or even to give a formal account of particular aspects of multi-party dialogue (such as dialogue control) and concerning only some issues (such as the *shared* context between interlocutors) [8].

In multi-party dialogue, several speakers interact with the system, that thus has to be able to assign each human speech turn to a speaker identifier (in other words, the system has to figure out not only what has been said, but also *by whom* that has been said). Therefore, the speech recognition component in the dialogue system has to *track* the speakers as they produce their turns, and this should happen as fast as possible, so that the total processing time, for a request from a user, is as reduced as possible.

Thus, in this chapter we propose an algorithm for tracking speakers in dialogue; the procedure consists in an "on the fly" unsupervised MLLR (Maximum Likelihood Linear Regression) adaptation of acoustic models to speakers, where we derive a decision tree based on speech recognition scores at utterance level. This decision tree is then used to cluster utterances into speaker identities. An important point to emphasize is that the clustering process is performed *on-line*, for each new user utterance, but relying on previous utterances as well. The novelty of the method proposed in this chapter resides in that we use only "classical" unsupervised MLLR adaptation, but through a careful handling of confidence scores and decision-making based on these. Moreover, the computational simplicity (essentially assuming a top-down left-right traversal of a decision tree) is suited for dialogue applications where real-time operation is an important constraint. Concerning this latter aspect, several strategies for reducing speaker tracking time are studied: from a rather "naïve" parallelization of recognition processes (for a speaker-independent system and several speaker-adapted systems), we further optimize the MLLR adaptation process *per se*, by combining the usage of phoneme-level regression classes [9] with a parallel running of the adaptation: MLLR is performed in parallel for the regression classes.

After a brief overview, in § 2.1, of related research, concerning mostly *off-line* indexation of audio recordings of multi-party meetings, where real-time constraints thus do not apply, we review, in § 2.2, the fundamentals of MLLR adaptation; then, in § 3.1 we present the baseline speaker tracking algorithm, along with motivational background from psycholinguistics and corpus study; in § 3.2 we propose several strategies for improving the runtime performance of the algorithm, via parallelization at several levels. Furthermore, in § 4 we discuss several experiments performed with different versions of the speaker tracking algorithm, in the context of a book reservation multi-party dialogue application, in French and Romanian languages. The last section concludes the paper and proposes further enhancements.

2 Background

2.1 Related Work

As for the current state of the art regarding speaker tracking, most of the work is related to speaker segmentation and/or indexing of offline multimedia content (or recorded meetings); in that case, the task is eased by several points: meetings usually take place indoor, speakers have rather fixed positions, their number is rather constant throughout the meeting [18]. Thus, one of the few previous works on multi-party dialogue segmentation assigns turn taking likelihoods to a language model that reflects the nature of the conversations [13]; two algorithms run in parallel for speaker and speech content estimation on TV sports news. Hence, dialogue issues are not directly considered, since *enough* data is available offline and runtime constraints do not apply; moreover, in multi-party dialogues, where speaker changes occur in an unpredictable manner (it is only progressively that speech turns become available), hence statistically modelling speaker changes is much more complicated, if not impossible.

Another strand of research boils down to performing both environment (i.e., noise features) adaptation and speaker adaptation and tracking, in pre-recorded meetings as well [18], [20], [21]. For example, in [20] and [21], an unsupervised speaker adaptation in noisy environments is performed, in order to segment recorded meetings, where usually several microphones (microphone arrays) exist and the relative positions of speakers with respect to the microphones can be exploited [12]. Usually, the approaches adopted in this context start from GMM (Gaussian Mixture Models)-based speaker identification systems, that are coupled with HMM (Hidden Markov Models)-based speech recognition systems [18], [20]. Concerning the microphone arrays approach, it usually relies on cross-correlations computed on signals coming from pairs of acoustic sensors [12]. However, none of these procedures apply to service-oriented dialogue applications, since they usually involve outdoor processing, where non-relevant speech signals exist as well and the geometry of the users positions with respect to the acoustic environment is not too much controllable [2], [11]. Moreover, there is another research strand that relies on multimodal input for speaker tracking, e.g. combining acoustics with vision [7].

However, the research closest to ours was pursued by Furui and colleagues [23], where one proposes an unsupervised, on line and incremental speaker adaptation methods that improves the performance of speech recognizers when there are *frequent* changes in speaker identities and each speaker produces a series of several utterances. Basically, the authors propose two speaker tracking methods, that are then applied to broadcast news transcription; first, an HMM based scheme is proposed, where the likelihood given by a speaker independent decoder is compared to the scores given by speaker adapted HMMs. The rationale behind this approach is that, for succeeding utterances from the same speaker, the speaker adapted decoder is expected to give a larger likelihood than the speaker independent HMM set; on the other hand, if the acoustic features differ from those of a previous (identified) speaker, then the speaker independent decoder is expected to yield a larger likelihood than the speaker adapted ones (unless the voice of the new speaker is very similar to the previous speaker's).

The adaptation is achieved using the MLLR method (see § 2.2), but the new (adapted) mean vectors of the Gaussian mixture components in the states of the HMMs are updated so that overtraining (in the case of sparse adaptation data) is avoided, by linearly interpolating the adapted mean vector with the original (unadapted) mean. Moreover, the algorithm also tackles the situation where phonemes remain unadapted, because they are not acoustically realized in the adaptation data stream; this latter point is handled using vector field smoothing, whereby unadapted mean vectors are transferred to a new vector space (corresponding to the adapted HMM states) by using an interpolated transfer vector [23]. Then, speaker tracking supposes simply recognizing an utterance with a speaker independent system and with a set of speaker adapted decoders and comparing the likelihoods yielded by these processes. The second algorithm relies on GMMs for discerning the speakers, since computation is thus reduced; nevertheless, the initial speaker independent HMMs are adapted to speakers' voices as well, in order to improve utterance recognition performance.

Since in the goal of Furui and colleagues' research [23] was to improve the segmentation of multi-party (e.g., broadcast news) conversations, the correct speaker tracking rate was not a relevant measure, hence results in this respect are not reported; only word error rate improvements using speaker tracking are shown.

2.2 Fundamentals of MLLR Speaker Adaptation

In a generic service-oriented spoken dialogue system the speech recognition component is usually instantiated as a medium or high vocabulary speaker *independent* HMM based Viterbi acoustic decoder, followed an *n*-gram based linguistic model [6]. This approach is legitimate for classical dual-party dialogues, where the machine has only one (human) interlocutor. However, in multi-party dialogue, as the system has to figure out also *who* has uttered a certain speech turn, besides *what* has been said in that turn, speaker tracking has to be performed. An approach to this resides in *adapting* a speaker independent HMM decoder to the voices of the particular speakers taking part in the dialogue and expecting higher recognition (probability) scores for the speaker adapted systems, with respect to the speaker independent decoder. This is motivated by studies that have shown that a speaker adapted system usually has a word error rate around two times lower than the word error rate obtained with the speaker independent system [1], [4], [6]. Obviously, these higher recognition scores and lower error rates are achieved for speech signals produced by the speaker to whom the system was adapted; hence, for each new dialogue, new adapted systems have to be built, using a rather low amount of data (the length of a few utterances) and proceeding in an incremental manner (adaptation is further pursued as the speaker produces more utterances in dialogue).

As any data-driven HMM parameter estimation method (i.e., learning method), HMM adaptation can be *supervised*, when adaptation data are already labeled with textual information, or *unsupervised*, when adaptation data are not labeled. Moreover, when all adaptation data are available at once, *static* adaptation is performed;

otherwise, if data become available along a time span, *incremental* adaptation is performed.

If static adaptation is performed in supervised manner, it can be done using the MLLR method, or the MAP (Maximum A Posteriori Probability) method. On the other hand, if incremental adaptation is performed, in supervised or unsupervised manner, it can be performed via the MLLR method [6]. Hence, we can follow two different approaches when adapting an HMM set to a speaker's voice:

- via MLLR method - for static or incremental adaptation, in supervised or unsupervised manner;
- using MAP criterion - only for static supervised adaptation.

Obviously, in multi-party dialogue applications where adaptation data becomes progressively available, incremental adaptation is the most appropriate choice, hence only the MLLR method can be used. In this case, the adaptation process essentially consists in computing a set of transforms which, applied to the HMMs due to be adapted, will reduce the mismatch between the HMMs and speaker dependent adaptation data. More specifically, MLLR is a model adaptation technique that estimates a set of linear transforms for the means and variances of the components in the Gaussian mixtures that model emissions at each transition between the states in the HMMs. The effect of these transforms stems from modifying the means and variances of these Gaussians, so that each state in the HMMs generates with maximum probability the adaptation data. However, it has been observed that, in practice, the most important performance improvements are obtained if only mean vectors are estimated, leaving the covariance matrices unchanged; modifying the latter parameters does not bring substantial improvements in recognition scores or error rates [9].

Denoting by $\mu = (\mu_1, ..., \mu_n)$ the n-dimensional mean vector of a component in a Gaussian mixture that models the output of one state in an HMM, and by $\overline{\mu}$, the re-estimated mean vector, using adaptation data stream s, the *transformation matrix* W is given by $\overline{\mu} = W \cdot \zeta$, where W is a $n \times (n + 1)$ matrix, and ζ is the *extended mean* vector: $\zeta = (\omega, \mu_1, ..., \mu_n)^T$, where the upper index T denotes the transposition operation, and $\omega \in \{0; 1\}$ is an *offset* term; usually, the value $\omega = 1$ is preferred [22], in order to induce a non-negligible offset on the initial mean vectors.

The transformation matrix W can be computed in a manner that is akin to the linear regression method [22]. We denote by $s = (s_1, ..., s_T)$ a set of acoustic observations (an adaptation data stream), where each s_i, $i = 1, ..., T$ is a multidimensional vector (its dimension is given by the number of acoustic parameters used for characterizing a frame of signal - for instance, Mel cepstra); we denote by s_t the observation vector at moment t, by m, the index of a Gaussian component in a mixture, by μ_{mj} the mean of the m_j-th component in the Gaussian mixture, by ζ_{mj} the extended mean vector of μ_{mj}, by Σ_{mj}, the covariance matrix for the Gaussian of index m_j, and by $L_{mj}(t)$ the *occupation* probability for the m_j-th component of a mixture, at time t (i.e., the probability that at time t mixture m_j models the output of the HMM, in the current state).

In order for the adaptation process to be robust enough with respect to data variability, it is sensible to compute distinct transformation matrices for different states in an HMM. However, computing a distinct transformation matrix for each state is often infeasible, due to data sparseness; this is why states can be *grouped*, so that they share a transformation matrix. One currently accepted criterion of state clustering (into *regression classes*) is the identity (or closeness) of the acoustic phenomena that these states account for [6], [9]: thus, the states that output, with maximum probability, the same phoneme type, are grouped in the same regression class, for which only one transformation matrix is built.

Thus considering that R Gaussian components of a mixture, forming a regression class denoted by the set of indexes $\{ m_1, ..., m_R \}$, are adapted by computing the transformation matrix W_m, it has been shown [9] that the transformation matrix can be obtain from the equation: $\sum_{t=1}^{T} \sum_{r=1}^{R} L_{mr}(t) \cdot \Sigma_{mr}^{-1} \cdot s_t \cdot \zeta_{mr}^{T} = \sum_{t=1}^{T} \sum_{r=1}^{R} L_{mr}(t) \cdot \Sigma_{mr}^{-1} \cdot W_m \cdot \zeta_{mr} \cdot \zeta_{mr}^{T}$. The occupation probabilities $L_{mr}(t)$ can be expressed as $L_{mr}(t) = P\ (\gamma_{mr}(t)\ |s,\ HMM)$, where $\gamma_{mr}(t)$ denotes the Gaussian of index m_r at time t, and HMM denotes the hidden Markov model currently adapted. This latter probability is usually computed using the *forward-backward* algorithm, from the adaptation data [6], [22]. Thus, the transformation matrix is computed in several steps: first, we denote the left-hand member of the equation above, by Z (since it does not depend on W_m); then, we define a new variable G_i with the elements $g_{jk}^{(i)} = \sum_{r=1}^{R} v_{ii}^{(r)} \cdot d_{jk}^{(r)}$, with $V^{(r)} = \sum_{t=1}^{T} L_{mr}(t) \cdot \Sigma_{mr}^{-1}$, and $D^{(r)} = \zeta_{mr} \cdot \zeta_{mr}^{T}$. Hence, the i-th row of W_m can be determined as $w_i^T = G_i^{-1} \cdot z_i^T$, where z_i is the i-th row of Z.

3 Speaker Tracking Algorithms

3.1 Baseline Adaptation Procedure

3.1.1 Outlook

The speaker tracking algorithm consists essentially in adapting a speaker-independent speech recognition system to each new utterance, then clustering these adapted systems into a more restrained set, denoting the speakers in multi-party conversation. The adaptation process is represented by an unsupervised MLLR adaptation of a set of speaker-independent HMMs, whereas the clustering is based on the top-down traversal of a decision tree involving the utterance-level log-likelihood scores obtained in speech recognition.

The inputs to the algorithm consist in:

- a set of speaker-independent trained HMMs (at a word, triphone or phoneme level); these are denoted by the system S_0;
- a set of acoustic features extracted from a test speech signal; such an utterance is denoted by ε_i, for the i-th user utterance in dialogue.

The output of the algorithm consists in an assignment of a speaker identifier to the input utterance. As for the intermediary information structures used, these consist in confidence scores obtained for each acoustic unit that occurs in an utterance; these scores are then averaged and the value obtained is denoted by σ_{0i}, for the i-th utterance and the system S_0. Another valuable set of intermediary data structures is represented by the MLLR transformation matrices [6], one matrix for each new adapted system. A system adapted to an utterance ε_i is obtained from S_0 via unsupervised MLLR using this utterance; hence, such a system consists in the original HMM set (S_0) together with the transformation matrix for the MLLR adaptation to ε_i, and is denoted by S_{ai}.

Sociolinguistic evidence prove that spontaneous multi-party dialogues tend to involve at most 5-6 speakers [2]; for a greater number of speakers, we tend to have several independent dialogues, although the interlocutors might still share the same environment (table, desk, etc.). In order to test these evidence, we have considered a corpus of multi-party dialogues; the data consists in three vaudevilles written in the 19th century by Eugène Labiche were considered (in French language): “La Cagnotte” (“The Jackpot”), “Les chemins de fer” (“The Railroads”), and “Le prix Martin” (“The Martin Prize”)[1].

Table 1. Characteristics of three vaudevilles

Play	“The Jackpot”	“The Railroads”	“The Martin Prize”
n°. of scenes	52	42	42
n°. of characters	17	18	8
n°. of main characters	6	6	4

Some relevant characteristics of these three plays are the number of scenes for these plays, their total number of characters, as well as the number of main characters in each play; we add that each scene has a number of speech turns varying from 2 to around 200, for a number of characters varying from 2 to 10. These characteristics are summarized in Table 1.

In order to provide a subtler characterization of this multi-party human-human dialogue corpus, we show in Table 2 the “raw” number of dialogues, with respect to the number of turns and dialogue partners. In Table 2 we can see that the distribution of the number of dialogues, with respect to the set of speakers and to their size is rather uneven: most of the dialogues have less than 50 speech turns, produced by less than 7 speakers. That is, our data presents an even distribution for dialogues of at most 50 turns, where at most 6 speakers participate. This situation is in accord with results from sociolinguistics, where, in social reunions, people tend to cluster in interacting groups of 4 to 6 individuals [19]. Moreover, dialogues tend not to be very long, namely, they usually contain less than 50 speech turns. However, there are a few longer dialogues, of around 80 speech turns,

[1] The electronic versions of these plays were downloaded from `http://fr.wikisource.org`.

where 3 or 4 speakers are involved. These elements provide, in our opinion, valuable guidelines concerning the limits of multi-party dialogues, in terms of number of turns and participants: the machine should thus handle mostly conversations where at most 6 speakers are involved (including itself), whereby at most 50 speech turns are produced.

One last remark finds its place here, namely that, summing over the dialogues in Table 2, we see that a number of 133 is obtained; however, summing over the scenes in Table 1, a number of 136 is obtained. Nevertheless, we have previously stated that each scene is assimilated to one multi-party dialogue situation. The difference between the two counts stems from the fact that some scenes (namely, 3) are *monologues*, hence not considered in Table 2.

Table 2. Number of dialogues, according to their size and number of interlocutors

N°. of characters / N°. of lines	2	3	4	5	6	7	8	9	10
2-10	**11**	**10**	**2**	**2**	0	0	0	0	0
11-20	**13**	**10**	**2**	**2**	**1**	0	0	0	0
21-30	**10**	**4**	**3**	**5**	**1**	0	0	0	0
31-40	**4**	**3**	**4**	**2**	**6**	**1**	0	1	0
41-50	**1**	**3**	**3**	**3**	**1**	**1**	0	0	0
51-60	**1**	**1**	0	0	**2**	**1**	0	1	0
61-70	0	0	**1**	0	1	**2**	0	0	0
71-80	0	2	**2**	0	0	0	**1**	0	0
81-90	0	0	0	**2**	0	0	**1**	0	0
91-100	0	0	0	0	0	0	0	0	0
101-110	0	**1**	0	0	**1**	0	0	0	**1**
111-120	0	0	0	0	0	0	0	0	0
121-130	0	0	0	0	**1**	0	0	0	0
131-140	0	0	0	0	0	0	0	0	0
141-150	0	0	0	0	0	0	0	0	0
151-160	0	0	0	0	0	**1**	0	0	0
161-170	0	0	0	0	0	0	0	0	0
171-180	0	0	0	0	0	0	0	0	0
181-190	0	0	0	0	0	0	0	0	0
191-200	0	0	0	0	0	**1**	0	0	0

Hence, the speaker tracking algorithm adopts different strategies, regarding whether this number (denoted by $\overline{L}$) of speakers has been reached or not: while the number of speakers detected is inferior to $\overline{L}$, the procedure will create a new adapted speech recognition system for each input utterance. When the number of speakers reached $\overline{L}$, the algorithm will perform several supplementary tests before creating a new adapted speech recognition system. However, $\overline{L}$ is an input parameter for the algorithm; Wizard-of-Oz dialogue simulations [3] or corpora investigations (as shown above) can assess the scale of the dialogue (in terms of maximum number of speakers) and, by consequence, provide an empirical, application-dependent value for $\overline{L}$.

3.1.2 Decision-Making

As stated before, the speaker tracking algorithm uses the information structures described above by constructing a fixed decision tree and then traversing it accordingly. The tree is specified offline, whereas its traversal depends on the confidence score obtained in recognizing the input utterances; thus, the procedure goes as follows (numbers indicate successive steps, while letters mark alternative paths):

1. start with the speaker-independent speech recognition system S_0, a total number of utterances $N \leftarrow 0$ and of speakers $L \leftarrow 0$; specify a maximum number of speakers $\overline{L}$ and an *offset* Δ (the number consecutive of input utterances where no new speaker is detected);
2. for an input utterance ε_1:

   2.1 perform unsupervised MLLR of S_0 on ε_1, obtaining the adapted system, S_{a1};
   2.2 perform speech recognition of ε_1, with both S_0 and S_{a1}; two recognition scores σ_{01} and σ_{a11} result, respectively; from the definition of MLLR, we should have that $\sigma_{a11} > \sigma_{01}$; we mark that ε_1 has been produced by the speaker l_1;
   2.3 $N \leftarrow N + 1$, $L \leftarrow L + 1$;

3. for a new utterance ε_2:

   3.1 perform unsupervised MLLR of S_0 on ε_2, obtaining the adapted system S_{a2};
   3.2 perform speech recognition of ε_2, with the three systems S_0, S_{a1} and S_{a2}; three recognition scores are obtained, respectively: σ_{02}, σ_{a12} and σ_{a22}; we can have one of the following possibilities:

      a) if $\sigma_{02} > \max(\sigma_{a12}, \sigma_{a22})$ then, from the definition of MLLR, we have an error;
      b) else, if $\sigma_{a12} > \max(\sigma_{02}, \sigma_{a22})$, then we have an error as well;
      c) else, if $\sigma_{a22} > \max(\sigma_{02}, \sigma_{a12})$, then ε_2 has been produced by a new speaker, l_2, so that $l_2 \neq l_1$; by consequence, $L \leftarrow L + 1$;
      d) else, $\sigma_{a22} \approx \sigma_{a12} > \sigma_{02}$; then ε_2 has been produced by the same speaker as ε_1, that is, l_1; by consequence, L remains unchanged and S_{a2} is discarded;

   3.3 $N \leftarrow N + 1$;

4. repeat step 3 until $L = \overline{L}$ or L remains unchanged for a number of utterances equal to Δ;

5. if $L = \overline{L}$ or L unchanged for Δ consecutive utterances, for a new utterance ε_m, assuming that we have 1 + W speech recognition systems, S_0, S_{a1}, ..., S_{aW} built as above, perform speech recognition on ε_m, with all the 1 + W systems, obtaining the scores: σ_{0m}, σ_{a1m}, ..., σ_{aWm}; we can have one of the following possibilities:
   (a) if $\sigma_{0m} > \max_{i = 1, ..., W} (\sigma_{aim})$, then ε_m has been produced by a new speaker, l_{L+1}, different from the already detected L speakers; in that case, we perform unsupervised MLLR of S_0 on ε_m, obtaining a new system $S_{a(W+1)m}$; we perform $L \leftarrow L + 1$ and $N \leftarrow N + 1$ as well (actually, $m = N + 1$);
   (b) else, if there exists an $i \in \{1, ..., W\}$ so that $\sigma_{aim} > \max(\sigma_{0m}, \sigma_{a1m}, ..., \sigma_{aWm})$, then ε_m has been produced by the emitter of a preceding utterance ε_i, with $i < m$; in this case, L remains unchanged and N gets incremented by one, to obtain m;
   (c) else, if there exists a k in $\{1, ..., W\}$ such that $\acute{\sigma}_{akm} \approx \sigma_{0m}$, then we have an error, from the definition of MLLR;
   (d) else, if there exist j and k in $\{1, ..., W\}$ so that $j \neq k$ and $\sigma_{ajm} \approx \sigma_{akm} > \max(\sigma_{0m}, \max_{t \neq j, k} (\sigma_{atm}))$; in this case:
      5.1. perform unsupervised MLLR of S_{aj} and S_{ak} on ε_m, obtaining the systems $\tilde{S}_{aj}$ and $\tilde{S}_{ak}$, respectively;
      5.2. perform speech recognition with $\tilde{S}_{aj}$ and $\tilde{S}_{ak}$ on the utterance ε_m; the scores $\tilde{\sigma}_{ajm}$ and, respectively, $\tilde{\sigma}_{akm}$ are obtained; in this point, two situations are possible:
         (a) if $\tilde{\sigma}_{ajm} \approx \tilde{\sigma}_{akm}$ and $\tilde{\sigma}_{ajm} \geq \sigma_{ajm}$ and $\tilde{\sigma}_{akm} \geq \sigma_{akm}$, then $S_{aj} \equiv S_{ak}$ and ε_m has been produced by the emitter of ε_j *and* ε_k; in this case, discard S_{ak} and $L \leftarrow L - 1$, $N \leftarrow N + 1$;
         (b) else, if $\tilde{\sigma}_{ajm} > \tilde{\sigma}_{akm}$ or $\tilde{\sigma}_{akm} > \tilde{\sigma}_{ajm}$, then denote by j_0 the index of the maximal score: $j_0 = \text{argmax}(\tilde{\sigma}_{ajm}, \tilde{\sigma}_{akm})$:
            (i) if $\tilde{\sigma}_{aj0m} \geq \sigma_{aj0m}$, then ε_m has been produced by the same speaker as ε_{j0} (the utterance used to obtained the system S_{aj0}); in this case, keep L unchanged and $N \leftarrow N + 1$;
            (ii) else, ε_m has been produced by a new speaker, which is neither the producer of ε_j, nor the producer of ε_k; in this case, $L \leftarrow L + 1$ and perform an unsupervised MLLR of S_0 on ε_m;
6. while there is an input utterance, go to step 5;
7. for i from 1 to N, return the identifier of the speaker that produced utterance ε_i.

In this algorithm, the decision tree is constituted by the "if" alternatives at steps 3.2, 5, 5.(d)5.2, and 5.2(b); the depths of the leaves are given by the nesting levels in the algorithm. The bottom-up traversal of the tree is inherently given by the nestings in the algorithm, whereas the left-right traversal is given by the order of the clauses: first, the loop in steps 3-4 is executed, then, the loop in steps 5-6. In

Figure this tree is represented, marking by "#" the first decision point, between the two speaker tracking strategies in steps 3-4, respectively 5-6; dotted arrows indicate the flow of the algorithm and the continuous lines mark alternative possibilities (the intersection of a set of such lines is a decision point). The rest of the symbols mimic those used in the specification of the algorithm; the tree should be read top-down, left-right.

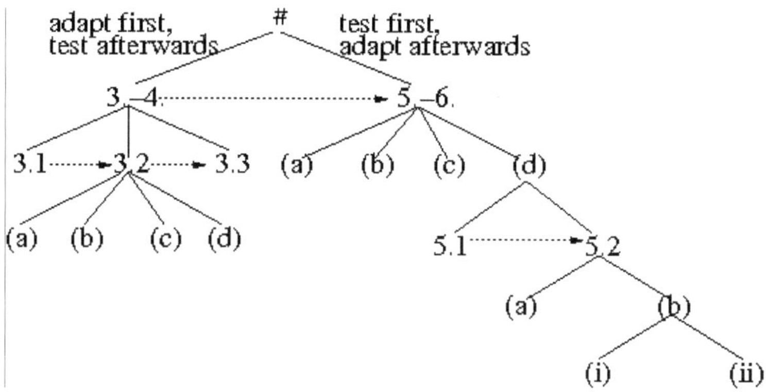

Fig. 1. Recognition score-based decision tree

As for the reliability of the algorithm, one objection might be that the variations in recognition scores can be induced by the variations in the content of the utterances used in adaptation or in recognition. However, this apparent problem is reduced by the fact that each comparison is performed between scores obtained on the *same* utterance, although the systems used for this might or might not have used the same utterance in training (or adaptation). If we had compared two scores obtained with two systems that had not used the same utterance set in training or adaptation, we could have had results "corrupted", i.e. the scores reflect rather the differences in utterances than the differences in speakers. The answer to this is that the scores are computed by averaging at the utterance level the scores obtained for each acoustic unit (e.g. word, triphone, phoneme); hence, the scores that are compared might depend only on the particular distribution of the acoustic units within the utterance, being independent of the length of that utterance. However, even in this case, if the speaker-independent system is well trained, the scores for the individual acoustic exhibit low variances from one unit to another. Therefore, if this variance is inferior to the difference between scores obtained on utterances from different speakers, then the problem is alleviated. A discussion in this respect is provided in § 4.2.

The parameter Δ (called "offset") represents the maximum number of consecutive utterances where no new speaker is detected, before the algorithm adopts the "less expensive" strategy, starting from step 5. Its value can be empirically chosen, but a reasonable value is just the number L of detected speakers.

Concerning the complexity of the algorithm, expressed in terms of the number of speech recognition processes (performed as Viterbi decoding [6]), the number of MLLR adaptation processes, the number N of utterances in dialogue, the "expected" number of speakers ($\overline{L}$) and the offset Δ, an estimate is provided here. Thus, considering the worst-case scenario where all the branches in the tree are visited and the values of L and N are as high as possible (i.e., $L = \overline{L}$ from a certain input utterance on, and N limited - between steps 3 and 5, to $\overline{L} - 1 + \Delta$, and $\Delta = \overline{L}$), and denoting by τ_{MLLR} the average time needed for a MLLR adaptation process, by τ_{ASR}, the average time needed for a speech recognition process, and by τ_{CMP}, the time required for a comparison, we obtain that the execution time of the algorithm has an expression of the form (α, β, α′ and β′ indicate constant non-zero real numbers): $T = \tau_{MLLR} \times (\alpha \overline{L} + \beta N) + (\tau_{ASR} + \tau_{CMP}) \times (\alpha' \overline{L}^2 + \beta' \overline{L} N)$.

Therefore, the algorithm is quadratic in $\overline{L}$ and linear in N, for a specified which is a constant for a running instance of the algorithm.

As for comparisons with previous work, we see that, according to Murani and colleagues [13], including speaker change information in the language model used in recognition improves recognition accuracy with around 2 % and speaker tracking performance with around 7 - 8 %; however, as already shown before, in multi-party dialogues such a model cannot be computed, because speakers' involvment in dialogues is not an already available information that could be used in training an n-gram. More interesting comparisons are possible with the work of Furui and colleagues [23]: for instance, although in our baseline speaker tracking algorithm we do not use phoneme-level regression classes, performance is still acceptable, because in our procedure we have a supplementary step: when recognition scores obtained with two or more speaker adapted decoders are identical (up to a slight difference, less than 5 % of the average value of the scores), then these systems are *further* adapted to the current input utterance and the variations of the recognition likelihoods are studied (step 5.(d)5.1).

3.2 Performance Improvements

3.2.1 Algorithm Parallelization Strategies

Several strategies can be pursued for improving the runtime performance of the speaker tracking process. One such way resides in building a GMM for each speaker, and relying on these models in order to discern among interlocutors; this has been pursued by Furui and colleagues [23], but has the main disadvantage of reducing the accuracy of the speaker models, if two or more speakers in dialogue have very similar voices. This is why we have followed a different approach in improving the runtime performance: we studied possibilities of *parallelizing* the speaker tracking algorithm at several levels. This is motivated by the fact that nowadays multi-core computers or even clusters become readily available.

Thus, a first and very important parallelization step (undergone, among others, also by Furui and colleagues [23]) consists in simultaneously performing speech recognition of an input acoustic stream, with the speaker-independent and the speaker-adapt*ed* systems. Hence, we obtain a relatively steady runtime performance improvement in that computation time remains relatively constant with respect to the number of speakers (hence, to the number of speaker-adapted systems) in dialogue; this obviously becomes more important as the number of speakers increases. The runtime gain results in that, instead of adding the recognition times for the speaker-independent and speaker-adapted systems, we divide the sum of these times by the number of processors available (and if this number is approximately equal to the number of speakers - usually, around 4-5 speakers, as pointed out in § 3.1, then the computation time practically does not increase when a new speaker gets involved in dialogue).

However, there is another point where performance can be improved, especially until the number of dialogue participants stabilizes, namely the actual MLLR adaptation process. Even if MLLR is not performed anymore, once the number of speakers in dialogue stabilizes (on average, to no more than 5 speakers), its runtime costs are important for the first speech turns in dialogue, where more and more speakers become involved. Actually, there are several dialogues where the number of speakers stabilizes towards the *end* of the conversation, thus MLLR could actually be performed at several points in dialogue, not only in its beginning. Moreover, a once involved speaker might leave the dialogue, but in some situations (e.g. when her/his voice was very similar to another, still active, speaker's voice), such an "on-leave" speaker has to be ruled out through further MLLR adaptation processes. This is why we have tried several MLLR parallelization strategies, finally adopting the most efficient one, as described in the next section.

3.2.2 Parallelizing the Adaptation Process

We have shown in previous studies [14], [15] that in parallelizing a data-driven HMM parameter estimation process, several strategies can be, in principle, adopted.

For example, we could try a *program-level* parallelization, where the actual sequential code that implements a parameter estimation procedure (e.g. Baum-Welch re-estimation, Viterbi alignment or MLLR estimation) is parallelized, either in an *algorithm-independent* way, via classical program optimization techniques such as *loop unrolling* or *multisampling*, or in an *algorithm-dependent* manner, by taking into account the inner working of the actual procedure (e.g. as in Ravishankhar's efficient Viterbi decoders in the Sphinx system [17]); in all these cases, all available data is used by each parallel program instance.

On the other hand, we can shift the focus from programs to data and thus follow a *data-level* approach, where the HMM parameter estimation procedure is parallelized by *distributing* the available data among several processors and then combining the results obtained; this can be realized either in an *algorithm-independent* way, where the details of the actual algorithm are not taken into account, but its (partial, since on partial data) results are combined or compared, as in [15], or in an *algorithm-dependent* manner, where the partial results are combined in a way that explicitly takes into account the inner workings of the algorithm, such as in HTK's parallel Baum-Welch HMM training [22]. In this strand, the algorithm-dependent (data-level) approaches have been shown to be the most efficient ones (as compared to the algorithm-independent data-level ones, that are, instead, more general).

Usually, in parallelizing such a procedure, we first investigate data-level approaches (first, algorithm-independent, then, algorithm-dependent), and only then we investigate program-level approaches (first, algorithm-dependent and, at last, algorithm-independent). This is motivated by the need to gradually modify a baseline, sequential procedure, that should be available as a starting point; this is of course necessary for proper bug tracking and development control.

Concerning MLLR parallelization, there is, to our knowledge, no reported research. This can be explained by two facts: first, MLLR is a rather new technique [9] and parallelizing it did not seem motivated by the (usually) *off-line* adaptation process; secondly, MLLR has been scarcely or not at all applied to actual multi-party dialogue systems, where real-time operation is of paramount importance. Hence, in trying to apply the strategies shown above, the most interesting ones are those that follow the data-level approach, since results are not affected by potential (hidden) parallelization bugs: the actual parameter estimation program remains *unchanged*, for example regarding a sequential baseline that has already been validated. We thus restricted our attention to data-level approaches to parallelizing MLLR: an algorithm-independent approach is not very useful, since, given that generally adaptation data is already rather scarce, the chances that saturation is achieved with less data (than that available) is very weak [9], which is not the case, for example, in Baum-Welch training, where indeed, data that do not further help the parameters converge can be thus detected and actually not used in training [15].

For the reasons above, we investigated the possibility of parallelizing MLLR in a data-level algorithm-*dependent* approach, where we cluster phoneme occurrences in a data stream, into regression classes, one class for each phoneme type that occurs in a (continuous) stream of adaptation data. Thus, a given stream s, of a certain speaker, contains the words $w_1^{(s)}$, ..., $w_N^{(s)}$, as recognized with the speaker-independent HMM system. Each word $w_i^{(s)}$ contains several phoneme occurrences (tokens), $p_1^{(is)}$, ..., $p_M^{(is)}$. Assuming that a word-level HMM modelling with a variable number of states per HMM (proportional to the number of phonemes in the word) is used, for each phoneme $p_j^{(is)}$ we have a number of states in the HMM of word $w_i^{(s)}$; otherwise, if phoneme-level HMM modelling is used, for each phoneme $p_j^{(is)}$ we have the model $HMM_{\varphi k}$ if $p_j^{(is)} \in c_{\varphi k}$, for phoneme type φ_k, $k = 1, ..., P$.

We compute a MLLR transformation matrix for each phonemic regression class; all phoneme tokens $p_j^{(is)}$ take values in the set $\{ \varphi_1, ..., \varphi_P\}$ of phoneme (types) in a given (e.g. Romanian or French) language. The MLLR adaptation process computes a transformation matrix for each Gaussian (mixture) in each HMM state. We *cluster* the states that belong to the same phoneme in an adaptation data stream. For adaptation data stream s we have $p_j^{(is)}$, $j = 1, ..., |w_i^{(s)}|$, $i = 1, ..., |s|$, where $|w_i^{(s)}|$ denotes the number of phoneme occurrences in word $w_i^{(s)}$, and $|s|$, the number of word occurrences in stream s, and for each $p_j^{(is)}$ we have three states in a word-level HMM.

Now, denoting by $c_{\varphi k}^{(s)}$ the set of phoneme tokens in stream s that are equal to the phoneme type φ_k ($c_{\varphi k}^{(s)} = \{ p_j^{(is)}: j \in \{ 1, ..., |w_i^{(s)}|\}, i \in \{1, ..., |s|\} \,|\, p_j^{(is)} \equiv \varphi_k \}$), we can compute the time taken by several adaptation configurations:

(a) *Baseline sequential*
In this situation, no regression classes are used, all adaptation data are used to estimate only one transformation matrix; data are expressed as $s = s_1, ..., s_K$, where $(s_i)_{i = 1, ..., K}$ is an acoustic (observation) vector (for example, of MFCC cepstral coefficients). Thus, we use $\{s_1, ..., s_K \}$ to perform MLLR for all (HMM) states of all phoneme tokens $p_j^{(is)}$ in stream s. Denoting by τ the time needed for computing one transformation matrix, using one acoustic frame of adaptation data, in this process we have a time of $K \times\tau$ for computing the unique transformation matrix for all the states in the composite HMM of stream s.

(b) *Sequential, with regression classes*
In this setting, a regression class is constructed for each phoneme type that occurs instantiated in stream s; in each such class we cluster all phoneme tokens in the words in s, that are identical, to a certain phoneme type. To make things clearer, we assume that each acoustic observation (vector) corresponds to the acoustic realization of one phoneme, and we denote by $s^{(k)}$ the set of acoustic frames that correspond to phoneme occurrences in $c_{\varphi k}^{(s)}$: $s^{(k)} = \{ s_{i1}^{(k)}, ..., s_{iQ}^{(k)}\}$, with $\{ i_1^{(k)}, ..., i_Q^{(k)} \} \subset \{ 1, ..., K \}$. Assuming that in stream s only $P_s^{'} \leq P$ phoneme types are acoustically realized, we have that $\sum_{k=1}^{P_s^{'}} |s^{(k)}| = K$, thus the total MLLR adaptation time is

$$\tau\times \sum_{k=1}^{P_s^{'}} |s^{(k)}| = \tau\times K,$$ as in the baseline sequential case.

(c) *Parallel, with regression classes*
In a parallel processing environment, we assume that we have Π processors that can work in parallel (for example, in a cluster computer[2], or in a multi-core architecture). In this case, Π adaptation processes can be performed in parallel. Assuming that, for an adaptation stream s, we need to perform $P_s^{'} \leq P$ adaptations, two cases arise:

1. $\Pi \geq P_s^{'}$: In this situation, each adaptation (using $|s^{(k)}|$ acoustic frames, for $k = 1, ..., P_s^{'}$) can be performed in parallel, hence the computing time is given by $\tau\times\max_{k \in \{ 1, ..., P_s^{'} \}} |s^{(k)}| < \tau\times K$. The latter inequality always holds if $P_s^{'} > 1$, that is, we have more than one phoneme acoustically realized in adaptation stream s. This is *usually* (but not always!) true in spoken dialogue (where we can have speech turns composed of interjections, e.g. "aaa!"). In those (relatively rare) cases where $P_s^{'} = 1$, obviously, the parallelization strategy described here does not bring any performance improvement, concerning the computation time.
2. $\Pi < P_s^{'}$: In this situation, only Π MLLR adaptation processes can be performed in parallel, at a time. Denoting by ρ the "process-to-processor ratio", $\rho = [P_s^{'} / \Pi]$, where $[x]$ denotes the integer part of x, we have a *sequence* of ρ sets of *parallel* Π adaptations; thus, the total adaptation time is lower than $\tau\times$

[2] We should however note that in a cluster computer the total processing time might be slightly higher, due to interprocessor communication delays.

$\left(\sum_{j=0}^{\rho-1} \max_{k \in \{j \cdot \Pi + 1, \ldots, (j+1) \cdot \Pi\}} |s^{(k)}| + \max_{k \in \{\rho \cdot \Pi + 1, \ldots, P_s'\}} |s^{(k)}|\right) < \tau \times K$. The total computing time is given by the latter estimation if each set of Π adaptation processes have to finish completely before a new set of adaptations start. Obviously, this is neither optimal, nor compulsory; indeed, the adaptation processes can be performed such that as soon as one transformation matrix computation has finished in a set of Π parallel adaptations, a new MLLR process can start. In this case, of course, there is an upper bound of the computation time on average lower than that given by the maximum operator on $|s^{(k)}|$.

Obviously, out of these two cases, the second one is more interesting for practical purposes, if we assume that a phoneset in a language averages 35 ($P \approx 35$), and if the actual number of phone types acoustically realized in a data stream s averages 15 ($P_s' \approx 15$), and that in a (relatively widespread) dual-core computer (or four-core server) we have a relatively low number of processors ($\Pi \in \{2; 4\}$).

Finally, we should remark that while the baseline speaker tracking strategy was unsupervised, since adaptation data did not need to be decoded before performing MLLR adaptation, the parallel MLLR algorithm needs the word level (and, hence, phoneme-level) transcription of adaptation data. Therefore, parallel MLLR is supervised, but this does not add any further computational cost, since every input utterance is decoded using the speaker independent HMM set, before any adaptation process takes place. Moreover, the usage of phoneme-level regression classes brings further improvements, in both recognition rate and speaker tracking performance, as shown in studies like [13] or [23].

4 Experiments

4.1 Speech Recognition System

The algorithm described in this chapter was applied in a continuous speech recognition system, designed for "virtual librarian" multi-party dialogue applications, in Romanian language. The system was trained at word level, using no language modelling information. Thus, 92 words, related to library services were used, along with a supplementary set of 16 cue words; for each word a left-right (Bakis) [6] HMM was trained, with a variable number of states for each word (equal to 2 (initial and final, non-emissive states) + 3 × the number of phonemes in the word); the output observations are modelled with one Gaussian for each emissive state.

Each word-level HMM was trained in a speaker-independent manner, using around 4 hours of recorded speech (in laboratory conditions: SNR ≥ 25 dB), containing these words, uttered in context. The acoustic characteristics of the training data are (i) acquisition: unidirectional head-set microphone; (ii) sampling frequency: 16 kHz; (iii) signal frame size: 300 samples; (iv) weighting window type: Hamming; (v) parameterization: 12 MFCC (mel frequency cepstrum coefficients) per frame, along with the energy and with the first and second-order derivatives of these features; this results in a total of 39 acoustic features per frame.

The training was performed in two steps, following a classical "isolated unit training" strategy, based on hand-labeled data (at a word level)[3]:

1. the parameters of the set of *prototype* word-level HMMs[4] are initialised through Viterbi alignment to the data [22];
2. the parameters of the initialised HMMs are re-estimated via the Baum-Welch procedure [6].

The system achieves a word-level accuracy of around 79 % when tested against spontaneously-uttered speech produced (in laboratory conditions, akin to those used in training) by locutors not used in the training process; this relative low percentage can be explained through the spontaneous nature of the test utterances, where the word boundaries are not easy to localise and, moreover, words exhibit incomplete or altered acoustic realisation.

We stress on the fact that language modelling was not used in the system; we did not even use phonemic acoustic modelling, since the speech recognition system can be and is used as a word spotter for dialogue purposes: bearing in mind that the ultimate goal of speech recognition in dialogue contexts is to *analyse* the utterance from a semantic and pragmatic point of view [3], spotting words that are relevant to the task (that convey semantic information) and cue words (that convey discourse and pragmatic information) is more effective and efficient that full-blown continuous speech recognition [10], [11]. However, in the experiments described in this chapter, the utterances (for training and testing) were chosen so that they contain only the words considered, and the system was used as a speech recognizer.

4.2 Baseline Speaker Tracking Performance

The speaker tracking procedure was tested using a maximum of four speakers in multi-party dialogues. The dialogues are driven by a set of scenarios involving several typical tasks: (i) enrollment of a new customer of the library; (ii) request for books on a certain topic or subject; (iii) request for a specific book; (iv) restitution of a book; (v) payment of fines due to delays in book restitution.

Table 3. Runtime variations in various configurations

$\Delta \backslash \overline{L}$	1	2	3	4
1	0.6674	0.7855	0.8843	0.9638
2	N/A	0.7975	0.8963	0.9759
3	N/A	N/A	0.9084	0.9879
4	N/A	N/A	N/A	1.0

[3] The labels include temporal information as well.

[4] The prototype HMMs are specified by the user and contain constraints on the size and topology of the models.

The multi-party dialogues are constructed so that every possible speaker order and "weight" (in terms of number of turns per speaker / number of turns / conversation) is achieved; for the moment, given the fact that the entire dialogue system is, for the time being, a work in progress [3], the conversations are only between humans, out of which one plays the part of the librarian. Thus, a number of around 400 conversations were used for testing the ability of the algorithm to map utterances to the appropriate speaker identifiers.

It is worth mentioning that the number of speakers (between two and four, the maximum, as stated above) and their utterances were not previously known to the system; moreover, the speakers used in testing were different from those used in training the speaker-independent speech recognizer described in § 4.1.

The first relevant performance figure in our experiment is given by evaluating the word recognition of a system that is adapted to a certain speaker via unsupervised MLLR, on an utterance of that speaker, versus the same measure obtained, in the same test conditions, with the speaker-independent system. Thus, the average word recognition rate on a set of utterances produced by a certain speaker (out of those used in testing) reaches more than 80 % for the adapted system, versus around 79 % for the speaker-independent system.

The most relevant performance measure of our algorithm is represented by the number of correct speaker identifier assignments for each utterance, divided by the total number of utterances (since each utterance has a speaker identifier associated with it). Moreover, we studied the variation of this ratio (denoted by ρ in this chapter) with respect to the parameters $\overline{L}$ and Δ, the "expected" number of speakers and the offset, respectively (see § 2 for details). Thus, $\overline{L}$ was varied from 1 to 4, and Δ, from 1 to $\overline{L}$ (see § 2.2 for the rationale behind the choice of Δ). A score of 81.2 % was obtained for ρ; this score is independent of $\overline{L}$ and Δ, which makes sense, since these parameters control the strategy adopted for speaker tracking in respect with runtime, rather than with the actual decisions being made. Thus, a more interesting evaluation concerning these parameters is related to the runtime of the algorithm.

Table 4. Confusion matrix for speaker identity assignments

	M_1	F_1	M_2	M_3
M_1	**1296**	8	112	184
F_1	88	**1408**	8	96
M_2	180	12	**1216**	192
M_3	144	4	172	**1280**

However, since in our experiments the algorithm was implemented as a series of Bash and Python scripts driving HTK (Hidden Markov Modelling Toolkit) tools [22], absolute runtime values are not relevant, preferring relative values instead, scaling them with respect to the runtime for $\Delta = \overline{L} = 4$; thus, we mark by 1.0 the runtime of the algorithm for the values specified just above for $\overline{L}$ and Δ (hence, this is the baseline case), and providing relative coefficients for the other combinations for these two parameters; details are given in Table 1, for a series of around 400 dialogues, each one counting between 4 and 20 utterances (speech turns).

From Table 3 we see that the fastest speaker tracking algorithm is actually obtained for the minimal values for $\overline{L}$ and Δ and the runtime increases with both these parameters; this proves that, after only the first utterance had been processed (i.e., a first speaker identity assigned to it, and the speech recognition system adapted to it), it is worth adopting the "lazy" speaker tracking strategy stated in steps 5 and 6 in the algorithm. That is, the algorithm could be simplified, in that steps 3 and 4 could be eliminated completely, running directly steps 5 and 6, from the second input utterance on (i.e., after step 2); this is more efficient because in these latter steps first testing is performed, and then, if necessary, further adaptation, whereas in steps 3 and 4, adaptation is performed first, then testing and, if necessary, adapted systems are discarded. However, for the first input utterance, it makes sense to perform first the adaptation, as in step 2.

Yet, a further refined analysis of the performances of the algorithm can be emphasized, namely by showing the confusion matrix concerning the assignment of speaker identities to utterances. Thus, denoting by M_1, M_2 and M_3 the three male speakers, and by F_1 the female speaker used in testing the algorithm, the confusion matrix is shown in Table 4, where the figures on the lines indicate the assignments performed by the algorithm.

From this confusion matrix, *precision* (defined as the number of correct speaker assignments, divided by the total number of speaker assignments) and *recall* (defined as the number of correct speaker assignments, divided by the total number of real speaker-to-utterance mappings) can be derived, for each speaker and, consequently, the corresponding F-measures (defined as the harmonic mean of precision and recall). These quantities are shown in Table 5 for each of the four test speakers considered. We can see that these quantities are evenly balanced, although, as we could expect, the best results are obtained for the (only) female speaker. This shows, on the one hand, that the performances of the algorithm are robust to speaker variations, and that the usage of only one female speaker introduces an artificial bias on the results. Therefore, the most relevant performance figures are those obtained for the male speakers. Moreover, we can see that there is a balance between precision and recall as well, which might be a hint that further tests are needed in order to see whether this is a feature of the algorithm, or of the test data.

Table 5. Performance measures for every test speaker

	Precision	Recall	F-measure
M_1	0.76	0.81	0.78
F_1	0.98	0.88	0.93
M_2	0.81	0.76	0.78
M_3	0.73	0.80	0.76

Concerning the values of the recognition scores obtained, the word-level recognition log-likelihood scores evolve around –2000 for adapted systems recognizing utterances produced by the speakers that the systems are adapted to, and around –3700 for the non-adapted speaker-independent system. On the other hand, the word-level log-likelihood score variances are in the range of around ±800, thus, less than the difference between the first two average scores. Therefore, tests indicate that the usage of log-likelihood scores is a reliable strategy for speaker tracking.

4.3 Performance Improvements

The performance optimizations that stem from parallelizing the speaker tracking algorithm yield manyfold effects: first, a reduction in the processing time is achieved (considering both parallel recognition of input utterances, with speaker independent and speaker adapted HMM sets, and parallel MLLR). Secondly, when multiple regression classes are considered, speaker tracking and speech recognition accuracies improve. In this section one will show some quantitative effects of the parallel speaker tracking algorithms.

First, runtime improvements regarding the speaker tracking algorithm where speech decoding processes run in parallel, versus the case where they run in sequence have already been discussed in previous research [13], [23], hence here we only say that, given the fact that these recognition processes bare a significant (average) weight (of around 75 %) in the total tracking time, for a given utterance, the performance gain is mostly important as the number of speakers increases, and the number of processing units available is at least equal to the the number of speakers.

Secondly, word error rate improvements of around 0.5 - 1.5 % have been reported in previous studies [23], for the case where phoneme-level regression classes are used in MLLR adaptation, versus the case where no regression class is used (i.e., only one transformation matrix is computed for all the states in the HMMs). Hence, in this chapter we will emphasize the less studied issue of the effects of MLLR parallelization on the total speaker tracking runtime, for each new input utterance. Thus, performance can be characterized by looking at two measures: first, the ratio between the time taken by MLLR adaptations, and the total speaker tracking time, for an input utterance; we denote this measure by ΔT_{MLLR}. Then, the ratio between the time taken by a parallelized MLLR process, and a sequential (normal) MLLR adaptation can be considered; we denote this measure by $\Delta T_{\parallel MLLR}$. Obviously, $\Delta T_{MLLR} \in [0; 1)$, since we can have situations where no adaptation is performed for a tracking process, and $\Delta t_{\parallel MLLR} \in (0; 1]$, since there are situations where parallel MLLR is not faster than sequential MLLR (e.g., for utterances composed of a sequence repeating the same phoneme type).

For the beginning of a dialogue, where speakers begin to take part in conversation, MLLR adaptation is more important (since, depending on the values of $\overline{L}$ and Δ (see § 4.2), for the utterances produced until the number of speakers stabilizes, MLLR adaptations take place roughly for any speaker tracking process). Afterwards, MLLR adaptations only take place if new speakers occur in dialogue, or if there are at least two speakers with very similar voices (therefore, supplementary adaptation processes are needed). As for the advantages that parallel MLLR

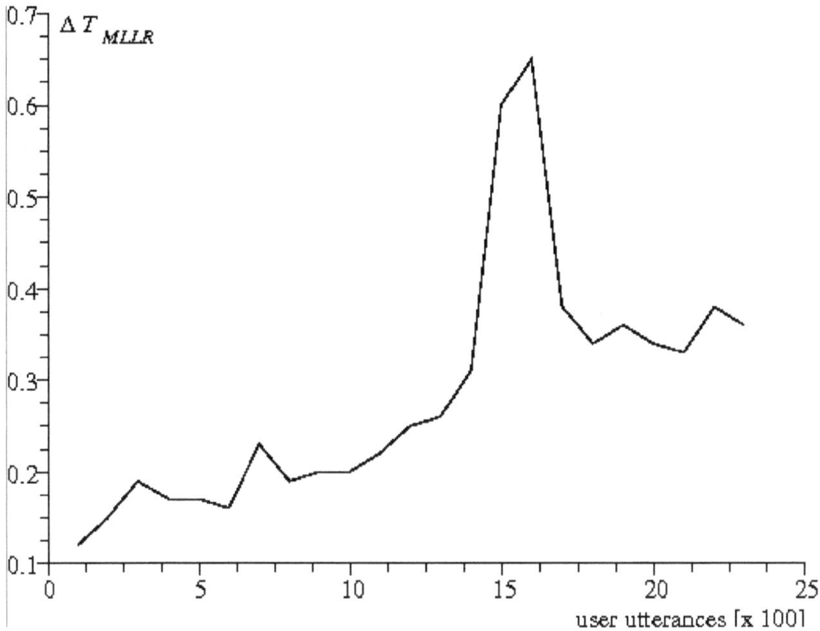

Fig. 2. Weight of MLLR time in a speaker tracking process

adaptation brings, with respect to sequential MLLR, they are more perspicuous if several phoneme types are instantiated in input utterances (according to decoders' outputs). On the other hand, if too many phoneme types are instantiated in an utterance, word error rate can degrade, due to insufficient adaptation data.

The variation of ΔT_{MLLR} is shown in Figure 2, in the context of speech recognition being ran in parallel, over $\Pi = 4$ processors; hence, the figure shows the weight of the (sequential) adaptation time in the total tracking time, for an utterances. On the horizontal axis user utterances (scaled by a factor of 100) are plotted, for the four users in the 400 dialogues considered in § 4.2; on the vertical axis, ΔT_{MLLR} is plotted. We can see that, on average, MLLR takes around 30 % of the total speaker tracking time, for an input utterance.

As for $\Delta T_{\parallel\, MLLR}$, that quantifies the runtime effects of parallelization on MLLR alone, it is shown in Figure 3; the same conventions and context as in Figure 2 are used. We can see that, when $\Pi = 4$ processing units are used in parallel, runtime reductions of more than 50 % are achieved. Theoretically, $\Delta T_{\parallel\, MLLR}$ should be around 25 %; however, due to uneven phoneme balance in utterances, and to inherent operating system-determined computational load, variable percentages are obtained.

The two figures, 2 and 3 altogether show that via MLLR parallelization an overall performance improvement in speaker tracking runtime of $\Delta T_{MLLR} \times (1-\Delta T_{\parallel\, MLLR})$ is obtained, that is, on average, around 19 %, for each input utterance.

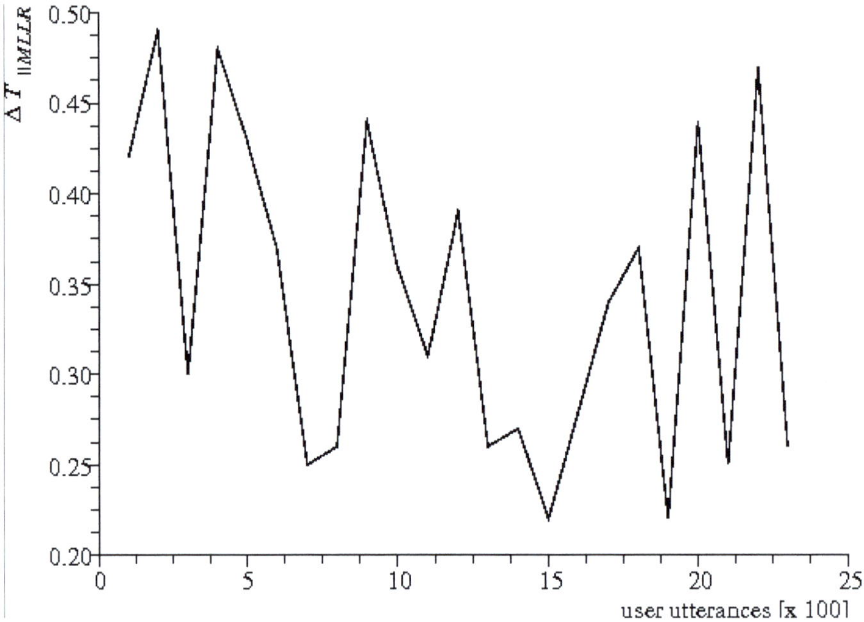

Fig. 3. Parallel MLLR runtime performance, for Π = 4 processors

5 Conclusions

We have presented a computationally simple strategy for speaker tracking in multi-party human-computer dialogue. The approach is based on the traversal of a decision tree that relies on speech recognition scores and unsupervised MLLR adaptation of a speech recognition system to input utterance so that these scores are maximized. Thus, an algorithm linear in the number of previous utterances in dialogue is obtained, that achieves a speaker tracking performance of around 80 % on spontaneous speech. The algorithm has been tested in the context of a virtual librarian dialogue application, in Romanian and French languages, and exhibits good runtime performance.

Moreover, several performance improvements were proposed, especially concerning the runtime of the speaker tracking algorithm; they basically rely on the idea of parallelizing the adaptation process. We have studied a parallel MLLR adaptation strategy that relies on simultaneous adaptation of HMM states grouped in regression classes, whereby MLLR runtime is reduced by a factor proportional to the number of processors available. Thus, taking into account the weight of roughly 25 - 30 % that MLLR adaptation has in the total speaker tracking time, for an input utterance, we remark that for a common four-core processor, total speaker tracking time is reduced by a factor of about 19 %, on average.

However, several issues remain to be done for tuning a speech recognizer into a front-end to multi-party dialogue systems; for instance, the algorithms proposed in this paper could be improved with contextual information (expressed at semantic, or even *discourse* levels [3]): a certain speaker is more likely to say certain things, or, viceversa, certain things were more likely said by a certain speaker. Thus, a statistical model of user conversational preferences could be trained, starting from discourse structures (rhetorical relations between utterances or sets of utterances) where the user's utterances integrate.

Another rather important technical detail that might be improved concerns the way we built the regression classes for the MLLR adaptation: we either did not use any regression class at all (by computing only one, global, transformation matrix for all the states in the HMM set), or we used phoneme-level regression classes (where for each phoneme type we compute a transformation matrix for all the states that correspond to that phoneme type). Neither of these two approaches is optimal, in the sense that the former is too coarse (which impends on the reliability of the transforms computed), whereas the latter is too fine-grained (which poses problems in case of data sparseness). Hence, a better approach would be to follow a strategy similar to that implemented in the HTK toolkit [22]: MLLR could use a regression class tree to group Gaussians in the HMM set, so that the set of transformation matrices to be computed can be chosen according to the amount and type (e.g. phonemes contained) of the adaptation data available. Thus, the tying of each transformation matrix across several mixture components makes it possible to adapt distributions for which no adaptation data was available. In our current setting, such distributions are not adapted unless relevant data (i.e., that contains acoustic realizations of the phoneme type that indicates the regression class referred) is available.

Acknowledgments. The research reported here was funded in part by the Romanian Government, under the National Research Authority ("CNCSIS") grant IDEI, no. 930/2007.

References

1. Barras, C.: Reconnaissance de la parole continue: adaptation au locuteur et contrôle temporel dans les modèles de Markov cachés., PHD Thesis, University of Paris VI, Paris (1996)
2. Branigan, H.: Perspectives on multi-party dialogue. Research on Language and Computation 4, 153–177 (2006)
3. Caelen, J., Xuereb, A.: Interaction et pragmatique - jeux de dialogue et de langage. Hermès Science, Paris (2007)
4. Christensen, H.: Speaker adaptation of hidden Markov models using maximum likelihood linear regression. MA Thesis, University of Aalborg, Denmark (1996)
5. Ginzburg, J., Fernandez, R.: Scaling up from dialogue to multilogue: Some principles and benchmarks. In: Proc. ACL, Michigan, pp. 231–238 (2005)
6. Huang, X., Acero, A., Hon, H.-W.: Spoken language processing: a guide to theory, algorithm and system development. Prentice Hall, New Jersey (2001)
7. Landragin, F.: Dialogue homme-machine multimodal. Hermès Science, Paris (2005)
8. Larsson, S., Traum, D.: Information state and dialogue management in the TRINDI Dialogue Move Engine Toolkit. Natural Language Engineering 6(3-4), 323–340 (2000)

9. Leggetter, C.J., Woodland, P.C.: Maximum likelihood linear regression for speaker adaptation of continuous density hidden Markov models. Computer Speech and Language 9, 171–185 (1995)
10. McTear, M.F.: Spoken language technology: Enabling the conversational user interface. ACM Computing Surveys 34(1), 90–169 (2002)
11. Minker, W., Bennacef, S.: Parole et dialogue homme-machine. CNRS Editions, Paris (2001)
12. Motlicek, P., Burget, L., Cernoký, J.: Non-parametric speaker turn segmentation of meeting data. In: Proc. Eurospeech, Lisbon (2005)
13. Murani, N., Kobayashi, T.: Dictation of multiparty conversation considering speaker individuality and turn taking. Systems and Computers in Japan 34(13), 103–111 (2003)
14. Popescu, V., Burileanu, C.: Parallel implementation of acoustic training procedures for continuous speech recognition. In: Burileanu, C. (ed.) Trends in speech technology. Romanian Academy Publishing House, Bucharest (2005)
15. Popescu, V., Burileanu, C., Rafaila, M., Calimanescu, R.: Parallel training algorithms for continuous speech recognition, implemented in a message passing framework. In: Proc. Eusipco, Florence (2006)
16. Popescu-Belis, A., Zufferey, S.: Contrasting the Automatic Identification of Two Discourse Markers in Multi-Party Dialogues. In: Proc. of SigDial, Antwerp (2007)
17. Ravishankhar, M.: Efficient algorithms for speech recognition. PHD thesis, Carnegie Mellon University, Pittsburg (1996)
18. Sato, S., Segi, H., Onoe, K., Miyasaka, E., Isono, H., Imai, T., Ando, A.: Acoustic model adaptation by selective training using two-stage clustering. Electronics and Communications in Japan 88(2), 41–51 (2004)
19. Trudgill, P.: Sociolinguistics: an introduction to language and society, 4th edn. Penguin Books, London (2001)
20. Yamada, M., Baba, A., Yoshizawa, S., Mera, Y., Lee, A., Saruwatari, H., Shikano, K.: Unsupervised acoustic model adaptation algorithm using MLLR in a noisy environment. Electronics and Communications in Japan 89(3), 48–58 (2005)
21. Yamada, S., Baba, A., Yoshizawa, S., Lee, A., Saruwatari, H., Shikano, K.: Unsupervised speaker adaptation for robust speech recognition in real environments. Electronics and Communications in Japan 88(8), 30–41 (2005)
22. Young, S., Evermann, G., Kershaw, D., Moore, G., Odell, J., Ollason, D., Povey, D., Valtchev, V., Woodland, P.: The HTK book. Cambridge University, United Kingdom (2005)
23. Zhang, Z., Furui, S., Ohtsuki, K.: On-line incremental speaker adaptat ion for broadcast news transcription. Speech Communication 37, 271–281 (2002)

The Fuzzy Interpolative Control for Passive Greenhouses

Marius M. Balas and Valentina E. Balas

Aurel Vlaicu University, 77 B-dul Revolutiei,
310130 Arad, Romania
balas@inext.ro

Abstract. The passive greenhouses are independent of any conventional energetic infrastructures (electricity, gas, hot water, etc.) They are relying exclusively on alternative energy sources: sun, wind, geo-thermal, etc. Their extensive use facilitates a massive ecological reconstruction of our planet, which could eventually reduce the CO_2 concentration in the atmosphere and the consequent global heating. The paper is approaching the passive greenhouses' control, proposing a fuzzy-interpolative controller with internal model. Simulations performed with a structural model are provided.

1 Introduction

A Passive Greenhouse (PG) is independent of any conventional energetic infrastructure (electricity, gas, hot water, etc.) The concept is inspired by the solar passive greenhouses which are using only the solar radiation and the natural ventilation. If provided with alternative energy sources: geo-thermal water, wind, photovoltaic, etc. the solar passive greenhouses become PGs. Assisted by Internet capabilities, PGs can be installed virtually anywhere on the surface of the Earth where phreatic water is disposable.

Our team is interested in PGs because they are offering the opportunity of a global scale ecological reconstruction that could eventually reduce the CO_2 concentration in the atmosphere and the consequent global heating. The reasoning scheme is the following: PGs are consuming great amounts of CO_2, comparable to the same surface of forest, thanks to the high density of the plants and the ideal growing conditions → PGs could feed more that five times more people than the same conventional agricultural surface → the unshackled agricultural surfaces resulted could be covered with forests, meadows, etc.

The paper is proposing a structural computer model that can assist the technical design and the economic analysis of PGs, as well as a dedicated fuzzy-interpolative controller that is using the previous model as an internal model.

2 The Ecological Reconstruction

The cause-effect relation between of the carbon dioxide concentration in the atmosphere and the global warming is admitted by the majority of the science community.

H.-N. Teodorescu, J. Watada, and L.C. Jain (Eds.): Intel. Sys. and Tech., SCI 217, pp. 219–231.
springerlink.com

In order to illustrate the global growth of the CO_2 concentration, we will quote only an example: its evolution recorded along 45 years in the Mauna Loa station [1], that shows a 19.4% increase in the mean annual concentration, from 315.98 parts per million by volume (ppmv) of dry air in 1959 to 377.38 ppmv in 2004. This represents an average annual increase of 1.4 ppmv per year.

Even if the global warming was not linked to the growth of the CO_2 concentration, this tendency is unhealthy and must be stopped as soon as possible. The processes that reduce the CO_2 concentration are known as *carbon offsets*. An obvious carbon offset strategy is based on the *reforestation*. Plants are storing carbon through photosynthesis, converting carbon dioxide and water into oxygen and plant matter. Because of their greater size comparing to cereals or other natural vegetation, trees are the best carbon sequesters, although the balance of the carbon is depending of the environmental factors. The carbon storing capacity of trees was measured by several researchers. For instance, for the Siberian typical trees, as the Siberian spruce (*Picea obovata*), the Siberian larch (*Larix sibirica*) and the weeping birch (*Betula pendula*), in summer, trees consumed daily 210 kg CO_2/ha (57 kg C/ha) in variable weather and 117 kg CO_2/ha (32 kg C/ha) in cloudy weather [2].

Our work is connected to greenhouses, so we will recall some facts about the CO_2 concentration in greenhouses [3], [4]. In normal conditions the CO_2 concentration is of 0.03%, respectively 0.3cm^3/l or 0.589mg CO_2/l at a temperature of 0ºC and a pressure of 760mmHg. This concentration presents daily and long term variations. The daily variations are produced by the plants themselves: during the day the CO_2 concentration is getting lower because of the photosynthesis, while during the nights it's getting higher, because of the respiration of the plants. The CO_2 assimilation depends of a set of internal and external factors. For the synthesis of one gram of glucose, the leafs of a generic plant have to absorb the CO_2 from 2500 l of air. The usual daily CO_2 consumption is 100-250kg/ha, equivalent to the reduction by 10-20% of the concentration. As one can observe, the CO_2 consumption in the greenhouses is very high, similar to a forest, due to the high density of the plants and to the ideal growing conditions.

The reforestation is only one aspect that is not solving the entire problem. The ideal solution would be, of course, the drastic reduction of the CO2 emissions. This can be hardly imaginable before the reaching of the industrial fusion energy technology. The reforestation could also cause some inconvenient effects (a higher retention of the solar energy, the extraction of the carbon of the soil, etc.) so we should better think to a *global ecological reconstruction*. Anyway, the major obstacle against the global ecological reconstruction is the huge agricultural surface needed to feed the human population [5].

The feeding of the people can be accomplished by two possible food chains:

a) A three trophic levels chain: plants → animals → humans

b) A two trophic levels chain: plants → humans

Much energy is lost into the environment at each transfer from one trophic level to another. That is why the food chain a) needs much more agricultural surfaces than the food chain b). The surface demanded by the food chain b) can be further reduced by using greenhouses. We consider that a sustainable solution to the CO_2 problem is the drastic reduction of the agricultural surface that would be possible if we will intensively use greenhouses. But using conventional greenhouses supplied with fuel burn energy would be nonsense, so the real answer to this challenge is the *passive greenhouse* [6], [7], [8].

3 The Passive Greenhouse and Its Computer Model

In some previous papers [7], [8], we have proposed a particular PG structure, presented in Fig. 1. This structure includes the most common alternative energy sources, and is valuing the experience acquired with the experimental greenhouse realized by the LSIS Laboratory of the Southern University of Toulon-Var, France [9], [10], etc. Our PG is aggregating three complementary energy sources: a heat pump extracting energy from cold phreatic water [11], [12], a wind generator [13], [14] and an orientable matrix of solar photovoltaic panels [15]. Another essential element is the DC accumulator that stocks the wind and the sun energies that are much less reliable than the geo-thermal one. The basic energy source is the heat pump, in the two water wells constructive version. The wind generator is providing the DC electric energy needed by the heat pump for recirculation and may also be connected to an electric emergency heating device. The solar panels are also charging the accumulator and in the same time they are shading the plants when solar radiation is excessive. When the sun is not too strong, the solar panels are parallel to the sun rays, and the greenhouse is directly heated and lighted by the sun. The solar panels might seem expensive at the first sight, but we found at least three reasons for including them into the system:

- Their price is continuously decreasing;
- They are replacing the actual rolling curtain system, that is shading the plants when the solar radiation is too strong;
- They are making possible the use of smaller heat pumps and wind generators.

The accumulator is recirculating the heat pump and supplying the GP's electric equipment, and the emergency electric heating device when necessary.

Generally speaking, alternative energy sources are expansive. In this case, we have three such items. The only way to make feasible such a structure is the dimensional optimization associated with a smart control. Each of the energy sources involved has created its own market; our problem is just the correct choice of the products. The nominal capacity of each element (the constructive parameters of the greenhouse and the powers of the heat pump, wind turbine and solar panels) must be carefully balanced, taking into account the climatic features of the location. Besides the internal temperature which is the key factor, an optimization problem can be targeted also to the investment costs. Because the PG system is fairly complex, we have built a simplified model, able to assist the optimization problems and the smart automate control.

Gas burning devices can also be installed, in order to cope with the occasional extreme cold weather and to increase the CO2 concentration in air, that is benefic for plants.

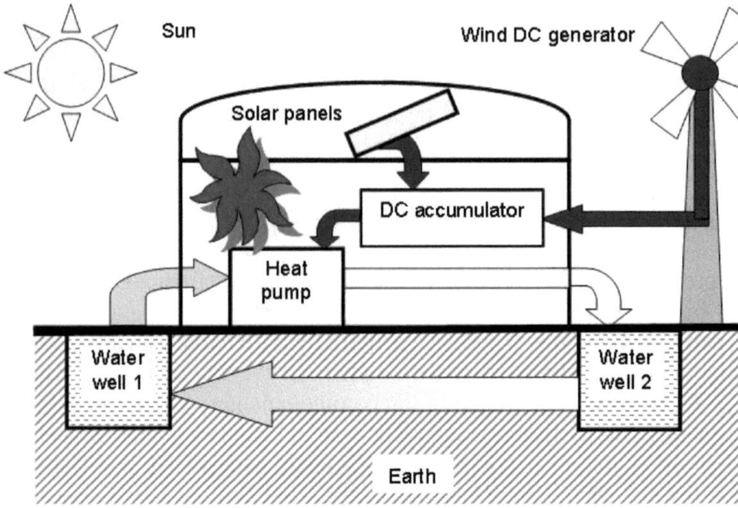

Fig. 1. The energetic passive greenhouse

The PG internal temperature $T_I(t)$ is decomposed into quantities representing the individual contribution of each energy source:

a) T(t) [^{0}C], the basic inside temperature due to the environment influence, realized by the heat flow through the walls and by natural or forced ventilation:

$$\frac{dT(t)}{dt} = [k_\alpha(t) + k_v \cdot F(t)] \cdot [T_E(t) - T_I(t)] \tag{1}$$

where $k_\alpha(t)$ [s^{-1}] is the coefficient of the heat flow through the walls, k_v [m^{-3}] the ventilation coefficient, F(t) [m^3/s] the ventilated air flow and T_E [^{0}C] the external temperature. $k_\alpha(t)$ is a nonlinear parameter, embedding several functional influences: constructive (shape, dimensions, material of the walls), of the wind, etc. The model used in this work is considering the two major physical effects characterizing the heat flow through the walls, the radiation and the convection, by two specific coefficients, $k_{\alpha R}$ [s^{-1}] and $k_{\alpha C}$ [m^{-1}]:

$$k_\alpha = k_{\alpha R} + k_{\alpha C} \cdot V_W \tag{2}$$

where V_W[m/s] is the speed of the wind.

The ventilation coefficient k_v, that is considered constant for the time being, could be also treated as a nonlinear variable, influenced by the shape and the dimension of the ventilation fans, the wind, etc.

b) The equation of the heat pump is:

$$\frac{dT_{HP}(t)}{dt} = k_{HP} \cdot P_{HP} \tag{3}$$

with $T_{HP}(t)$ the temperature amount added to T(t) by the heat pump, $k_{HP}[^0C \cdot s^{-1} \cdot W^{-1}]$ the heat pump coefficient, and $P_{HP}[W]$ the power of the heat pump.

c) The equation of the wind generator is:

$$\frac{dT_W(t)}{dt} = k_W \cdot P_W \tag{4}$$

where $T_W(t)$ is the temperature amount added to T(t) by the wind generator if connected to the electric heating device, $k_W[^0C/m]$ the wind coefficient and P_W the power supplied by the wind generator. We are considering a generic wind generator modeled by the following equation [14]:

$$P_W(t) = 0.5 \cdot \eta_W(V_W) \cdot \rho \cdot \pi \cdot r \cdot V_W^3 \tag{5}$$

where η_W is the efficiency coefficient of the wind generator and r the radius of its helix.

d) The equation of the solar energy is:

$$\frac{dT_S(t)}{dt} = k_W \cdot L_S \tag{6}$$

where $T_S(t)$ is the temperature amount added to T(t) by the sun, $k_S[^0C \cdot m^2 \cdot s^{-1} \cdot W^{-1}]$ the greenhouse effect coefficient and $L_S[W/m^2]$ the intensity of the solar radiation [18].

e) The resulting inside temperature is given by the equation:

$$\frac{dT_I(t)}{dt} = \frac{dT(t)}{dt} + \frac{dT_{HP}(t)}{dt} + \frac{dT_W(t)}{dt} + \frac{dT_S(t)}{dt} \tag{7}$$

As one can observe, the model is composed by first order subsystems with time varying parameters. The main advantage of this approach is the compatibility to the low level programming (ASM, μC, DSP, C#, etc.), that is fundamentally facilitating the internal model techniques and the remote control via Internet, which is essential for our concept. A continuous time Simulink-Matlab version of this model was developed (see Fig. 2).

The tuning of the model is based on the adaptation of the identification results obtained in previous papers concerning the Toulon greenhouse: [16] for k_{PC} and k_W, [17] for $k_{\alpha R}$ and $k_{\alpha C}$, and [18] for k_{ES}. In the case of k_V, experimental data are missing, so a plausible value is allocated. The numeric values of the parameters applied are:

$$\begin{cases} k_{\alpha R} = 0.001207 \ s^{-1} \\ k_{\alpha C} = 0.000036 \ m^{-1} \\ k_V = 0.005 \ m^{-3} \\ k_{PC} = (250 \cdot 4560)^{-1} = 0.87712 \cdot 10^{-6} \ W \cdot {}^oC \cdot s^{-1} \\ k_W = (250 \cdot 4560)^{-1} = 0.87712 \cdot 10^{-6} \ W \cdot {}^oC \cdot s^{-1} \\ k_{ES} = 50.87719 \cdot 10^{-6} \ m^2 \cdot {}^oC \cdot s^{-1} \cdot W^{-1} \end{cases} \tag{8}$$

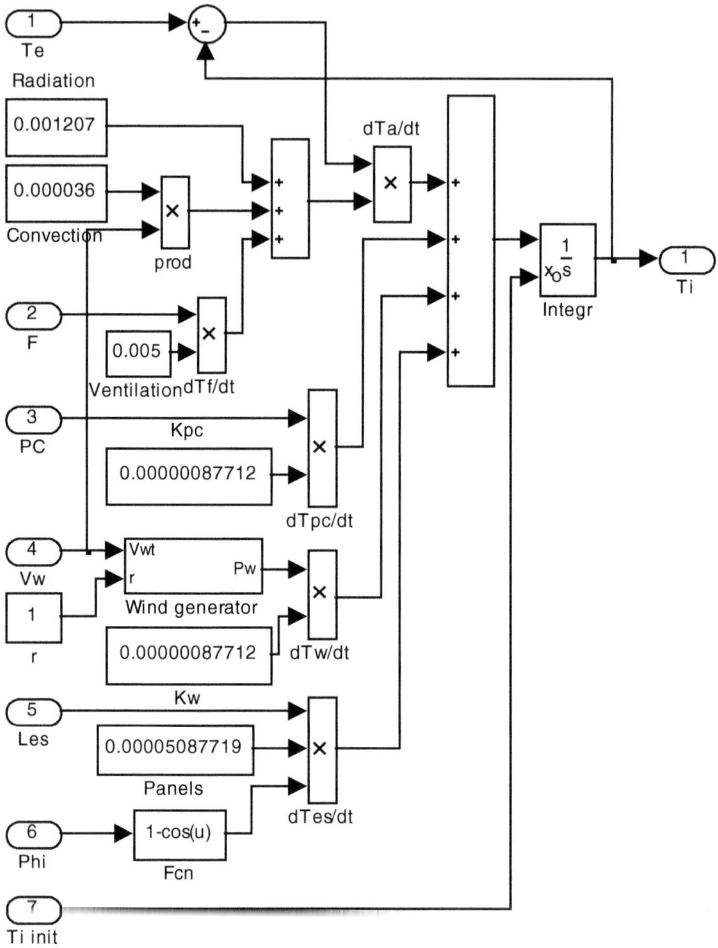

Fig. 2. The Simulink-Matlab passive greenhouse model

4 A Fuzzy-Interpolative Controller for Passive Greenhouses

When designing the PG's control algorithm we must have in mind the characteristics of the three energy sources, phreatic water, wind, and sun. The heat pump is recommended to operate in steady regimes. The wind energy is inconstant, but in this case, since we want only to accumulate its energy, no particular operating constraints are needed. The solar panels are connected to the accumulator and they have two positions: **1** when they are capturing the solar radiation and shading the plants (perpendicular to the solar radiation) and **0** when they are disconnected and letting the sun light the plants (parallel to the solar radiation). Connecting solar panels to the heating device is nonsense, because when the sun is directly heating the greenhouse, no electric heating is necessary. If the accumulator is fully charged and the sun is too strong, the panels may shade the plants and be disconnected.

This kind of application (highly nonlinear, but with no particular accuracy constraints), may be conveniently controlled with expert systems. Our approach will rely on the fuzzy-interpolative expert systems that may be implemented by look-up tables with linear interpolations, or with any other interpolative software or hardware networks [19], etc. The fuzzy approach was previously applied for the natural ventilated buildings [20].

The PG controller system is MIMO (Multiple Inputs Multiple Outputs). We are presenting the fuzzy inputs and the Boolean outputs in a minimal version.

a) Inputs

- **εT_I** the control error of T_I: imposed_T_I - T_I, with three linguistic labels: *negative*, *zero* and *positive*;
- **dT** the difference T_E - T_I with three linguistic labels: *negative*, *zero* and *positive*;
- **V_W** with two linguistic labels: *weak* and *strong*;
- **L_S** with two linguistic labels: *weak* and *strong*;
- **A_{CC}** with two linguistic labels: *loaded* and *unloaded.*

b) Outputs

- **HP**: the heat pomp switching device;
- **F**: the ventilation fan switching device;
- **W**: the wind generator switching device;
- **WAC**: the wind generator – accumulator switching device;
- **WH**: the wind generator – heating equipment switching device;
- **SP**: the solar panels switching device;
- **SPAC**: the solar panels – accumulator switching device;
- **SPAH**: the solar panels – heating equipment switching device;
- **ACH**: the accumulator - heating equipment switching device.

The operation consists in commutations of the energy sources: turns on, turns off, connections to the accumulator, connections to the heating device, etc. The sequential control actions are:

- **HP→1**: the heat pump is warming the greenhouse; this is the generic situation when weather is colder that the imposed temperature;
- **HP→0**: the heat pump is turned off, when weather is warm;
- **F→1**: the greenhouse is naturally ventilated;
- **F→0**: the ventilation fan is closed;
- **W→1**: the wind turbine is turned on;
- **W→0**: the wind turbine is turned off;
- **WAC→1**: the wind energy is accumulated;
- **WH→1**: the wind energy is heating PG;
- **SPAC→1**: the solar panels are connected to the accumulator;

- **SPAC→ 0**: the solar panels are not connected to the accumulator; they are oriented parallel to the solar radiation;
- **SPAH→ 1:** the solar panels are connected to the accumulator;
- **ACH→ 1**: the accumulator is connected to the heating equipment;

The PG system depends on three energy sources, but only the heat pump has a significant power, while the others are inconstant and, very often, weak. Since the heat pump has a notable inertia, we expect that the PG's control will be slower than the usual greenhouses'. That is why we must rely on PD controllers that have a certain predictive capacity. In order to avoid the undesired complication of the rule base we will replace the derivatives of all the inputs by a single derivative, cumulating all the dynamic effects: $\mathbf{cT_i}$ the estimated change of the internal temperature, with three linguistic labels: *negative, zero,* and *positive*.

The estimation is done with the help of the model, over a certain time horizon [21], [22]. In other words, the model is used as a predictor of the PG's dynamic behavior, supporting the control rules' design. The model is receiving the measured parameters of the real greenhouse and performs a simulation of the next sampling period. Since the model is very simple, its integration is very fast. Based on the estimated tendency, we are able to write anticipative rules, enhancing the desired tendencies and rejecting the bad ones.

The kernel of the rule base is containing the next rules:

1. If **dT** is *positive* then **HP → 1**
2. If **dT** is *positive* and $\boldsymbol{\varepsilon T_I}$ is *negative* then **F → 0**
3. If **dT** is *not positive* and $\boldsymbol{\varepsilon T_I}$ is *positive* then **F → 1**
4. If $\mathbf{V_W}$ is *strong* and $\mathbf{A_{CC}}$ is *unloaded* then **WAC → 1**
5. If $\mathbf{L_S}$ is *strong* and $\mathbf{A_{CC}}$ is *unloaded* then **SPAC → 1**
6. If $\mathbf{L_S}$ is *weak* then **SPAC → 0**
7. If **dT** is *positive* and $\boldsymbol{\varepsilon T_I}$ is *negative* and $\mathbf{V_W}$ is *strong* and $\mathbf{cT_i}$ is *negative* then **WH → 1**
8. If **dT** is *positive* and $\boldsymbol{\varepsilon T_I}$ is *positive* and $\mathbf{A_{CC}}$ is *loaded* and $\mathbf{L_S}$ is *weak* and $\mathbf{cT_i}$ is *negative* then **ACH → 1**

The rules may be linguistically described as follows:

1. When outside is colder than the imposed temperature the heat pump is turned on.
2. When outside is cold and inside is colder than desired, the ventilation is stopped.
3. When outside is not cold and inside is too warm the ventilation is working.
4. The wind turbine is loading the accumulator when the wind is strong and the accumulator is not fully loaded.
5. The solar panels are loading the accumulator when this one is unloaded.
6. When the solar light is not too strong the panels must let the light get to the plants.
7. If outside is cold, as well as inside, and the wind is strong, the wind turbine is directly heating the greenhouse. This is an emergency situation.
8. If outside is cold, as well as inside, and the accumulator is loaded, the accumulator is heating the greenhouse.

5 Simulation Results

Our preliminary simulations are indicating that although the PG system is very slow, it is basically controllable by our controller. This type of fuzzy-interpolative controllers is extremely flexible, with numerous adjustment possibilities. The adjustments may be done at the expert system level (control rules), as well as at the fuzzification level (number of linguistic labels, position and shape of the membership functions).

Next, we present three simulations, oriented towards different objectives.

a) A predictor of the internal temperature

A simulation performed over a certain time horizon, using the current input data, is extrapolating the current evolution of the system. This way, we can achieve different tasks, such as the following two:

a1) The prediction of dangerous situations (basically freezing or overheating):

Let us consider the following parameters: P_{HP} = 15 kW, T_E = 0 °C, initial T_I = 20 °C, L_S = 10 W/m^2 (a cold night, with no ventilation). A 60 s simulation is producing a 0.02 °C decrease of the temperature. A two hours simulation is indicating that T_I will stabilize at 8 °C. If the greenhouse's crop is consisting in tomatoes, we can accept this situation, taking into account the fact that the plants can tolerate such short time cool periods.

Imagine now that the PG's walls are damaged, resulting a 0.5 m^3/s air flow. After two hours, T_I will stabilize at 3.5 °C, which is not tolerable. The decreasing of T_I in 60 s is this case 0.056 °C. Since this value is significantly exceeding the normal 0.02 °C decrease, we are able to diagnose the damaged walls after just one minute, and we can produce immediately the necessary interventions.

a2) The assistance at the dimensioning of the energy sources:

Consider the following situation: we want to install a PG in a region where the lowest winter temperature is -20 °C, the mean winter temperature is -0 °C, and the mean wind speed is 5 m/s during nights. Which heat pump should we choose, if we want to maintain a minimum T_I of 10 °C?

A two hours simulation for T_E = -20 °C is indicating that we need a total power of 47 kW. The same simulation for T_E = -0 °C is indicating only 16 kW. Since the extreme temperatures are very rare, we can choose a 12 kW heat pump, and provide the remaining 4 kW with the solar panels. In order to cope with the extreme temperature we can think at an emergency gas burning device.

b) Testing the controller

The PG controller can be tested by performing simulations under different scenarios.

Consider the case of a very bright day, when the greenhouse effect would produce overheating. Fig. 3 is presenting the Toulon experimental greenhouse data recorded in 2004.02.09 between 7 a.m. and 19 p.m. One observes that T_I is reaching 37 °C.

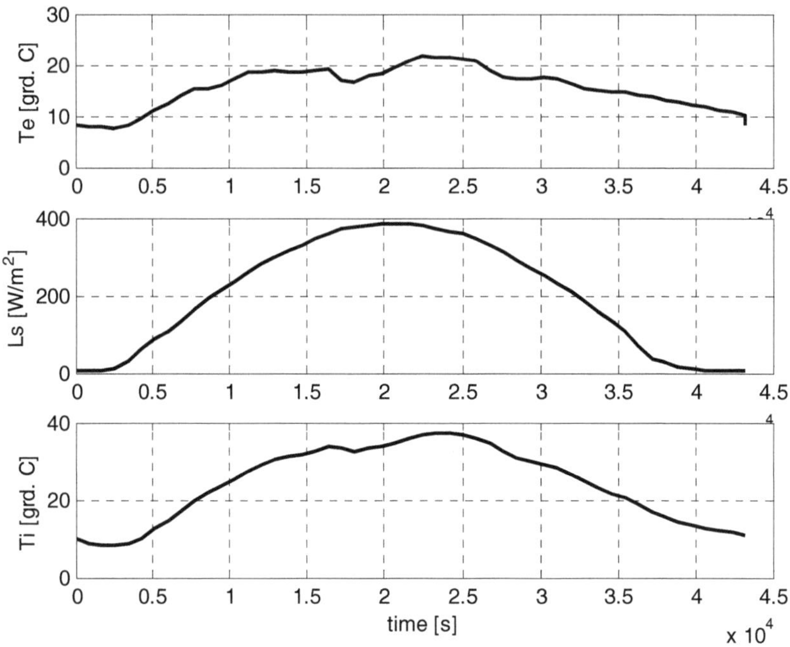

Fig. 3. The greenhouse effect with no control actions

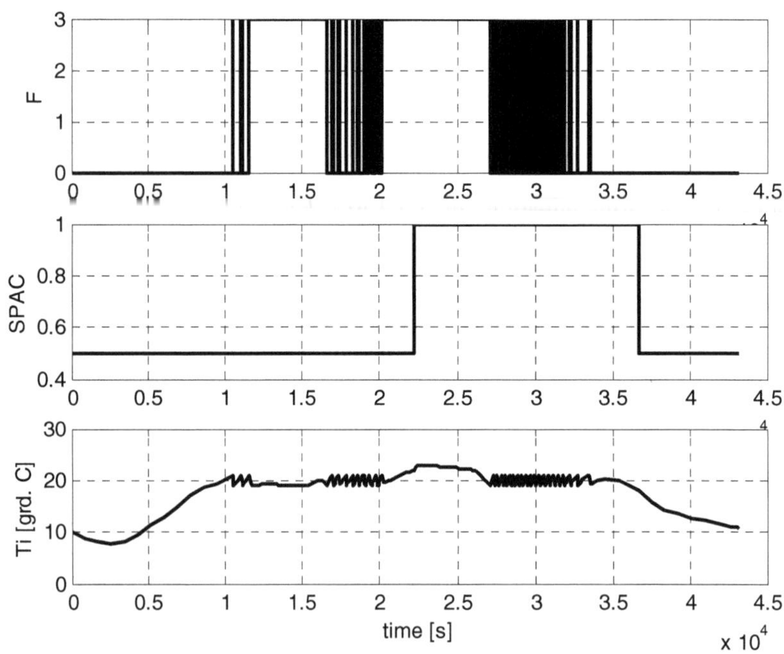

Fig. 4. The performance of the PG controller

We need to avoid overheating and to maintain T_I around 20 °C. Figure 4 is presenting the PG controller action. The control actions are aiming to reduce T_I by:

- natural ventilation (F = 0 for the ventilation fan closed and F = 1 for the ventilation fan open.) We considered a mean ventilated air flow of 3 m^3/s, which is typical for the Toulon greenhouse in the Mistral period.
- the shading of the plants. When SPAC = 1 the plants are shaded and the solar panels are charging the accumulator. When SPAC = 0.3 the panels are parallel to the solar radiation.

Notice that during the first half of the day, when $T_I < 20$ °C, the natural ventilation is able to cool the air by itself. When T_I is close or is exceeding 20 °C the shading of the plants becomes necessary.

6 Conclusions

A large scale ecologic reconstruction of the Earth surface is possible with the help of the energetic passive greenhouses that make possible a great reduction of the agricultural surfaces. The unshackled resulting surface can be reconverted as natural environment, with benefic effects on the atmosphere, including the reduction of the CO_2 concentration and the reduction of the global glasshouse effect.

We are proposing a specific passive greenhouse structure, aggregating three renewable complementary energy sources: the sun light (photovoltaic panels), a cold water heat pump and a DC wind generator. The disposable energy that is not necessarily to heat the greenhouse is stored into DC accumulators. The wind generator and the solar panels have the purpose to ensure the water recirculation needed by the heat pump and the supplying of the control, driving and communication systems of the greenhouse. Since passive greenhouses are independent of infrastructures (excepting the roads), they can be positioned virtually anywhere one can find underground cold water resources. Each element of this greenhouse can be easily found on the market. The key points of this solution are the optimal design of each element's nominal capacity that must be matched to the climatic features of the location and the smart control of the aggregated installation.

A deterministic model that is facilitating an optimal design of the passive greenhouse is presented. The model is implemented in Simulink-Matlab. A dedicated multi inputs multi outputs fuzzy-interpolative controller is introduced and tested with the help of the model.

Our further work will be focused on the implementation and the refining of the computer model, as well as on the construction of an experimental passive greenhouse. The control rule base will be improved after significant experimental tests. Smart control elements will be added, for instance by introducing anticipative rules that may avoid limit situations, save energy and adapt PG to the environment conditions. The remote control via Internet will be also investigated and developed.

References

1. Keeling, C.D., Whorf, T.P.: Atmospheric carbon dioxide record from Mauna Loa. Carbon Dioxide Research Group, Scripps Institution of Oceanography, University of California, http://cdiac.ornl.gov/trends/co2/sio-mlo.htm
2. Tuzhilkina, V.V.: Carbon dioxide exchange in the photosynthetic apparatus of trees in a mature spruce phytocenosis of the northern taiga subzone. Ekologiya 2, 95–101 (2006)
3. Horgos, A.: Legumicultură specială. Agroprint, Timisoara (2003)
4. Voican, V., Lacatus, V.: Cultura in sere si solarii. Ceres, Bucharest (1998)
5. Borlaug, N.E.: Feeding the World in the 21st Century: The Role of Agricultural Science and Technology, Speech given at Tuskegee Univ. (April 2001)
6. Bellows, B.: Solar greenhouses. ATTRA National Sustainable Agriculture Information Service, Fayetteville (2003), http://www.attra.org/attra-pub/solar-gh.html
7. Balas, M.M., Cociuba, N., Musca, C.: The energetic passive greenhouses. Analele Universitatii "Aurel Vlaicu" din Arad, 524–529 (2004)
8. Balas, V.E., Balas, M.M., Putin-Racovita, M.V.: Passive Greenhouses and Ecological Reconstruction. In: 12th IEEE International Conference on Intelligent Engineering Systems INES 2008, Miami (2008)
9. Lafont, F., Balmat, J.F.: Modélisation floue itérative d'une serre agricole. Actes des Rencontres Francophones sur la Logique Floue et ses Applications (LFA), 281–288 (2001)
10. Bouchouicha, M., Lafont, F., Balmat, J.F.: Neural networks, Fuzzy logic and Genetic algorithms for greenhouse identification. In: 2nd International Conference – Tunisian Conference on Electro-Technical and Automatic Control JTEA, pp. 356–362 (2002)
11. Bouchouicha, M., Lafont, F., Balmat, J.F.: Ochsner Waermepumpen, http://www.ochsner.com
12. Olivier, C.: Ground source heat pump in France in the residential. International Summer School on Direct Application of Geothermal Energy, Skopje (2001), http://www.geo-thermie.de/tag-ungkongresse/vortrag-sprogramm_igd_2001
13. Olivier, C.: Wind Turbine Design Cost and Scaling Model. Technical Report NREL/TP-500-40566 (2006), http://www.nrel.gov/docs/fy07osti/40566.pdf
14. Olivier, C.: Wind Energy Manual. Iowa Energy Center, http://www.energy.iastate.edu/Renewable/wind/wem/windpower.htm
15. Bradford, T.: Solar Revolution: The Economic Transformation of the Global Energy Industry. MIT Press, Cambridge (2006)
16. Balas, M.M., Duplaix, J., Balas, S.: Modeling the heat flow of the greenhouses. In: The IEEE International Workshop on Soft Computing Applications SOFA 2005, Szeged-Hungary & Arad-Romania, pp. 37–43 (2005)
17. Balas, M.M., Duplaix, J., Bouchouicha, M., Balas, S.V.: Modeling the Wind's Influence over the Heat Flow of the Greenhouses. Journal of Intelligent & Fuzzy Systems 19(1), 29–40 (2008)
18. Balas, M.M., Balas, V.E.: Modeling Passive Greenhouses - The Sun's Influence. In: IEEE International Conf. On Intelligent Engineering Systems INES 2008, Miami, pp. 71–75 (2008)

19. Kóczy, L.T., Balas, M.M., Ciugudean, M., Balas, V.E., Botzheim, J.: On the Interpolative Side of the Fuzzy Sets. In: IEEE Sofa 2005, Szeged-Arad, pp. 17–23 (2005)
20. Dounis, A.I., Bruant, M., Santamouris, M., Guaracino, G., Michel, P.: Comparison of Conventional and Fuzzy Control of Indoor Air Quality in Buildings. Journal of Intelligent & Fuzzy Systems 4(2), 131–140 (1996)
21. Dounis, A.I., Bruant, M., Santamouris, M., Guaracino, G., Michel, P.: Solar Greenhouse Project. Dutch Governmental Research Program on Ecology, Economy and Technology, http://nt1.aenf.wau.nl/mrs/Projects/SolarGreenhouse/index.html
22. van Ooteghem, R.J.C.: Optimal Control Design for a Solar Greenhouse, Ph.D. thesis Wage-ningen University (2007), http://library.wur.nl/wda/dissertations/dis4110.pdf

13
A Complex GPS Safety System for Airplanes

Dan-Marius Dobrea and Cosmin Huțan

"Gh. Asachi" Technical University of Iasi, Faculty of Electronics, Communications and Information Technology, Bd. Carol I, No. 11, 700506 Iasi, Romania
mdobrea@etc.tuiasi.ro, cos_hc@yahoo.com

Abstract. There are many applications where the exact position and dynamics of different objects are needed in real time. We propose a system that is able to locate simultaneously several "objects" and to present them, in real time, on a map. The system is dedicated mainly to airports for tracking maintenance cars and persons (in this last case the system works as a personal locator device) and to avoid disasters that could happen on the runway. Several results and aspects of the system are investigated and presented.

1 Introduction

The GPS and the GSM are two of the mature technologies existing on the market, with a large number of commercial applications. The GPS technology is mainly used to obtain the absolute coordinate and position of an "object". However, in some applications, the GPS technology is applied for obtaining a time stamp and time synchronization of different processes moreover, for triggering different events. In car position, control and navigation systems [1], [2], [3], [4], [5], [6], physics [7], [8], aircrafts [9], railway transportation [5], [10], user position [11], telecommunications [12], geoscience [8], [13], [14], [15], [16], automotive safety systems [17], etc. the GPS technologies are wildly used.

On December 30, 2007, at the International Airport "Henri Coandă", Romania a Boeing 737 plane with 117 passengers and having 200 Km/h hit, during the take-off time, a maintenance car. Fortunately, no one was killed or injured in this incident, but the plane was severely damaged and the car was completed destroyed. Also, in another incident, which happened in 1987, a MD-80 airplane landed in heavy fog at Helsinki Airport and hit a maintenance car on the runway [18]. In both incidents, even if on the airports there were strict procedures regarding the landing and the take-off it seems that these procedures were not enough for avoiding accidents.

At this moment the standard method used to track the objects, vehicles and aircrafts is the surface movement radar in on the large airports. But the surface movement radar has several disadvantages and as it was presented previously it was unable to prevent this types of accidents. Moreover the surface movement radars are very expensive and, because of this, small airports like the one in Thessaloniki, Greece can not afford ground surface radar [20] and use camera

H.-N. Teodorescu, J. Watada, and L.C. Jain (Eds.): Intel. Sys. and Tech., SCI 217, pp. 233–247.
springerlink.com

systems to prevent different types of accidents. But these methods are prone to error and unusable in bad weather and low visibility conditions. One of the drawback of the surface movement radar is gave by the presence of building and other aircrafts that mask and blind the radar. This problem can be solved easily using a large number, but this number of radar antennas is limited by the health risks and by the electromagnetic radiation interference they produce.

Based on the effect of resulting disturbances of the Earth's magnetic field due to the quantity of ferromagnetic metal existing in the aircrafts, the researchers and engineers have built a magnetic sensor able to be used in order to avoid runaway crashes [20]. But the big disadvantage of this sensor is given by the cover range – 50 meters. For a big airport like Frankfurt hundreds of these sensors must be placed [20]. But this sensor can be used only in the key point to complete the information offered by the surface movement radar [20].

To avoid the collisions between airplanes and other ground objects we believe that the air control staff must to have and to operate a system that should be able: to locate simultaneously a large number of different types of targets operating on the plane railway and to depict them, in real time, on a map. Such a system will contribute directly to the safety and efficiency of the air traffic services.

This paper presents a complete solution, software and hardware, for the above problem. Also, the obtained results are presented.

The remainder of this paper is organized as follows: Section 2 outlines the system concepts and organization. Section 3 presents the mobile platforms. Section 4 presents the master applications. Section 5 presents the results and finally, section 6 gives the conclusion.

2 The Airplane Safety System, Concepts and Organizations

The proposed system is based on the existence of several mobile locator devices able to acquire continuously the position of an "object" and to send it to the master application. These mobile devices should be placed on all vehicles used in the airports (maintenance cars, tow tractors, etc.).

The master application receives the positions from all mobile locator devices placed on the vehicles and equipments and represents, in real time, these positions on a map. Based on this information, the air control staff obtains a clear image of the positions and dynamics of the vehicles and equipments situated in the airport airside areas (airside areas include all areas accessible to the aircrafts). Having the information provided by the system formed by the mobiles locator platforms and the master application and knowing the airplane position, the planes will receive the take-off or landing clearance only when the procedure will support this decision (no reported vehicles and/or equipments located by the airplane safety system on the plane runway).

3 The Mobile Locator System

The mobile locator system is built having the Freescale MCF5213 processor as the heart of the system. The Freescale MCF5213 is a microcontroller on 32 bits having a Version 2 ColdFire variable-length RISC processor core.

The software, running on the microcontroller acquires continuously the GPS position and sends this information, through a GSM connection, to the master application. The software acquires the position from a RCB-LJ ultra-low power GPS receiver produced by the uBlox company. The GPS receiver is based on the ANTARIS® GPS Engine that was jointly developed by Atmel and uBlox. This core provides: a) excellent navigation performance under dynamic conditions, in areas with limited sky view (like urban and canyons), b) high sensitivity (acquisition -140 dBm, tracking -149 dBm, by using an active antenna) for a weak signal and c) support of DGPS (Differential GPS) and multiple SBAS (Satellite Based Augmentation Systems) systems like WAAS (Wide Area Augmentation System) and EGNOS (European Geostationary Navigation Overlay Services).

The position obtained from the GPS systems is sent through the GSM cellular network. The GSM module is a Fastrack M1306B cellular Plug & Play Wireless CPU module with GSM/GPRS connectivity for machine to machine applications.

For determining the accuracy of the acquired position the DOP (Dilution of Precision) parameter was used. The DOP parameter is a unitless value that indicates when the satellite geometry provides the most accurate results. This parameter can be determined for the horizontal position – horizontal dilution of precision (HDOP) – and for the vertical position – vertical dilution of precision (VDOP).

But, the most commonly used DOP parameter is the position dilution of precision (PDOP). PDOP is a combination of HDOP and VDOP. PDOP parameter is the mathematical representation of the quality of the navigation solution; mainly, this quality is based on the geometry of the satellites on the sky (required to calculate the position) and on the receiver's mask angle of the antenna (the mask angle determines the minimum elevation angle below which the receiver will no longer use a satellite / the satellites in its computations). The number of the visible satellites and their relative positions in the sky mainly control the PDOP; however, the PDOP can be affected (made larger) by signal obstruction due to the terrain, foliage, building, vehicle structure, etc.

A PDOP value of 1 indicates an optimum satellite constellation and the highest quality data. Meanwhile, a PDOP values in excess of 8 are considered poor. For example, a point calculated with a PDOP of 30.0 may be placed by more than 150 m from its true location [19].The mobile locator system has two working modes. The first one, named **tuning mode**, is used in order to set up the confidence threshold level in the master application. Due to the position error generated by: the geometry of the satellites used in position calculation, by the signal path obstruction (by buildings, foliage, covers, snow, etc.), the multi-path effects, the ionospheric and tropospheric effects, etc. around each plane runway a safety zone must be imposed.

If an "object" is placed on the airplane runway or in the safety zone, the airplanes will not receive the take-off or the landing clearance. In this case the risk of an impact is considered high. Mainly, because the position's error is also determined by the receiver himself – due to the antenna shortcomings (poor gain of the GPS antenna, poor directivity of the GPS antenna, poor matching between antenna and cable impedance, poor noise performance of the receiver's input stage

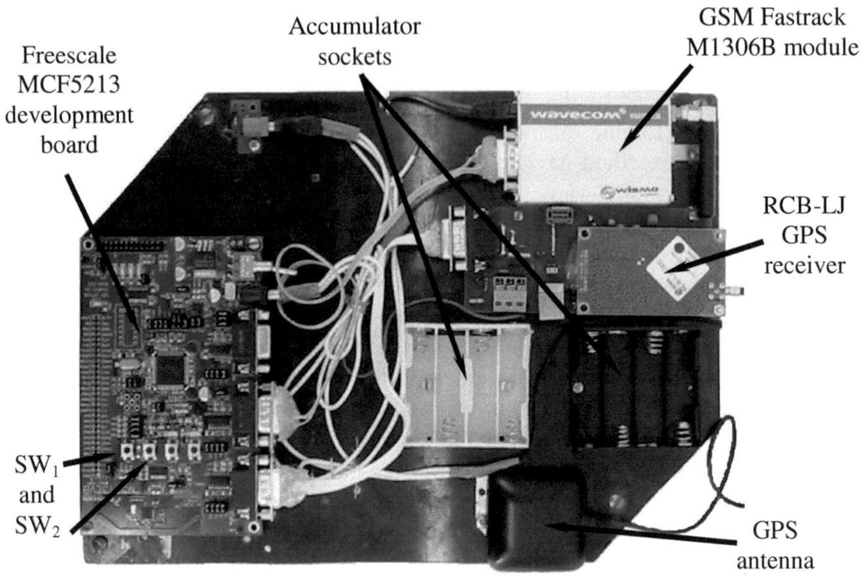

Fig. 1. The mobile locator prototype board

or the antenna amplifier), to the electrical environment (jamming from the external signals, jamming from the signals generated by the receiver itself), to the presence at the GPS module level of different satellite based augmentation systems (WAAS and EGNOS), etc. –, the confidence threshold level is different for different GPS receivers. In the tuning mode the airplane safety system determines the confidence threshold level around the plane runway for a specific PDOP value. In this mode the mobile locator systems send the GPS position for all the PDOP values. Making a statistical analysis and correlating the true position with the determined position the confidence threshold level is determined for the RCB-LJ GPS module presented above.

The second working mode is used in order to track the mobile locator system's positions. This mode was named **tracking mode**. In this operating mode, from time to time the tracking module sent its coordinate and its unique ID code. The time period between the mobile locator coordinate communications can be set from 10 second up to several thousands of seconds (e.g. 5 minute is a usual time interval that was used in system tests and validation). For maximum accuracy, the GPS receiver is set in Continuous Tracking Mode (CTM). Our GPS module can be interrogated up to 4 times in a second. If the PDOP parameter is smaller, then a predefined threshold determined in the tuning process of the entire system, the position will be sent to the master application; if the PDOP is grater then the same threshold a new set of coordinates position will be acquired. If after several readings from the GPS module the performance reflected through the PDOP parameter does not improve the mobile locator system will send the coordinate together an error message and with the PDOP value. The communication between

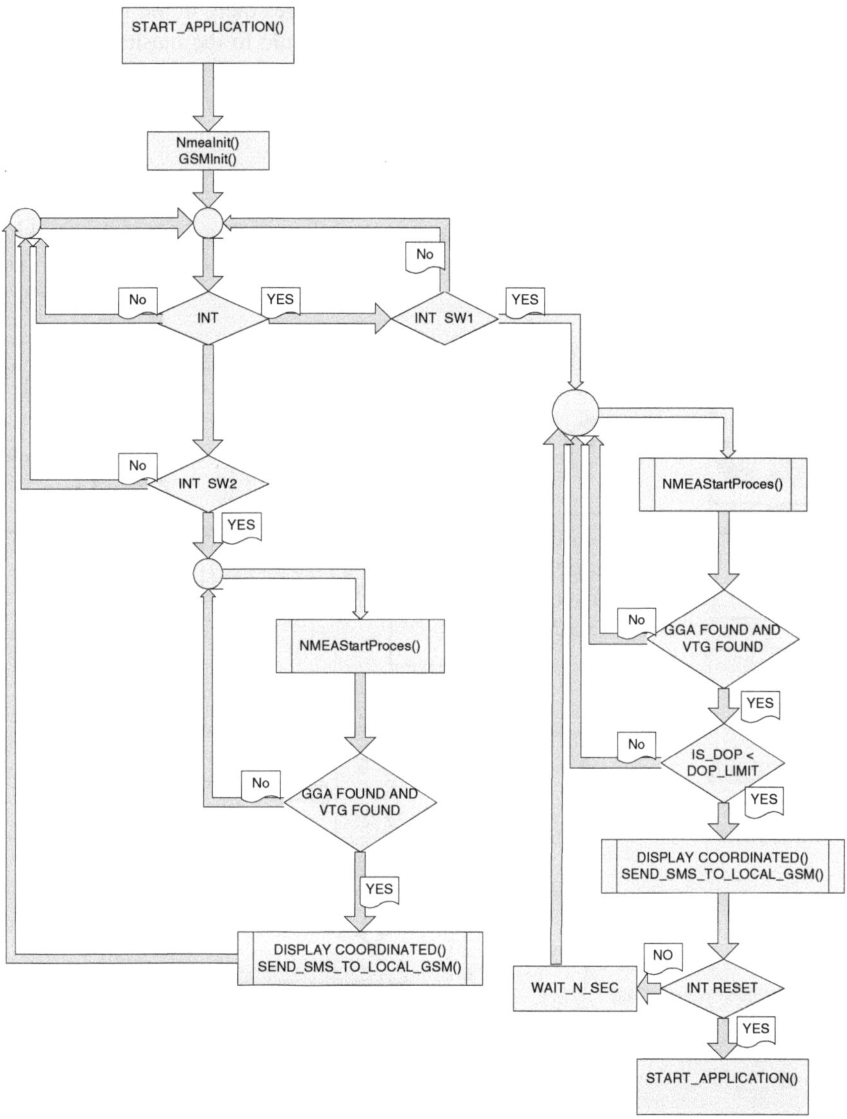

Fig. 2. Software diagram from the mobile locator system

the GPS module and the microcontroller is a serial one, based on the NMEA 0183 standard protocol. The NMEA (National Marine Electronics Association) protocol is an ASCII based standard data communication protocol used by the GPS receivers.

The working modes are selected based on the state of the two external switches, SW_1 and SW_2, Figure 1. The application waits until the GGA message is received from the GPS module. If the PDOP is smaller then a predefined threshold (PDOP parameter is encapsulated in the GSA NMEA message) then, in the next step, the position is extracted and sent through the GSM module to the master application (see Figure 2, the software diagram); after this, the cycle presented above is repeated.

4 The Master Application

The master application has two working modes: the map mode and the tracking mode. The data flow for the master application is presented in Figure 3. The

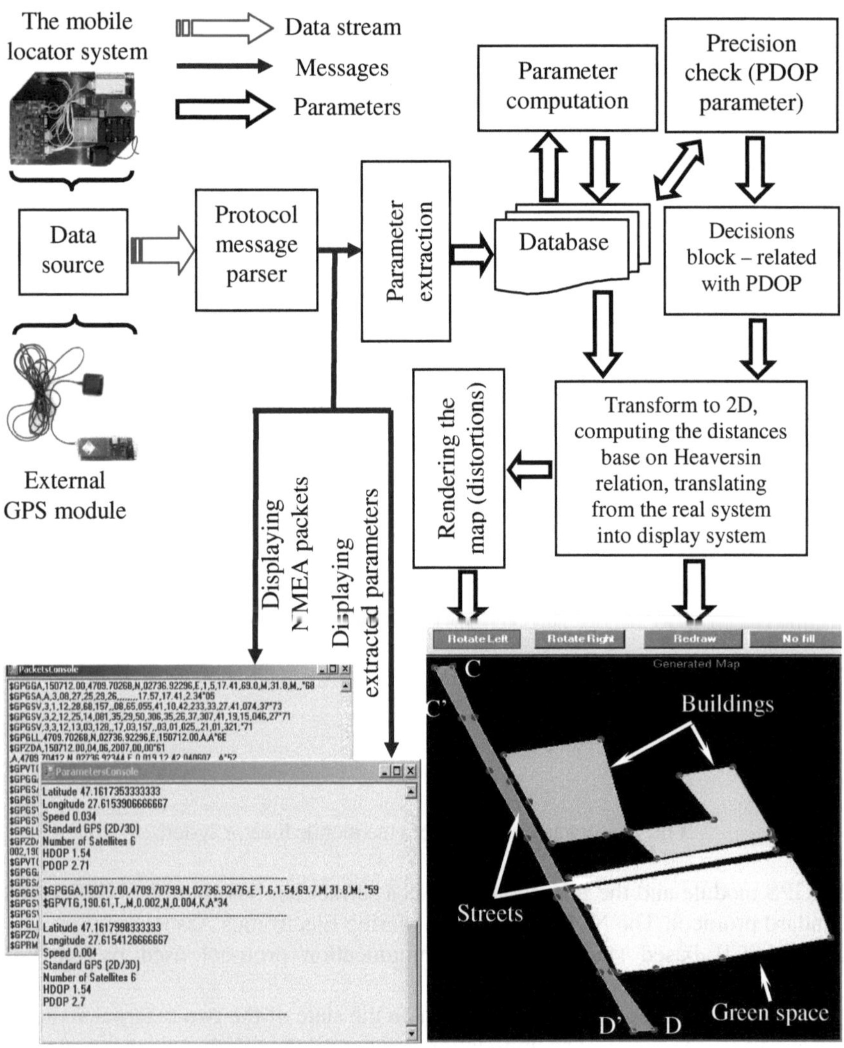

Fig. 3. The data flow for the master application

master application was written in C# (Visual Studio 2005) and the SQL supports the data base.

In the map mode, the master application communicates with the GPS receiver connected to the serial port. In this mode, the system is able to acquire the position of the different points and to store this information; finally, based on these points the map is drawn, see Figure 3. The map is sketch in real time (in the same time with the point acquisition).

$$\phi = \arccos\left(\sin\phi_1 \sin\phi_2 + \cos\phi_1 \cos\phi_2 \cos(\lambda_1 - \lambda_2)\right) \tag{1}$$

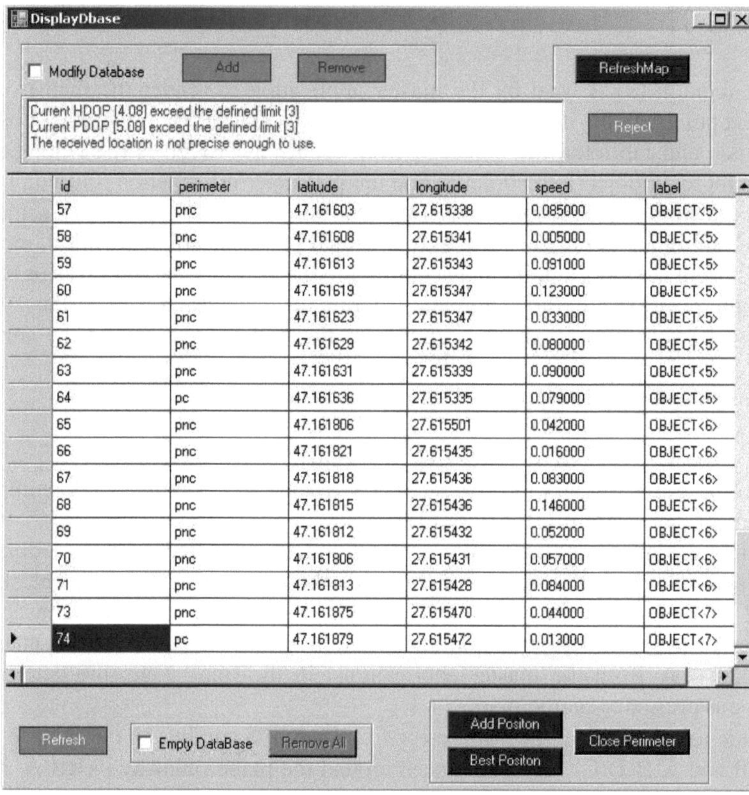

Fig. 4. Data base edit window

For two points with coordinates $\{\phi_1, \lambda_1\}$ and $\{\phi_2, \lambda_2\}$, (longitude and latitude), the easiest way to determine the angle between the two radius (that have as an end point the center of the earth and the second endpoint one of the two points previously presented) is:

Due to the errors that the relation (1) introduces, mainly the rounding errors, this relation is infrequently used in navigation. The Haversine relation is more accurate and, in consequence, it is used in a larger number of applications:

$$\phi = 2\arcsin\left\{\sqrt{\sin^2\left(\frac{\phi_2 - \phi_1}{2}\right) + \cos\phi_1 \cos\phi_2 \sin^2\left(\frac{\lambda_1 - \lambda_2}{2}\right)}\right\} \tag{2}$$

Even if the relation (2) is more accurate than the relation (1) for a larger type of distances it also induces some large errors, especially from the points placed on opposite diameters. For these reasons a more complicated relation, (3), is used for the all types of distances:

$$\phi = \arctan\left\{\frac{\sqrt{[\cos\phi_2 \sin\Delta\lambda]^2 + [\cos\phi_1 \sin\phi_2 - \sin\phi_1 \cos\phi_2 \cos\Delta\lambda]^2}}{\sin\phi_1 \sin\phi_2 + \cos\phi_1 \cos\phi_2 \cos\Delta\lambda}\right\} \tag{3}$$

The relation (3) is used by the master application to render the map from its acquired points. In the database are stored, for each single point; the point position (longitude and latitude), the object id from each point (it gives the belonging of each point to an object), the point id and the perimeter information (closed or not). The acquisition of the coordinates for each point can be done using a manual method and in an automatic way. In the manual procedure of acquisition, the user of the system acquires a point and according to the PDOP value he saves or rejects the point position. In the automatic coordinate acquisition mode, the system acquires 10 values in 10 seconds and saves the best values, having the smallest PDOP value.

Finally the distance, represented in meters, is obtained as:

$$d = R \cdot \phi \tag{4}$$

In (4) R is Earth radius (approximately 6378 Km) and ϕ is the value computed with one of the relations (1), (2), or (3).

In the tracking mode, the master application has connected a GSM module to a serial port. The mobile locator system sends the position of the different objects tagged by them from time to time. This time interval between two consecutive sessions of position determination and position communication can be set in the automatic way from the master applications. In the final stage, the determined position is presented on the map.

Due to the obtained position error from the GPS module, a confidence zone (ACDB and A'B'D'C') must be placed around the plane runaway (ABB'A' zone) in order to be sure that no type of collision will take place, see Figure 5. The confidence zone was determined using the maximum distance error related to the mean position and is equivalent with the AC distances from Figure 5. The mean position was determined using a large series of coordinate positions recorded in conditions as closed as possible to real conditions.

There are now only two difficulties for having a correct representation of the runaway collision zone. First, the runaway is represented through the coordinates (longitude and latitude) and the confidence zones are determined using the estimated distance error of the GPS module. Finally, all information is stored in database in coordinate (longitude and latitude) form.

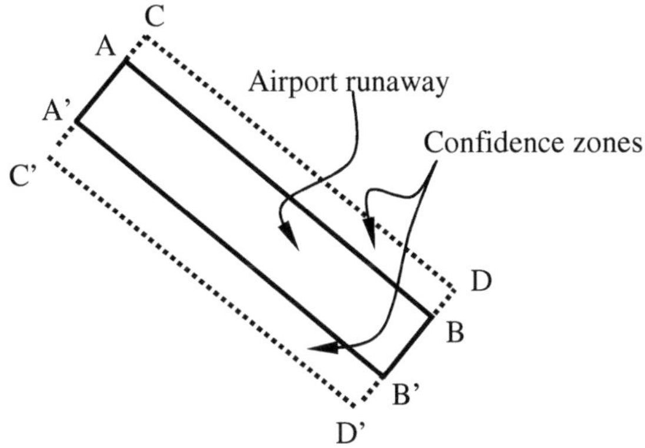

Fig. 5. The plane runaway collision zone: CDD'C'

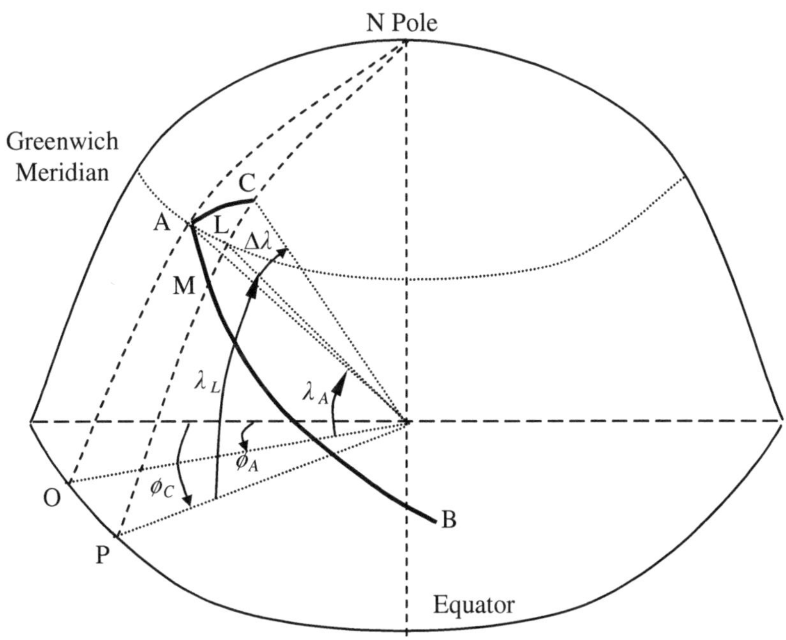

Fig. 6. The real plane runaway edge, AB, and the collision zone determination base on AC

The second problem is related to the Earth curvature and it must be taken into account in order to obtain an excellent representation. The long range distances (as the distance between two cities like Paris and Moscow) are more difficult to determine exactly from the coordinates (longitude and latitude) than the short ones and, as a result, the computational error is greater than in case of the short distances. From the geographically point of view, long range distances involve following a curved line which is not like the approximately straight line used in a normal case. Practically this problem is solved by breaking the curved line into several straight segments. In our case, all the distances are short (e.g. the airport runaway is around 3.5 Km) and, for these type of distances, the obtained errors are very small and the problem can be translated to be solved in a flat surface.

In order to find out the C point coordinates we must to add $\Delta\lambda$ and ϕ_C - ϕ_A, see Figure 6, to the A point coordinates.

The slope made by the airplane runaway with Equator can be determined based on:

$$\alpha = \mathrm{mod}\left\{ a\tan\left[\frac{\sin(\phi_A - \phi_B)\cdot\cos(\lambda_B)}{\cos(\lambda_A)\cdot\sin(\lambda_B) - \sin(\lambda_A)\cdot\cos(\lambda_B)\cdot\cos(\phi_A - \phi_B)} \right], 2\pi \right\} \quad (5)$$

Knowing α, the CL and AL segments distances can be determined directly based on AC segment distance. Having CL and knowing the Earth radius, $\Delta\lambda$ results directly from (4). Using basic geometric relations, ϕ_C - ϕ_A is easily determined, Figure 7. In similar ways, the other points D, D' and C' are determined.

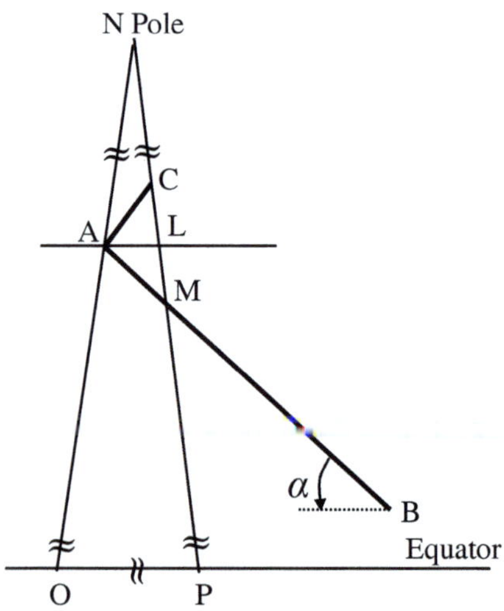

Fig. 7. The geometrical problem for collision zone coordinates determination

In this implementation of the master application, the vehicles and all the associated other devices and equipments used or associated with these vehicles, tracked by the mobile locator systems, are only placed in real time on the map. In this stage of the system development, there an automatic warning module to notify the human operator when a vehicle is placed on the plane runaway is not implemented.

5 Results

In testing the system, three mobile locator systems were used and one master application. All the mobile locator systems were identical. The master application was installed on a laptop PC. First, the map was generated, see Figure 3. The tests were conducted in a perimeter of around 600m x 200m. The tests were done for static and, also, for dynamic mobile locator systems, in conditions as close as possible to the real ones (those on the airports) – we are referring here to the electrical, the environmental and the weather conditions. The airplane safety system showed its ability to track and present the positions for all the mobile locator systems.

Table 1. A statistical analysis of the results obtained for the case when there are no obstacle to obstruct the GPS

	Coordinate		Distance from the mean position [m]	PDOP
	Latitude [deg]	Longitude [deg]		
Minimum	**45.451501**	**28.043954**	**0.258944**	**1.7**
Maximum	**45.451552**	**28.044004**	**3.614854**	**1.7**
Average	**45.451523**	**28.043977**	**2.229548**	**1.7**
Deviation	**0.000020**	**0.000014**	**0.988741**	**0**
No. samples	**226**			

The standard existing software tools were used for determining the confidence threshold level for the master application. First, one of the mobile locator systems was placed on the perimeter, in three different situations. Second, the master application continuously received the position information and saved the results into a data base. All this time the mobile locator system was set to work in tuning mode. Finally, the data were analyzed. The results of the statistical analysis are presented in Table 1, Table 2 and Table 3. The data given in Table 1 and Table 2 were acquired on very good weather conditions, on a sunny day without clouds, water vapors or smoke. Unlike these, the data presented in Table 3 were acquired on a raining day. As a conclusion, these last distance errors are affected by the weather conditions and they cover the worst case situation.

The analyzed situations correspond to three different cases. In the first analysis, the GPS receiver was placed in such a position that it had direct line of sight with

Table 2. A statistical analysis of the results obtained for the case when the GPS view to the sky is obstructed by a wall for one side

	Coordinate		Distance from the mean position [m]	PDOP
	Latitude [deg]	Longitude [deg]		
Minimum	45.451279	28.044010	0.271503	3.5
Maximum	45.451343	28.044157	6.637724	3.9
Average	45.451313	28.044080	3.271895	3.8
Deviation	0.000019	0.000042	2.091876	0.1
No. samples	299			

Table 3. A statistical analysis of the results obtained for the case when the GPS view to the sky is obstructed by two walls (for right and back side)

	Coordinate		Distance from the mean position [m]	PDOP
	Latitude [deg]	Longitude [deg]		
Minimum	45.450934	28.044003	0.222265935	1.8
Maximum	45.451199	28.044291	17.23316686	5.8
Average	45.451075	28.044135	6.008740634	4.6
Deviation	0.000048	0.000070	4.65934139	1.2
No. samples	150			

all the satellites on the sky. For this case, the results are presented in Table 1. Even if in other applications this situation is not so frequently encountered, in our case (in the environments of airports airside areas, characterized by large open space) this situation represents the normality. From the Table 1 we can conclude that a confidence threshold level of 4 m is enough.

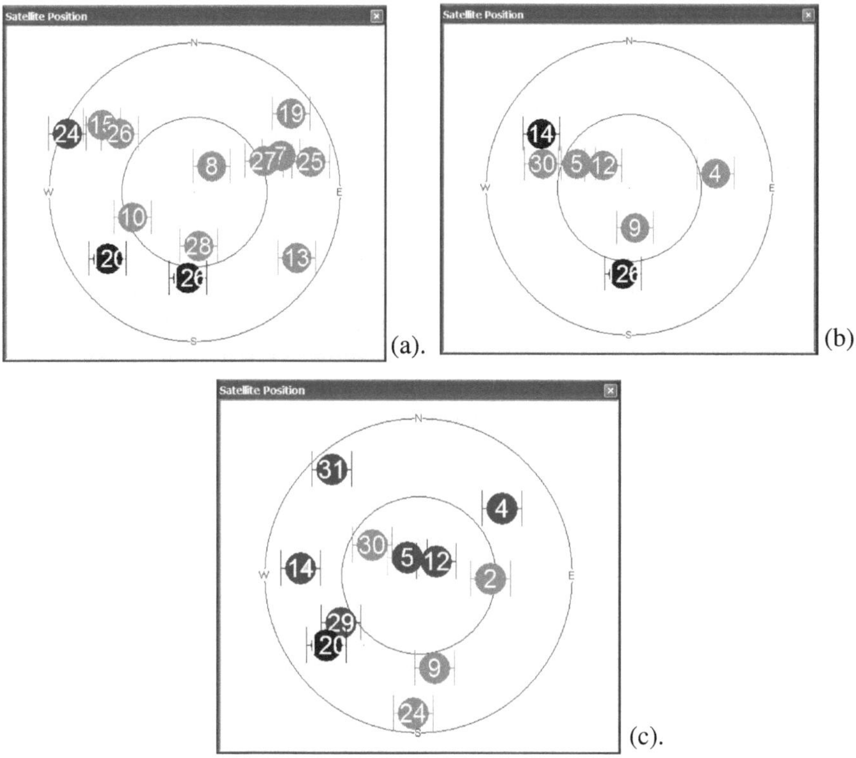

Fig. 8. Some satellites patterns on the sky for: (a). PDOP = 1.7, (b). PDOP = 3.5 and (c). PDOP = 5.3

The second analysis models the situation when a building obstructs the direct line of the site to the satellites placed in only one direction on the sky. This simple situation models a multi-path environment. In this type of environment not only some satellites are masked but from a part of the satellites the GPS will receive a direct path waves, while, in addition, other radio waves will be reflected by the buildings. For this case, a confidence threshold level of 7 m is more than enough.

When the direct satellites line of sight is blocked for two directions, the error increases and the confidence threshold level must to be set at almost 20 m. This situation is a very infrequent one on the airports but it should be taken into consideration.

In Figure 8 the PDOP parameters values for different satellites position on the sky are presented (with green - the active satellites, with blue - the satellites with limited connectivity, with red - the satellites with low signal) related with the situation presented in the Table 1, Table 2 and Table 3. The examples presented in Figure 8 were chosen from a variety of PDOP values starting with good PDOP value, 1.7, that generates a good GPS performance and ending with a quite high PDOP value, that generates a degrade of the GPS performance.

6 Conclusions

The paper presents a complete solution for an airplane safety system, able to avoid incidences that could happen during landing or take-off between an airplane and the different types of vehicles used in the airports or all other devices and equipments used or associated with these vehicles.

The main idea of the system is based on the existence of several mobile locator devices able to acquire continuously the position of an "object" and to send it to the master application. The master application receives the positions from all mobile locator devices placed on the vehicles and equipments that could be present on the plane runway and then it represents them, in real time, on a map. Based on this information, the air control staff obtains a clear image of the positions and dynamics of the vehicles and equipments situated in the airport airside areas. As a result of the conducted tests, the airplane safety system proved its ability to track and present, without any problems, the positions for all the mobile locator systems.

Acknowledgements. We, authors, are grateful to the Silica an AVNET Company for the donation of the MCF5213 systems. We take the opportunity to thank again, to the same company, Silica an AVNET Company, for its first generous donation to our Faculty of Electronics, Telecommunication and Information Technology that opened, for us and for our students, the world of the powerful 32 bits microcontrollers.

References

1. Sharaf, R., Noureldin, A.: Sensor Integration for Satellite-Based Vehicular Navigation Using Neural Networks. IEEE Transactions on Neural Networks 18(2), 589–594 (2007)
2. Bevly, D.M., Parkinson, B.: Cascaded Kalman Filters for Accurate Estimation of Multiple Biases, Dead-Reckoning Navigation, and Full State Feedback Control of Ground Vehicles. IEEE Transactions on Control Systems Technology 15(2), 199–208 (2007)
3. Hong, S., Lee, M.H., Kwon, S.H., Chun, H.H.: A car test for the estimation of GPS/INS alignment errors. IEEE Transactions on Intelligent Transportation Systems 5(3), 208–218 (2004)
4. Huang, J., Tan, H.S.: A Low-Order DGPS-Based Vehicle Positioning System Under Urban Environment 11(5), 567–575 (2006)
5. Seong, Y.C., Choi, W.S.: Robust positioning technique in low-cost DR/GPS for land navigation. IEEE Transactions on Instrumentation and Measurement 55(4), 1132–1142 (2006)
6. Obradovic, D., Lenz, H., Schupfner, M.: Fusion of Sensor Data in Siemens Car Navigation System. IEEE Transactions on Vehicular Technology 56(1), 43–50 (2007)
7. Berns, H.G., Burnett, T.H., Gran, R., Wilkes, R.J.: GPS time synchronization in school-network cosmic ray detectors. IEEE Transactions on Nuclear Science 51(3), 848–853 (2004)

8. Nilsson, T., Gradinarsky, L.: Water Vapor Tomography Using GPS Phase Observations: Simulation Results. IEEE Transactions on Geoscience and Remote Sensing 44(10), 2927–2941 (2006)
9. Williamson, W.R., Abdel-Hafez, M.F., Rhee, I., Song, E.J., Wolfe, J.D., Chichka, D.F., Speyer, J.L.: An Instrumentation System Applied to Formation Flight. IEEE Transactions on Control Systems Technology 15(1), 75–85 (2007)
10. Bertran, E., Delgado-Penin, J.A.: On the use of GPS receivers in railway environments. IEEE Transactions on Vehicular Technology 53(5), 1452–1460 (2004)
11. Kuusniemi, H., Wieser, A., Lachapelle, G., Takala, J.: User-level reliability monitoring in urban personal satellite-navigation. IEEE Transactions on Aerospace and Electronic Systems 43(4), 1305–1318 (2007)
12. Spiegel, S.J., Kovacs, I.I.G.: An efficient integration of GPS and WCDMA radio front-ends. IEEE Transactions on Microwave Theory and Techniques 52(4), 1125–1131 (2004)
13. Gleason, S., Hodgart, S., Yiping, S., Gommenginger, C., Mackin, S., Adjrad, M., Unwin, M.: Detection and Processing of bistatically reflected GPS signals from low Earth orbit for the purpose of ocean remote sensing. IEEE Transactions on Geoscience and Remote Sensing 43(6), 1229–1241 (2005)
14. Rivas, M.B., Martin-Neira, M.: Coherent GPS reflections from the sea surface. IEEE Geoscience and Remote Sensing Letters 3(1), 28–31 (2006)
15. Zuffada, C., Fung, A., Parker, J., Okolicanyi, M., Huang, E.: Polarization properties of the GPS signal scattered off a wind-driven ocean. IEEE Transactions on Antennas and Propagation 52(1), 172–188 (2004)
16. Grant, M.S., Acton, S.T., Katzberg, S.J.: Terrain Moisture Classification Using GPS Surface-Reflected Signals. IEEE Geoscience and Remote Sensing Letters 4(1), 41–45 (2007)
17. Tan, H.S., Huang, J.: DGPS-Based Vehicle-to-Vehicle Cooperative Collision Warning: Engineering Feasibility Viewpoints. IEEE Transactions on Intelligent Transportation Systems 7(4), 415–428 (2006)
18. Woldring, M.: The Human Factor After an Incident. Skayway the Eurocontrol Magazine 6(24), 11–13 (2002)
19. Christen, T.: ANTARIS TIM-Lx GPS Modules System Integration Manual/Reference Design, Switzerland (2005)
20. Dumiak, M.: Magnetic Field Sensor Could Help Halt Runaway Crashes. IEEE Spectrum 45(4), 9–10 (2008)

14

Exploring the Use of 3D Collaborative Interfaces for E-Learning

Gavin McArdle

School of Computer Science & Informatics
University College Dublin
Belfield, Dublin 4, Ireland
gavin.mcardle@ucd.ie

Abstract. Today, Learning Management Systems are the most popular technique for delivering learning material to students electronically. Despite the success and popularity of such systems, there is evidence which highlights that courses which rely solely on Learning Management Systems, have a significantly higher dropout rate than that experienced by courses operating in a classroom environment. Factors such as an absence of interaction with tutors and other students, combined with a lack of stimulation caused by unappealing user interfaces can contribute to the high attrition rates. This chapter examines these issues and explores possible solutions. Particular emphasis is placed on the use of Three Dimensional onscreen Graphical User Interfaces to stimulate users which can be combined with multi-user and synchronous communication techniques to facilitate meaningful interaction. Research indicates that there is also a need to include the social aspects of classroom-based teaching within e-learning environments, an issue which has not been fully explored in the literature. This chapter describes our own system, called Collaborative Learning Environments with Virtual Reality, the design of which includes tools for social interaction between students. To date no major evaluation of Three Dimensional interfaces for e-learning has been conducted. This chapter describes the approach and positive results of an evaluation conducted to determine the usability of CLEV-R.

1 Introduction

Learning Managements Systems (LMSs) are the e-learning technology of choice for a large number of universities who utilise such systems as an accompaniment to traditional face-to-face teaching and as a mechanism for distance learning. Table 1 lists some of the predominant LMSs in vogue at present. A LMS 'enables an institution to develop electronic learning materials for students, to offer these courses electronically to students, to test and evaluate the students electronically, and to generate electronically student databases in which student results and progress can be charted' [1]. Additionally, several LMSs provide tools to assist lecturers in the preparation of learning material and quizzes. LMSs rely heavily on

H.-N. Teodorescu, J. Watada, and L.C. Jain (Eds.): Intel. Sys. and Tech., SCI 217, pp. 249–270.
springerlink.com

text-based web pages to deliver learning material; however, tutors can also use images and multimedia embedded in the web page to accompany the text. One of the strengths of LMSs is their ability to recognise different users and provide varying levels of privilege accordingly. For example, lecturers have access to more resources, such as students' grades and contact information, than teaching assistants might. Administrative tools within LMSs can be used to provide assistance for lecturers. One powerful feature offered by LMSs is their ability to monitor students' actions and report information on students' progress back to the lecturer, often graphically using charts. Communication within LMSs tends to be asynchronous, taking the form of message boards, email distribution lists and online discussion forums [1]. Further functionality of the LMSs includes tools to provide access to exams and assignments. Auxiliary services such as announcements and bulletin boards are also provided. Additionally, many LMSs provide modules that allow the system to be incorporated with the university infrastructure such as the library and administration facilities.

Table 1. Common Learning Management Systems

Platform Name	Web Address
Blackboard	www.blackboard.com
WebCT	www.webct.com
Top Class	www.wbtsystems.com
Moodle	www.moodle.org

While the success of such LMSs is in no doubt, research indicates that courses, which rely solely on mainstream e-learning applications such as LMSs, have a higher dropout rate than their classroom-based counterparts [2]. Studies show boredom, ennui and a lack of motivation are contributing factors to the high attrition rates within online courses [3]. The use of web pages and text-based interfaces involves the student reading large passages of text, a task which many find boring and not very stimulating [4]. One of the major drawbacks with existing LMSs is the lack of support for instant communication. The absence of real-time interaction precludes timely discussions between learner and instructor which can lead to feelings of isolation and loneliness for students [5]. Collaborating with one's peers is an important element of learning in the real world [6]. It permits students to develop skills for dealing with people and teaches them about cooperation and teamwork. The asynchronous communication methods offered in mainstream e-learning applications are insufficient for organising group projects and so group work is often excluded from e-learning courses. While interaction is important for collaboration, so too is social interaction between students [7]. Students often build friendships with their classmates in the real world. This interaction with others plays a key role in the personal development of students and their formation of social skills. The asynchronous communication methods offered within traditional LMSs do not easily permit a natural flow of conversation and can hinder social interaction among students. Consequently they do not feel they have a social presence within the learning environment or experience a strong sense of a community, both of which can lead students to withdraw from their course of study before completing it [3].

The issues within mainstream e-learning applications, discussed above, can be broadly divided into two main areas, one concerning the lack of a stimulating interface and the other concerning an absence of synchronous communication. The focus of our research is to address these shortcomings. This chapter looks at possible solutions in the form of Three Dimensional (3D) Graphical User Interfaces (GUIs) combined with multi-user and synchronous communication techniques. A number of systems, which use these technologies within other domains, are examined in Section 2, before we discuss the use of such technologies in the e-learning situation. In Section 3, we examine Collaborative Learning Environments with Virtual Reality (CLEV-R), an e-learning interface that we have developed, for use by university students, to resolve the shortcomings of LMSs. Unlike similar systems, CLEV-R actively promotes social interaction through the inclusion of specialised social facilities. Prior to our work, no large-scale evaluation of 3D GUIs within the e-learning domain had been carried out. We conducted a usability survey to assess if 3D interfaces are useable within the e-learning domain and gauged students' perception of them. The evaluation processes along with the results are presented in Section 4 of this chapter, before a discussion and conclusion in Section 5.

2 Related Work

This chapter proposes the use of onscreen 3D graphics as a solution to the issues with LMSs. Such techniques are in widespread use in other domains; both as stand alone computer programs and web-based applications. They are used in a variety of contexts and for several types of task where they engage and captivate the user. This section firstly explores the use of 3D interfaces to access and interact with underlying information, and then as the focus of this chapter is on e-learning applications, several 3D interfaces developed for training in specific tasks are examined. Our research interests lie specifically in the development of Collaborative Learning Environments (CLEs), which offer a general e-learning interface via a 3D environment. A review of the work in this area is presented before details of our own research and how it differs is discussed.

2.1 3D Interfaces for Information Access and E-Commerce

3D graphics are an established means of entertainment. This is evident from the succession of first-person 3D computer games where players take on the role of a character within virtual worlds and must navigate around them to solve a series of tasks in order to progress to the next stage of the game. The appeal of such games continues to grow as users find the 3D platform stimulating and engaging. Recently such games have begun to provide mechanisms for interacting with others and this especially appeals to players of such games [8]. The 3D paradigm has been extended to form GUIs in a number of diverse areas. In all cases the use of a 3D GUI provides an intuitive means of accessing underlying information and data.

For example, the authors of [9] describe their efforts at making a 3D library of antique books available to users. Similarly, the authors of [10] use the computer game engine from Quake II to model the library at the University of Karlsruhe in Germany. Other research focuses on using 3D GUIs to assist with administrative chores on a computer. The onscreen 3D window management systems described in [11, 12] use a 3D paradigm to allow computer users to keep track of the windows they currently have open on their computer screen. This is extended further to the 3D file management system developed by the authors of [13], who believe visual cues are import in recalling the location of files in 3D environments. Rather than the traditional folder-based archiving method, they use the metaphor of mountains and landscapes as a means for computer users to archive files.

In addition to using 3D GUIs for viewing and retrieving information, their use as a tool for e-commerce has also been examined. The authors of [14] developed an online store called HAVS (Human Animated Virtual Shop), which locates products based on users' searches and then places them in a 3D environment which mimics a shop. The authors of [15] have designed a virtual environment modelling a traditional supermarket where virtual products are placed on virtual shelves within a 3D space. As is the case in HAVs, users navigate through the store and select items they wish to purchase. It is argued that this is a more natural way for consumers to shop online as it is more familiar to shoppers than lists of available items [15]. E-Tailor is a VR online clothing boutique [16], where the consumers can browse the products available through a 3D interface. However, uniquely in E-Tailor, they can also visualise how the clothes would appear on them. This is achieved by providing measurements, which the system then uses to create a lifelike 3D mannequin. The clothes are then resized to fit the model so the shopper can appreciate how they may look in real life.

2.2 3D Interfaces and Simulators for Training

This paper is primarily concerned with the use of 3D for e-learning and there are many examples of its use within this area. The ability of 3D technologies to model the real world makes them an ideal tool to use in training simulations. The scientific, military and increasingly the medical communities all use simulators to train staff.

The military use simulations as a means of training recruits and as a medium for rehearsals for battle. The initial simulators were flight simulators, which provided a mock up of a plane's cockpit. Nowadays, other simulators replicate complete environments, examples include a virtual environment which recreates the interior of a naval vessel to serve as a training area for cadets [17]. An embodied agent within the environment, called STEVE, demonstrates the operation of the machines to cadets. The environment can also be used for team training by encouraging cadets to work together to solve tasks [18]. More recently simulations have been used to provide leadership skills to lieutenants. Mission Rehearsal Exercise (MRE) is a training system designed to help members of the army develop leadership and decision-making skills [19]. Agents, which appear as characters similar to STEVE, populate the 3D environment and can interact with the users of the system in a natural way [20].

The use of VR technologies also has a lot to offer the medical domain as a training tool. For example, computer based simulators can be used for training medical students. Simulators are particularly beneficial for training surgeons in Minimally Invasive Surgery (MIS) techniques such as laparoscopic procedures. The authors of [21] have created a computer-generated 3D anatomical model of a liver which a surgeon can interact with in real time using a screen and specialised haptic tools. The use of haptic tools gives a sensation of forced feedback to the surgeon identical to that in real surgery and adds to the realism of the simulation. More recently, this technique has been applied to intestinal surgery [22]. Again onscreen displays are combined with haptic tools to provide feedback to the surgeon.

Such simulations as those described above for military and medical training are ideal for exposing inexperienced students to situations they may encounter in the real world via the safe environment of the 3D interface. Research has also been carried out to develop training situations within virtual laboratories. The authors of [23, 24] describe the development of the Virtual Chemistry Laboratory which is an accurate 3D recreation of a physical chemistry laboratory. The facility allows undergraduate university students to become familiar with the layout of laboratories and the equipment provided within them. The implementation of a training simulator called VIRAD, which is used for training radio pharmacy students, is discussed in [25]. Like the Virtual Chemistry Laboratory, a simulation of a radio pharmacy is created using onscreen 3D graphics. Communication tools provide facilities to permit a number of radio pharmacists to collaborate and work together despite being at different geographical locations.

The use of such 3D interfaces and simulators for learning and training provides an excellent means for carrying out particular tasks in an interactive and realistic manner. However, recreating accurate 3D models of real-world situations is time consuming and often results in a model that can only be used within one specific training scenario. Simulators lack the ability to provide a general interface to e-learning and generally cannot be used for more than one learning activity. In the next section, several examples of interfaces which offer a general 3D environment and can be used for teaching different subjects are discussed.

2.3 3D Collaborative Learning Environments

3D CLEs can act as an interactive interface for accessing e-learning material. All such CLEs include multi-user tools to support interaction among learners and tutors. Various modes of real-time communication facilities are also provided to support synchronous learning and collaboration in a number of different learning scenarios. Several CLEs are described in this section before details of our research and how it differs is discussed.

The Virtual European Schools (VES) project uses 3D graphics to provide a desktop immersive environment for school children to interact with [26, 27]. The goal of the research is to encourage the use of computers in the classroom. The environment consists of a number of themed rooms; each one is tailored with learning content for a particular school subject. These themed rooms provide information about the subject in the form of slide shows, animations and links to external sources of information. The 3D environment within VES contains a number of features which allow users to interact with each other and collaborate. As with all the systems discussed in this section, more than one person can access the environment concurrently and users are aware of one another via their onscreen persona; in this case an animated 3D character. Users of this system can communicate with each other using text-chat services that are akin to the services offered by Yahoo! [28] and MSN Messenger [29].

VES is aimed at school children; however, the technologies which it utilises can also be used in other contexts. For example, they can be used to support group learning where it is not feasible for all participants to be present at the same physical location at the same time. The Intelligent Distributed Virtual Training Environment also known as INVITE is one such system [30]. INVITE offers a 3D multi-user environment which can be used for on-the-job training of employees. The multi-user aspects of the system set it apart from traditional video conferencing techniques, as it attempts to make people feel as if they are working as part of a group rather than alone in front of a computer. The project focuses on the importance of creating a sense of presence within a learning environment. Again, onscreen personas which are created using 3D characters are used to achieve this. The authors of [31] describe the technologies required for such a system. The design allows synchronous viewing of content within the 3D environment through a presentation table. Users can see pictures, presentations, 3D objects and pre-recorded videos simultaneously and collaboration is provided through application sharing.

Like INVITE, the Education Virtual Environments (EVE) system is a web-based, multi-user environment for training [32, 33]. The system addresses two main challenges. Firstly, the technological challenge to develop a learning environment that resembles the real world and that provides functionality to enhance the users' experience. Secondly, the pedagogical challenge of making an educational model that contributes in the most efficient way to distribute knowledge. EVE has 2 modes of interaction, a personal desk space and a collaborative training area. The personal desk space is an asynchronous area where a student can use a 2D interface to access and review course material, view and compose personal messages and manage their profile. The training area is a 3D environment and resembles a classroom. This area is used for synchronous learning; it contains a presentation table and a whiteboard. Features such as application sharing, brainstorming and text and audio communication are supported to allow students to work together. As in other cases, each user within the system is represented on the computer screen by an avatar.

Second Life [34] and Active Worlds [35] are web-based VR environments in which communities of users can socialise, chat and build 3D VR environments in a vast virtual space. Thousands of unique worlds for shopping, chatting and playing games have been created. The technology of Active Worlds has been made available for use by educational institutions to develop collaborative 3D environments for learning [36]. Within this online community known as Active Worlds Educational Universe (AWEDU), educators can build their own space using a library of customisable objects and can then place relevant learning material in it. Through these environments students are able to explore new concepts and can communicate using text-chat. As in Active Worlds, users are represented in the onscreen environment by characters that help them feel immersed in the educational space. The AWEDU environment is extremely versatile and may be used for a number of types of learning. For example, the authors of [37] present its use as a form of distance education within a university. A description of a class held within Appedtec, an Active Worlds environment designed for training teachers on the integration of technology into the classroom is provided in [38]. Further examples of a number of cyber campuses, which have been developed in AWEDU, can be found in [39]. There are also many examples of the use of Second Life as a tool for collaborative e-learning. For example, Second Life has been used to create a science museum, in which users can learn about the planets in a virtual planetarium. Animations have been added to the environment to increase the users' understanding of events [40]. Other research focuses on examining techniques to incorporate Second Life with existing LMSs such as Moodle [41], where the 3D interface of Second Life can be used to provide and interactive means of accessing the information maintained within the LMS.

The e-learning platforms described in this section address the needs of students by providing an interactive medium for accessing learning content. As shown in Table 2, they address the issues with existing LMSs to varying extents. One aspect which is lacking in many of these systems is the provision of dedicated tools for socialising online. In particular, the file sharing functionality, which is one of the key elements of social networking websites, is restricted for use by tutors in many of these systems. This lack of social interaction with others comes amid the many research studies which argue for the inclusion and need for such functionality within the e-learning domain [7].

Table 2. Comparison of features of current CLEs and CLEV-R

	VES	INVITE	EVE	AWEDU	Second Life	CLEV-R
Multi-user	X	X	X	X	X	X
Avatars	X	X	X	X	X	X
Text Communication	X	X	X	X	X	X
Voice Communication		X	X			X
Web-cam Feeds						X
Tutor File Uploading		X	X	X	X	X
Student File Uploading		X	X			X
Defined Social Areas				X	X	X
Tutor-Led Activities	X	X	X	X	X	X
Student-Led Activities		X				X

While social facilities are provided in Second Life and AWEDU, they are predominantly social environments which have been adapted for e-learning. Therefore they are not dedicated e-learning systems and instead use ad-hoc methods for delivering learning content. The quality of the learning tools thus depends on the creative abilities of the person building the course. Furthermore, VES, Second Life and AWEDU do not ordinarily provide voice communication between students and so interaction within these systems is limited. In our system, CLEV-R, the need for social tools is not neglected and instead forms a major element of the design. This is achieved through the addition of specialised tools which encourage and foster social interaction. While these facilities allow natural communication between students within the 3D environment, further dedicated functionality also permits students to share and discuss photos and videos with each other. As highlighted in Table 2, student-led collaboration activities can take place which allow students to share files and participate in group work. Real-time voice and webcam communication is also supported through the 3D environment of CLEV-R which further enhances the learners' experience. The next section provides a description of CLEV-R.

3 CLEV-R

Similar to some of the CLEs described above, our own system, CLEV-R, uses virtual reality, multimedia and instant communication methods to present an online learning experience to users. CLEV-R provides a multi-user 3D environment which multiple users can access simultaneously. The 3D environment mimics a real university and is divided into a number of different areas including lecture rooms, meeting rooms and social areas [42]. In order to create a feeling of presence and an awareness of others, each user is represented within the 3D environment by a humanoid avatar. Interaction with both the environment and other users is achieved using a number of specialised features provided via the interface. A number of problems with existing LMSs were highlighted in Section 1, and while the CLEs described in Section 2 offer a solution to several of these issues, they neglect to include services for social interaction among students. Due to the importance of interaction and especially social interaction, CLEV-R offers dedicated facilities to support informal discussion and media sharing among students in order to create a sense of community.

CLEV-R is a web-based application, accessible via an Internet Browser. When a user access CLEV-R, they are presented with a web page which, as seen in Figure 1, is split into two distinct sections. The upper area displays the 3D environment while the lower section consists of a 2D GUI. The multi-user 3D environment supports synchronous learning and interaction between students. This interaction is further enhanced by the 2D GUI which provides accesses to the real-time communication facilities offered by CLEV-R. The following sections describe CLEV-R and in particular the features it provides for learning, collaborating and socialising online.

3.1 3D Interface

The 3D interface provides an onscreen environment which mimics a university setting and so contains the facilities offered in a real-world university. For example, a lecture room, a library, meeting rooms and social areas are all provided. Each room and area contains appropriate features to support its particular function.

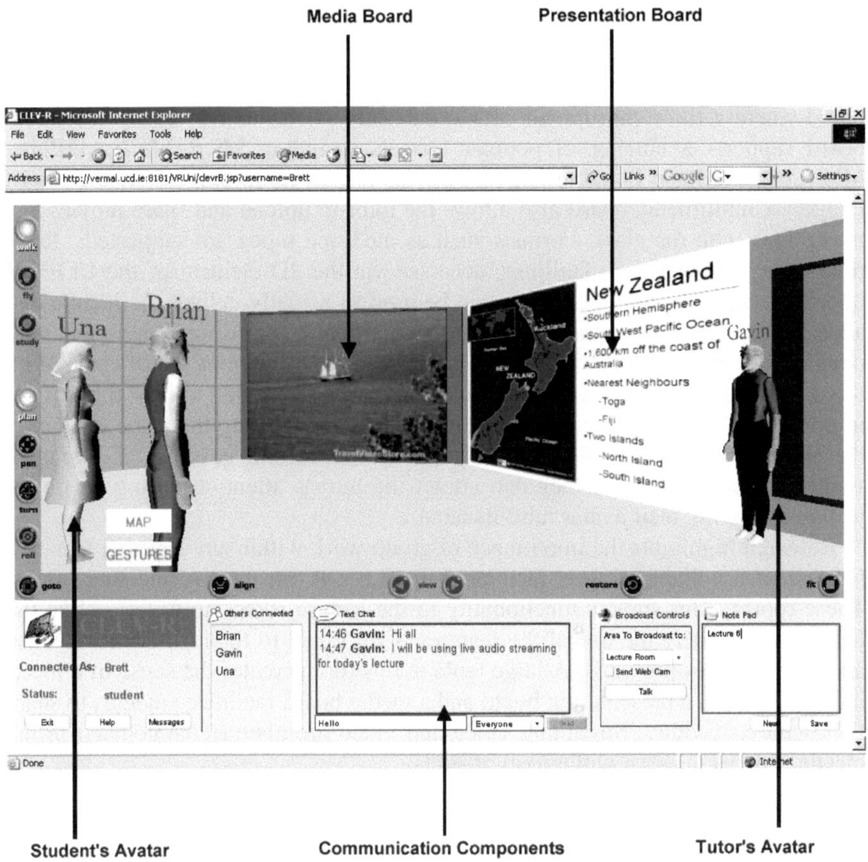

Fig. 1. Students Partaking in a Synchronous Lecture in CLEV-R

The users of the system, both tutors and students, are represented in the 3D environment by a character, selected when they first access CLEV-R. Each character is unique with different styles and colours of clothing and hair. Once a user is logged into CLEV-R their character, including their movements, become visible to all other users of the system. These avatars play an important role in creating an awareness of others. In addition to displaying a walking action while moving, the avatars are equipped with further gestures such as raising their hand. While these features serve as a useful function, they also add to the sense of presence which users experience. The functionality of the 3D environment can be broken down into the facilities concerned with learning and those which facilitate social interaction.

3.2 Delivering Learning Content

All learning content is presented to users through the 3D university environment. CLEV-R caters for several different learning scenarios. Firstly, the traditional lecture style of learning is supported; a tutor provides course material and delivers it to the class in the lecture room of the 3D environment. During live online lectures the students and a tutor are logged into the system simultaneously and congregate in the lecture room (Figure 1). The lecture room contains a presentation board where a tutor can upload course notes for others to see. The presentation board supports a number of popular formats including Microsoft PowerPoint, Microsoft Word and Portable Digital Format, as well as several different types of image. A multimedia board also allows the tutor to upload and share movies and music files with the class. Formats such as mp3 and mpeg are supported. Real-time audio and web-cam facilities, accessed via the 2D element of the CLEV-R interface, which is discussed later, can be used to actually deliver the lecture and address the class. Supplementary to the media board, the lecture room also contains a video board which can display a live feed from the tutor's web-cam. With this functionality a tutor can deliver a lecture in a similar way to how they would in a real-world situation with all students seeing and hearing the same content at the same time. As each avatar is equipped with certain gestures, if a student wishes to ask a question, they can attract the tutor's attention during an online lecture by having their avatar raise its hand.

Research highlights the importance of group work within any learning situation [6] and such collaboration is facilitated in CLEV-R via the two meeting rooms. These rooms offer similar functionality to the lecture room, however, unlike the lecture room where the use of the features is restricted to tutors, students can use all of the features provided. A large table in this room creates the sense of a meeting space. Both a presentation board and a media board facilitate students to share files with each other. Streaming voice and video are also accommodated in the meeting rooms to assist with group projects.

Many of the current LMSs offer facilities for individual, self-paced learning and CLEV-R also supports this, however the methods for accessing learning material via CLEV-R are more interactive and motivating for the students than via the text-based LMSs. Individual learning is supported through a virtual library feature of the 3D environment. Within the library, a bookcase contains a directory of all the lecture material uploaded by the tutor. When a student selects a book, the corresponding notes are displayed on a desk in the library where the student can study them or download them to their own computer. Other services such as links to external knowledge sources including online dictionaries and encyclopaedias are also provided in order to assist students with their studies.

3.3 Interactive Socialising Facilities

The inclusion of facilities to support social interaction among students is paramount within the e-learning situation [7]. It is this lack of interaction with others which leads to feelings of isolation and loneliness [5]. In the 3D environment of CLEV-R social interaction is supported in a number of ways. Firstly all of the rooms in the 3D environment are arranged around an open plan common area. Students can use this facility before and after learning activities to interact with each other.

Voice communication is supported so students can talk directly to each other in this informal setting. The common area is supplemented with a coffee area; here students can congregate around some coffee tables and use the voice communication to chat with each other. Unlike the common area, which is a general forum for discussions, only those present at a particular coffee table can partake in the conversation therefore making it more private.

The social networking websites which have become extremely popular offer users the ability to share files, generally media files, with each other. In order to facilitate this form of interaction within CLEV-R, dedicated social rooms have been developed. As in all locations of the 3D environment, students can use the voice and text communication tools in these rooms. In addition, a presentation board, and media board, similar to those in the lecture room, allow students to share multimedia files. The media board also facilitates the display of video content from the YouTube [43] website. As with all content displayed in the 3D environment, students can view it simultaneously and are aware of the presence of others.

3.4 General Features

The goal of the CLEV-R Interface is to support learning and socialising while also addressing the shortcomings of existing e-learning systems. In order to achieve this it is also necessary to have a number of support features to assist users while they are interacting with CLEV-R. In order to aid navigation, an onscreen interactive map is provided; this map provides a plan of the layout of the environment so users can locate different areas. The map also tracks other users and makes locating them simple. Each room within the 3D environment is equipped with interactive audio assistance. When a user requires help, a pre-recorded audio file can be played which details the function of the particular room and how to use the tools which it contains. A further feature, which supports students in their studies, is the provision of an announcement space within the common area of the virtual university. Students and tutors can place announcements in this area for others to see.

3.5 2D Graphical User Interface

The 2D element of the GUI, shown in Figure 2 supports communication within CLEV-R. The GUI is divided into a number of sub-sections. The user's own name and status are displayed in the first section while a list of all others currently connected to the system is also displayed. The avatars within the 3D environment give an indication as to who is in your immediate vicinity; however the list of other connected users on the 2D GUI lists everyone currently using the system. A text-chat component allows users to exchange text messages with all other users. Alternatively users can select an individual to converse with via private text messages. When a new message is received, a box surrounding the text-chat component flashes several times to alert the user to the presence of the new message. The GUI also hosts the controls for the audio and web-cam features, which allow users to broadcast their voice and web-cam images directly into the 3D environment. As seen in Figure 2, a dropdown menu allows users to select a particular location in the 3D environment, they then simply press and hold the 'talk' button to broadcast

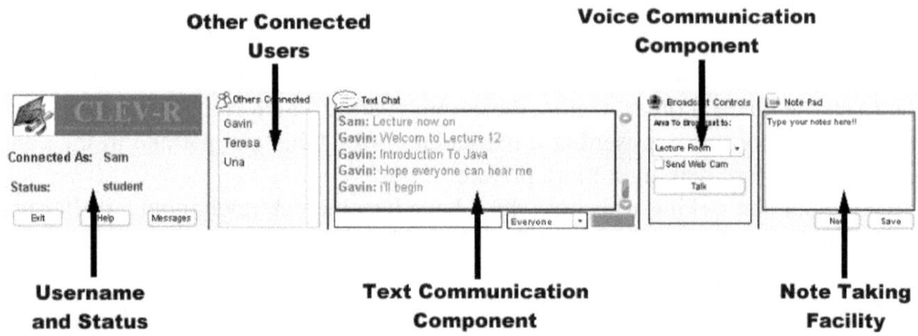

Fig. 2. The 2D Element of the CLEV-R Graphical User Interface

their voice into that area. Any users in that location of the 3D environment will automatically hear the broadcast. Web-cam images are shared in the same fashion. The GUI also includes an area where students can take notes, save them and access them at a later date via the library in the 3D environment.

4 Evaluation

Two usability studies were carried out using CLEV-R. The first, details of which can be found in [44], was carried out as implementation of the system neared completion. Its principle purpose was to determine any usability issues with the functionality of CLEV-R. Several technical issues were uncovered during this trial; however they were resolved prior to the commencement of a further, and larger, usability test involving CLEV-R, which is described below. Usability testing involves studying the interface and its performance under real-world conditions and gathering feedback from both the system itself and the users [45]. In this section we describe the sample of users who took part in the trial, the approach that we adopted to evaluate CLEV-R and the results from a series of standard usability questionnaires administered after the user trial. The user trial served two functions, firstly, to determine any usability issues with the CLEV-R interface and ensure that the 3D paradigm creates a useable GUI for use within e-learning, secondly, to gauge the reactions of users towards this new interface and particularly the social elements of CLEV-R. Details of the evaluation along with the results are presented in the next sections.

4.1 Test-Subjects

Twenty volunteers took part in the user trial. This number has been shown to be more than sufficient for evaluating the usability of systems with the questionnaires we chose to administer [46]. User profiling determined the sample consisted of 16 postgraduate students from varied disciplines, 1 undergraduate business studies student, 1 teacher and 3 employed professionals. 15 were male and 5 were female with an average age of 26.27. CLEV-R is designed for use with university students and so this sample is a good representation of our target users. Each test-subject used computers on a regular basis for a variety of activities including email, word processing and browsing the Internet. 60% of the male subjects had played first-person computer games in the past. From the sample, 12 of the volunteers had used other e-learning systems in the past and were satisfied with their experience. All participants took on the role of students during the user trial.

4.2 Approach

Four tasks, involving Social, Learning and Collaborating activities, with scenarios to simulate the use of CLEV-R in the real world were devised. The pretence of a geography course was used to provide sample course material for use within CLEV-R. Each of the tasks is outlined in Table 3 and more detailed descriptions of the four tasks are provided below.

Task 1 - Social Interaction

The first task consisted of a standard ice-breaking game, employed to make people feel comfortable interacting with others. The technique is extensively used within the area of business-training and has also been more recently extended to the online domain [47]. The classic guessing game: 'Who am I?' was chosen for this task. The game involves one participant selecting a notable person from history and the other participants must ask a series of question with yes or no answers in order to discover the identity of the historic figure. The task involves social interaction and also a certain amount of collaboration. This task requires the use of the audio and text communication tools available via the 2D interface. In this task the participants were also required to gather in a Meeting Room within the 3D environment and select a country from a list displayed on the presentation board in that room. The purpose of this element of the task was so that each participant could complete a short project on this country and present their findings to the other test-subjects during task 3.

Table 3. An Outline of the Tasks Involved in the User Trial

Task No.	Type of Task	Location	Task
1	Social Interaction	Social Room	Partake in Ice-Breaking Game
2	Online Learning	Lecture Room	Attend Synchronous Lecture
3	Collaboration	Meeting Room	Present a Project to the others
4	Social Interaction	Social Room	Share Media with Each Other

Task 2 - Online Learning

The second task involved the students attending a lecture in the Lecture Room. Content for the lecture consisted of facts regarding different countries. Interactivity was introduced through the use of multimedia material including Microsoft PowerPoint slides, movies and music files. The lecturer, a person with prior experience of using CLEV-R, also used the audio communication and web-cam features and encouraged participation from the test-subjects. Following the lecture, the participants had to locate the Library, find the appropriate set of notes for the lecture they had just attended and save them to their computer.

Task 3 - Collaboration

This task involved the students presenting the findings from the project which they chose during the first task. Each participant presented their work to the other students. This involved uploading a Microsoft PowerPoint file to the presentation board in one of the Meeting Rooms and using the audio communication facilities to talk about their element of the project. The other members of the group were then encouraged to ask questions, again using the audio communication.

Task 4 - Social Interaction

Task 4 was another social task. There were no particular instructions for the participants during this task and instead they were encouraged to interact and socialise with each other. This task highlighted the social tools within CLEV-R. Many of the participants shared videos and photos with each other in the social areas. The task gave the participants free reign with the system to explore any usability issues which might arise and provided an opportunity to see how students might use the system in a real-world situation.

During the evaluation, the entire sample of volunteers did not take part in the trial simultaneously; instead, the evaluation was conducted four times with five test-subjects and a lecturer taking part on each occasion. The evaluation was conducted in parallel to a user study involving mCLEV-R (mobile CLEV-R), which is a lightweight accompaniment to CLEV-R for use on mobile devices. A number of the users were required to carry out the tasks on both the mobile and desktop versions of the system in order to ascertain the effectiveness of mCLEV-R. Details of mCLEV-R and its evaluation can be found in [48].

4.3 Evaluation Questionnaires

Questionnaires were presented to the test-subjects once they completed all of the 4 tasks. The Computer System Usability Questionnaire (CSUQ) [49] was the main instrument used to obtain feedback. This is a standard usability questionnaire that assesses the appeal and usability of the interface. It consists of 19 questions to determine the overall level of user satisfaction with a computer system. Each question is a statement and the respondent replies by selecting a value on a 7-point likert-scale anchored at Strongly Agree (1) and Strongly Disagree (7). The questionnaire has shown to be a reliable measure of overall satisfaction with an interface with the scale having a coefficient alpha exceeding .89, indicating acceptable scale reliability. While the questionnaire can be used as an overall measure of satisfaction, it can also be subdivided into 3 sub-scales; *System Usefulness* (to assess the participants' overall satisfaction level towards the usefulness of the system), *Information Quality* (to assess the participants' overall satisfaction towards the level of help and support provided with the system) and *Interface Quality* (to determine the participants' overall satisfaction level towards the quality of the interface and the tools it contains).

A further questionnaire was administered in conjunction with the CSUQ. Seventeen questions, taken from a number of standard questionnaires relating to participants sense of presence in virtual environments [50, 51, 52, 53, 54], were combined with a series of 13 questions specifically related to CLEV-R. These questions gauged the sense of awareness, the sense of presence and the level social interaction experienced by the test-subjects. The feedback obtained from this questionnaire and the CSUQ are presented in the next section.

4.4 Results and Discussion

The three internal subscales of the CSUQ, referring to *System Usefulness*, *Information Quality* and *Interface Quality* are examined in the following sections before details of the results pertaining to the social awareness and presence elements of the questionnaire are presented.

System Usefulness

The first eight items on the CSUQ assess *System Usefulness*. The results are shown in Table 4. The overall trend regarding the usefulness of the system is a positive one; all results are positioned at the positive end of the 7-point axis, i.e. nearer 1. All of the results can be condensed to give an average score of 2.18 (on the 7-point likert scale). This indicates a high level of satisfaction among the subjects regarding the usefulness of CLEV-R. The ease of use was rated highly among the participants with the majority of them perceiving CLEV-R as simple to use. The study also revealed that the majority of participants found it was easy to learn how to use CLEV-R and felt they could become productive quickly while using the system. Despite many of the test-subjects being considered novice users with limited knowledge of operating in an onscreen 3D environment, 90% of them felt comfortable while using CLEV-R.

Table 4. Results for items 1 – 8 on the CSUQ referring to System Usefulness

Question	Average Response	Question	Average Response
Overall I am satisfied with how easy it is to use this system.	1.9	I am able to efficiently complete my work quickly using this system.	2.85
It is simple to use this system.	2.1	I feel comfortable using this system.	2
I can effectively complete my work using this system.	2.1	It is easy to learn to use this system.	2
I am able to complete my work using this system.	2.55	I believe I became productive quickly using this system.	2.15

Information Quality (Level of Help Support Provided)

Items 9 - 16 of the CSUQ can be used as a means of assessing the participants' satisfaction with the quality of the information associated with the system. In the case of CLEV-R it is important to recognise that we are not assessing the actual learning material and so *Information Quality* refers to the quality of help files and feedback provided within the system. When combined, the results for the seven questions of this element of the questionnaire give an average score of 2.60. This is a good response indicating an overall high level of satisfaction with this element of CLEV-R. The system scored well in relation to the information provided and how it is organised, with 85% of test-subjects agreeing the information provided is easy to understand, however, handling errors and recovering from mistakes are aspects which received the most negative responses from participants. For example, 85% of those that responded were indifferent or did not agree that CLEV-R gives error messages that clearly state how to fix a problem. While providing suitable support structures for users is an important aspect of developing a computer system, the focus of CLEV-R is to provide a 3D interface for e-learning. Developing a more sophisticated structure for offering assistance would enhance CLEV-R and potentially improve the feedback relating to the Information Quality of the system.

Interface Quality

The third metric obtained from the CSUQ provides a score for the Interface Quality. Three questions are used to give an average *Interface Quality* score of 2.02 with a standard deviation of 0.73. This is an excellent score and shows that participants were impressed with the interface. There were no negative answers returned for any of the items in this section of the questionnaire. All of the participants agreed that they liked using the interface and 90% found it pleasant to use. CLEV-R uses 3D interfaces, which is a new paradigm for e-learning and so assessing the quality of the interface is particularly important. The results obtained are encouraging. *Interface Quality* scored the highest from the three individual metrics in the CSUQ. The results indicate that CLEV-R offers a high quality interface that is supported by the functionality expected within an e-learning application.

Overall Score for Computer System Usability Questionnaire

The results obtained from the CSUQ can be condensed to give a final metric for an overall user satisfaction score. This is achieved by using the results from the above 3 metrics and combining them with an additional question regarding overall satisfaction. The mean of all the scores returned for the complete questionnaire is 2.21 (on the 7-point likert scale) with a standard deviation of 0.85. On the basis of this response, we can conclude that the participants found using CLEV-R to be an overall satisfying experience and the interfaces of CLEV-R are useable in an e-learning situation.

Social Awareness and Presence

A second questionnaire, consisting of a series of questions to gauge the participants' reaction to their sense of awareness and presence in the 3D environment

Table 5. Results Relating to the Sense of Presence and Engagement with CLEV-R

Question	Average Response	Question	Average Response
I was aware of the actions of other participants.	2.25	I experienced a strong sense of collaboration with other participants in this environment.	2.05
I could easily recognises other people in the 3D environment.	2.25	The presence of others in this environment engaged me in my learning experience.	2.15
I felt part of a group.	1.75	I had a strong sense of belonging.	2.65

was also administered after the user trial. The same 7-point likert scale that was used in the CSUQ was used for this questionnaire. Details of the test-subjects' average responses in relation to their awareness of others are shown in Table 5.

The responses are all skewed towards the positive end of the 7-point scale. 85% of the participants agreed that they were immediately aware of the presence of others in the 3D environment and 90% could easily recognise others, while almost all participants were aware of others' actions. Importantly, 85% of test-subjects agreed the presence of others engaged them in the learning experience, while 75% felt the presence of others actually enhanced their learning experience. The awareness of others contributed to 75% of the participants agreeing that they had a sense of belonging during the user trial and 80% of test-subjects agreed that they felt part of a group. Despite being in physically different locations, almost all of the respondents felt as if they were in the same room as the other participants. The results indicate that 95% of the test-subjects had a strong sense of collaboration with others in the 3D environment which is evidence that the facilities provided are an effective means of collaborating with others.

We were also interested in receiving feedback from the participants on their overall impressions of CLEV-R and their experience of using the 3D interface for e-learning. The most important results are presented in Table 6. The table shows that enjoyment levels during the trial were high, with none of the participants expressing dissatisfaction. These results are echoed by the fact that all test-subjects had their interest maintained in the virtual environment with an average response score of 1.8 on the 7-point scale. The sample consisted of twelve people who had prior experience of traditional e-learning systems. A lack of motivation is also cited as a failing of conventional e-learning systems, however, as the results indicate, this is not a factor in CLEV-R with 90% of the test-subjects agreeing that the learning material is presented in a motivating way.

Table 6. Results Relating to Enjoyment Levels while Using CLEV-R

Question	Average Response	Question	Average Response
Overall enjoyment level in this environment was high.	1.65	The material was presented in a motivating way.	2.45
I felt comfortable using the environment.	2.1	The virtual environment maintained my interest.	1.8

5 Conclusion and Discussion

Text Based LMSs are in widespread use within universities, where they are used to deliver learning material to students. They are versatile and can be used as a standalone distance-learning tool or integrated with face-to-face teaching. Despite their popularity, there is evidence which indicates that courses, which rely solely on LMSs, have higher drop out rates than their classroom based equivalents. The failure of students to complete such online courses is attributed to several factors including lack of stimulation caused by unattractive and unappealing user interfaces as well an absence of real-time interaction with tutors and other students. 3D interfaces offer a mechanism for interacting with underlying data in a stimulating way. Such interfaces are in use in various domains where complete onscreen 3D environments have been generated to act as interfaces for computer games, information retrieval and e-commerce.

While onscreen 3D simulators have proved themselves as an effective training tool, their use is often restricted to one particular training scenario and limited to a single context of use. 3D CLEs offer a solution to the issues with LMSs. They provide an interactive interface which can be used for various training and learning scenarios. Multi-user tools allow students to interact with each other while a range of facilities offer support for synchronous communication between users, for both learning and collaborating. A number of such CLEs have been developed and while they offer excellent support for learning many neglect the students' need for social interaction. This is not the case with CLEV-R, a system that we have developed. CLEV-R is a 3D environment which mimics a university and contains specialised areas which are dedicated to help students to interact informally with each other. Prior to our research, no major evaluation study concerning the usability of such environments has been conducted. We devised a scenario based field study which involved students interacting with CLEV-R under simulated real-world conditions. Feedback was then obtained from the test-subjects in the form of questionnaires. The main focus of the user trial was to establish the usability of the system; however we were also particularly interested in feedback from students regarding the social aspects of the system, since these are unique to CLEV-R.

Overall, the results, presented in Section 4, show that the test-subjects saw the system as usable. Furthermore the students involved in the study believed they could become productive quickly using CLEV-R and also that it was easy to learn how to use the interface. These results are significant because they show that 3D interfaces, which are a new paradigm within e-learning, are useable by students and something which they find acceptable and feel comfortable interacting with. The actual interface was also rated highly by the test-subjects who found it pleasant and all enjoyed using it.

In addition to questioning the participants regarding the usability of CLEV-R, feedback about the social elements of CLEV-R was also obtained. One of the key attributes of this type of multi-user environment is the sense of both presence and social awareness, which it can create, along with the ability to collaborate with others. These characteristics are often absent from traditional e-learning systems and so we wanted to measure to what extent they are delivered through CLEV-R. The results imply that the use of avatars within CLEV-R is sufficient to create a sense of awareness between the users and allow them to recognise each other easily within the 3D environment. The test-subjects felt the presence of others within the environment engaged them and this is evidence that students' awareness of others is important in the e-learning domain and further highlights the shortcomings of existing e-learning platforms, which fail to offer this level of awareness and interaction. These responses indicate that the features provided in CLEV-R are suitable for interacting with others and creating a sense of community. Furthermore, when users think back to the environment, they see it more 'as somewhere they visited rather than images they saw', indicating that the environment engaged them.

CLEV-R is a prototype which serves as a proof-of-concept; however, the results from the user trial indicate that the 3D paradigm has something to offer the e-learning domain. The ability of 3D CLEs to address the shortcomings of existing LMSs while also allowing them to be incorporated with social tools makes them a powerful medium for learning.

Acknowledgments

Research presented in this paper was funded by a Strategic Research Cluster grant (07/SRC/I1168) by Science Foundation Ireland under the National Development Plan. The authors gratefully acknowledge this support.

References

1. Paulsen, M.: Experiences with Learning Management Systems in 113 European Institutions. Educational Technology & Society 6(4), 134–148 (2003)
2. Owston, R.: A Meta-Evaluation of Six Case Studies of Web-based Learning. Presented at the Annual Meeting of the American Educational Research Association (AERA), New Orleans, Louisiana, USA (2000)
3. Serwatka, J.: Improving Retention in Distance Learning Classes. International Journal of Instructional Technology and Distance Learning 2(1), 59–64 (2005)
4. Anaraki, F.: Developing an Effective and Efficient eLearning Platform. International Journal of The Computer, the Internet and Management 12(2), 57–63 (2004)

5. Kamel Boulos, M.N., Taylor, A.D., Breton, A.: A Synchronous Communication Experiment within an Online Distance Learning Program: A Case Study. Telemedicine Journal and e-Health 11(5), 583–593 (2005)
6. Kitchen, D., McDougall, D.: Collaborative Learning on the Internet. Educational Technology Systems 27, 245–258 (1998)
7. Laister, J., Kober, S.: Social Aspects of Collaborative Learning in Virtual Learning Environments. In: Proceedings of the Networked Learning Conference, Sheffield, UK, March 26-28 (2002)
8. Gee, J.: What Video Games have to Teach us about Learning and Literacy. Computers in Entertainment (CIE) 1(1) (2003)
9. Cubaud, P., Thiria, C., Topol, A.: Experimenting a 3D Interface for the Access to a Digital Library. In: DL 1998: Proceedings of the 3rd ACM Conference on Digital Libraries, Pittsburgh, Pennsylvania, USA, June 23-26, pp. 281–382. ACM Press, New York (1998)
10. Christoffel, M., Schmitt, B.: Accessing Libraries as Easy as a Game. In: Börner, K., Chen, C. (eds.) Visual Interfaces to Digital Libraries. LNCS, vol. 2539, pp. 25–38. Springer, Heidelberg (2002)
11. Robertson, G., van Dantzich, M., Robbins, D., Czerwinski, M., Hinckley, K., Risden, K., Thiel, D., Gorokhovsky, V.: The Task Gallery: A 3D Window Manager. In: CHI 2000: Proceedings of the SIGCHI Conference on Human Factors in Computing Systems, The Hague, The Netherlands, April 1-6, pp. 494–501 (2000)
12. Cockburn, A., McKenzie, B.: Evaluating the Effectiveness of Spatial Memory in 2D and 3D Physical and Virtual Environments. In: CHI 2002: Proceedings of the SIGCHI Conference on Human Factors in Computing Systems, Minneapolis, Minnesota, USA, April 20-25, pp. 203–210 (2002)
13. Altom, T., Buher, M., Downey, M., Faiola, A.: Using 3D Landscapes to Navigate File Systems: The Mountain View Interface. In: Proceedings of the 8th International Conference on Information Visualisation, London, UK, July 14-16, pp. 645–649. IEEE, Los Alamitos (2004)
14. Sanna, A., Zunino, C., Lamberti, F.: HAVS: A Human Animated VRML-Based Virtual Shop for E-Commerce. In: Proceedings of theWorld Multi-Conference on Systemics, Cybernetics and Informatics (SCI 2002), Orlando, Florida, USA, July 15-19, 2002, pp. 24–29 (2002)
15. Chittaro, L., Ranon, R.: New Directions for the Design of Virtual Reality Interfaces to E-Commerce Sites. In: Proceedings of the 5th International Conference on Advanced Visual Interfaces, Trento, Italy, May 22-24, 2002, pp. 308–315 (2002)
16. Cordier, F., Seo, H., Magnenat-Thalmann, N.: Made-To-Measure Technologies for an Online Clothing Store. IEEE Computer Graphics and Applications 23(1), 38–48 (2003)
17. Rickel, J., Johnson, W.: Intelligent Tutoring in Virtual Reality: A Preliminary Report. In: Proceedings of the 8th World Conference on Artificial Intelligence in Education, Kobe, Japan, August 19-22, 1997, pp. 294–301 (1997)
18. Rickel, J., Johnson, W.: Virtual Humans for Team Training in Virtual Reality. In: Proceedings of the 9th International Conference on Artificial Intelligence in Education', Le Mans, France, July 19-23, pp. 578–585 (1999)
19. Hill, R., Gratch, J., Marsella, S., Rickel, J., Swartout, W., Traum, D.: Virtual Humans in the Mission Rehearsal Exercise System. K"unstliche Intelligenz 4(3), 5–10 (2003)

20. Rickel, J., Marsella, S., Gratch, J., Hill, R., Traum, D., Swartout, W.: Toward a New Generation of Virtual Humans for Interactive Experiences. Intelligent Systems 17(4), 32–38 (2002)
21. Ayache, N., Cotin, S., Delingette, H., Clement, J., Marescaux, J., Nord, M.: Simulation of Endoscopic Surgery. Journal of Minimally Invasive Therapy and Allied Technologies (MITAT) 7(2), 71–77 (1998)
22. Raghupathi, L., Grisoni, L., Faure, F., Marchal, D., Cani, M., Chaillou, C.: An Intestinal Surgery Simulator: Real-Time Collision Processing and Visualization. IEEE Transactions on Visualization and Computer Graphics 10(6), 708–718 (2004)
23. Dalgarno, B.: The Potential of 3D Virtual Learning Environments: A Constructivist Analysis. Electronic Journal of Instructional Science and Technology 5, 90–95 (2002)
24. Dalgarno, B., Bishop, A., Bedgood Jr., D.: The Potential of Virtual Laboratories for Distance Education Science Teaching: Reflections from the Development and Evaluation of a Virtual Chemistry Laboratory. In: Proceedings of UniServe Science: Improving Learning Outcomes Symposium, Sydney, Australia, October 2-3, vol. 3 (2003)
25. Alexiou, A., Bouras, C., Giannaka, E., Kapoulas, V., Nani, M., Tsiatsos, T.: Using VR Technology to Support E-Learning: The 3D Virtual Radiopharmacy Laboratory. In: Proceedings of the 24th International Conference on Distributed Computing SystemsWorkshops, Hachioji, Tokyo, Japan, March 23-24, pp. 268–273 (2004)
26. Bouras, C., Fotakis, D., Kapoulas, V., Koubek, A., Mayer, H., Rehatschek, H.: Virtual European School-VES. In: Proceedings of the IEEE International Conference on Multimedia Computing and Systems, Florence, Italy, June 7-11, pp. 1055–1057 (1999)
27. Bouras, C., Philopoulos, A., Tsiatsos, T.: E-Learning through Distributed Virtual Environments. Journal of Network and Computer Applications 24(3), 175–199 (2001)
28. Yahoo Messenger, `http://www.messenger.yahoo.com` (accessed May, 2008)
29. MSN Messenger, `http://www.messenger.msn.com` (accessed, May 2008)
30. Bouras, C., Hornig, G., Triantafillou, V., Tsiatsos, T.: Architectures Supporting E-Learning through Collaborative Virtual Environments: The Case of INVITE. In: Proceedings of the IEEE International Conference on Advanced Learning Technologies, Adison,Wisconsin, USA, August 6-8, pp. 13–16 (2001)
31. Bouras, C., Triantafillou, V., Tsiatsos, T.: Aspects of Collaborative Learning Environment Using Distributed Virtual Environments. In: Proceedings of ED-MEDIA 2001 (World Conference on Educational Multimedia, Hypermedia & Telecommunications), Tampere, Finland, June 25-30, pp. 73–178 (2001)
32. Bouras, C., Giannaka, E., Tsiatsos, T.: Virtual Collaboration Spaces: The EVE Community. In: Proceedings of the 2003 Symposium on Applications and the Internet, Orlando, Florida, USA, January 27-31, pp. 48–55 (2003)
33. Bouras, C., Tsiatsos, T.: Educational Virtual Environments: Design Rationale and Architecture. Multimedia Tools and Applications 29(2), 153–173 (2006)
34. Harkin, J.: Get a (Second) Life. Financial Times (November 17, 2006)
35. Hudson-Smith, A.: 30 Days in Active Worlds: Community, Design and Terrorism in a Virtual World. In: The Social Life of Avatars: Presence and Interaction in Shared Virtual Environments, pp. 77–89. Springer, New York (2002)
36. Dickey, M.: Three-Dimensional Virtual Worlds and Distance Learning: Two Case Studies of ActiveWorlds as a Medium for Distance Education. British Journal of Educational Technology 36(3), 439–451 (2005)
37. Dickey, M.: 3D VirtualWorlds: An Emerging Technology for Traditional and Distance Learning. In: Proceedings of the Ohio Learning Network; The Convergence of Learning and Technology-Windows on the Future, Easton, Ohio, USA, March 3-4 (2003)

38. Riedl, R., Barrett, T., Rowe, J., Smith, R., Vinson, W.: Sequence Independent Structure in Distance Learning. In: Proceedings of the Conference on Computers and Learning, Coventry, UK, April 2-4 (2001)
39. Prasolova-Førland, E., Sourin, A., Sourina, O.: Cybercampuses: Design Issues and Future Directions. The Visual Computer 22(12), 1015–1028 (2006)
40. Doherty, P., Rothfarb, R.: Building an Interactive Science Museum in Second Life. In: Proceedings of the Second Life Education Workshop at the Second Life Community Convention, San Francisco, California, USA, August 18-20, pp. 19–24 (2006)
41. Kemp, J., Livingstone, D.: Putting a Second Life Metaverse Skin on Learning Management Systems. In: Proceedings of the Second Life Education Workshop at the Second Life Community Convention, San Francisco, California, USA, August 18-20, pp. 13–18 (2006)
42. Monahan, T., McArdle, G., Bertolotto, M.: Virtual Reality for Collaborative E-Learning. Journal of Computers and Education, Elsevier Science (2007)
43. Youtube, `http://www.youtube.com` (accessed, May 2008)
44. McArdle, G., Monahan, T.: M. Bertolotto. Interactive interfaces for Presenting Online Courses: An Evaluation study. In: Proceedings of the 16th European Conference on Information Systems (ECIS), Galway, Ireland, June 9-11, vol. 11 (2008)
45. Nielsen, J.: Usability Engineering. Morgan Kaufmann Publishers Inc., San Francisco (1993)
46. Tullis, T., Stetson, J.: A Comparison of Questionnaires for Assessing Website Usability. In: Proceedings of Usability Professionals Association Conference, Minneapolis, Minnesota, USA, June 7-11 (2004)
47. Clear, T., Daniels, M.: A Cyber-Icebreaker for an Effective Virtual Group? In: Proceedings of the 6th Annual Conference on Innovation and Technology in Computer Science Education, Canterbury, UK, June 25-27, pp. 121–124 (2001)
48. Monahan, T.: The Integration of Mobile Devices into Interactive and Collaborative Online Learning Environments. Ph.D. Thesis submitted to the National University of Ireland, University College Dublin (2007)
49. Lewis, J.: IBM Computer Usability Satisfaction Questionnaires: Psychometric Evaluation and Instructions for Use. International Journal of Human-Computer Interaction 7(1), 57–78 (1995)
50. Schroeder, R., Huxor, A., Smith, A.: Activeworlds: Geography and Social Interaction in Virtual Reality. Futures 33, 569–587 (2001)
51. Biocca, F., Harms, C., Gregg, J.: The Networked Minds Measure of Social Presence: Pilot Test of the Factor Structure and Concurrent Validity. In: Proceedings of the 4th Annual International Workshop on Presence, Philadelphia, Pennsylvania, USA, May 21-23 (2001)
52. Slater, M., Usoh, M., Steed, A.: Depth of Presence in Virtual Environments. Presence: Teleoperators and Virtual Environments 3(2), 130–144 (1994)
53. Witmer, B., Singer, M.: Measuring Presence in Virtual Environments - A Presence Questionnaire. Presence: Teleoperators and Virtual Environments 7(3), 225–240 (1998)
54. Gerhard, M., Moore, D., Hobbs, D.: Continuous Presence in Collaborative Virtual Environments: Towards a Hybrid Avatar-Agent Model for User Representation. In: Proceedings of the 3rd International Workshop on Intelligent Virtual Agents Madrid, Spain, September 10-11, pp. 137–155 (2001)

15 An Overview of Open Projects in Contemporary E-Learning: A Moodle Case Study

Eduard Mihailescu

Technical University "Gheorghe Asachi "of Iasi
Faculty of Electronics and Telecommunications
meduard@etc.tuiasi.ro

Abstract. The technical core of an e-leaning project is the LMS (Learning Management System) that is being used. This chapter reviews several e-learning platforms and discusses the importance of open source e-learning platforms and analyzes the total costs of implementation/educational output ratio. The result of the assessment shows that the open platform Moodle outperforms the majority of other platforms and it is used in a wide variety of e-learning projects at different academic levels, both college and university. Finally, we describe a Moodle LMS case study, the eCNDS (Computer Network and Distributed Systems E-Laboratory) at The Faculty of Electronics and Telecommunications from Technical University "Gheorghe Asachi" of Iasi.

1 Introduction

1.1 The Evolution Path for Web-Based Education

It is becoming increasingly common at high and higher education establishments to sustain e-learning activities. The range of use extends from simple deployment of electronic courses instead of hard-copies to more complex forms of education. Research around online learning and the use of educational technology is also gaining more attention. More than 40 academic journals specialising in these topics [1] are published on a regular basis.

Recent research [2] have issued a three-generation scheme for the web based teaching (WBT) and learning history, taking into account the endorsing tutorial visions and the resulting social patterns and outcomes. The source [2] analyzes content, communications and assessment for this classification. The first generation of e-learning is characterized as "closed environment content (the manuscript), communication is mostly face-to-face, discussion forums are used for posting, and assignments are made by quizzes and graded by the teacher [2]". The second generation of e-learning is described as "open learning environment content (cases to be selected), blended online and face-to-face communication, discussion forums are for agreeing on procedures and targets, assignments are blended own-made and teacher-made [2]". The third generation of e-learning

H.-N. Teodorescu, J. Watada, and L.C. Jain (Eds.): Intel. Sys. and Tech., SCI 217, pp. 271–281.
springerlink.com

occurs nowadays and is mainly described as "open learning environment (cases to be found), blended online and face-to-face communication, discussion forums are for exchanging and commenting standpoints, and assignments are quiz, peer review, points won through voting and from trainer [2]".

1.2 Open Source E-Learning Platforms at a Glance

The popularity of open source software has risen in recent years into the world of online learning. These platforms cover most of the tools common in electronic means education: learning management systems (LMS), course authoring tools, tools to create media elements such as animations, audio and video clips and browsers and players to present academic content.

Some of the important benefits provided by open platforms are: a) low initial costs, due to the fact that open source software is free to download. Nevertheless, the hardware you need to run the e-learning system is not free. We deal with significant hardware, maintenance, desktop management and backup costs of running el-earning software on traditional servers and workstations. Some further development is necessary before the product would be adopted and will also involve costs, but still it remains a non-expansive approach; b) flexibility and customizability: the GPL (General Public Licence) licensing provides opportunities for localized integration of the LMS with other systems, thus there is a large possibility that one can modify the software as needed to make it fit better; c) multi-platform capabilities: most of the open source applications run on multiple platforms including Windows and Linux; d) consistence to e-learning standards: interoperability is a high priority for many open source developers, so the adherence to standards (SCORM and others) is closely followed [3].

2 Related Work

2.1 Terminology

Several of the common concepts used in e-learning (e.g. learning object, learning resource, learning activity, learning unit etc.) are briefly described. We shall use the definitions provided by the IEEE LTSC (IEEE's Learning Technology Standards Committee) [4]. A Learning Object is "any entity, digital or non-digital, which can be used, re-used or referenced during technology supported learning" [4]. LOs are different types of computer files: text, audio, video, presentation, web page and others; "a Learning Service is an activity carried by the e-learning platform for instance the collaborative services, the communication services: email, instant messenger, chat rooms, audio-video conferences [4]" ; a Learning Resource "can be any learning object or service [4]"; a Learning Unit (LU) "is an abstract representation of a course, a lesson, a workshop, or any other formal or informal learning or teaching event [4]". Thus, a learning unit is clustering the pedagogical goals that can be accomplished or evaluated using the unit. Larger learning units can be aggregated from smaller learning units or other learning activities and related learning resources.

Using the definitions above, we conclude that the e-learning is an educational activity carried out by the learner or the teacher on an organized basis, which aims to achieve the curricula and syllabi layered in learning objects LOs.

2.2 LMS, Core Component of E-Learning System

As emphasized, the Learning Management System (LMS) is a key constituent in any e-learning system [5]. Sometimes, they are referred as Virtual Learning Environments (VLE). For integration and standardization purposes, some basic features that a LMS should comply to are described.

a) Administration: the LMS must enable administrators to manage user registrations and profiles, define roles, set curricula, chart certification paths, assign tutors, author courses, and manage content. Administrators need to have complete access to the training database, enabling them to create standard and customized reports on individual and group performance. Reports should be scalable to include the entire workforce. The system should also be able to build schedules for learners, instructors, and classrooms. Most important by all, these features should be manageable using user-friendly administration screens; b) content integration: the LMS must provide native support to a wide range of third-party courseware; c) Adherence to standards: the LMS has to comply to major e-learning standards, AICC (The Aviation Industry CBT Committee) [6] and SCORM. Support for standards means that the LMS can import and manage content and courseware that complies with standards regardless of the authoring system that produced it; d) Assessment: although this is not compulsory for a LMS platform, it is desirable to have an evaluation engine that enables authoring within the product and includes assessments as part of each course.

Further, the goal is to identify whether the open e-learning platforms are suitable or not for the wide usage in academic environment.

3 Assessment of LMS Open Platforms

According to the literature [7], 9 e-learning platforms cover most of open-source projects. They are presented in alphabetical order: aTutor [8], Claroline [9], Dokeos [10], dotLRN [11], ILIAS [12], LON-CAPA [13], Moodle [14], OpenUSS [15], Sakai [16] and Spaghettilearning [17]. Criteria for description are as follows: management of LOs (Learning Objects), delivered services, extended services, learning activities and learning unit/objectives. A brief description of the main characteristics for several of these platforms is provided.

3.1 aTutor [8]

The open platform aTutor [8] stands creation of SCORM compatible LOs. It provides services like chat, discussion forums. Also there are possibilities to extend the services with various other modules, such as collaborative Web conferencing environments, audio and/or video conferencing, blog, course administrative tools, time tables, bulletin boards, course handouts, grades, class lists, surveys and evaluations, events calendar. The learning activities embedded in the system are: self- directed work with learning materials, simulations, usage of multimedia presentations, tutorials, listening of audio or watching of video lecture, participation in discussions. Assessment options activities are confined to individual and collective testing. The learning unit is the course.

3.2 Claroline [9]

Claroline [9] does not support creation of SCORM compatible learning objects, but could import LO developed through 3-party software tools. Description of the learning objectives is passive and can be used in description of the course. The delivered services are chat, discussion forum, collaborative or individual assignments tools, events calendar and notice board. The assessment of the learning activities is confined at self-assessment level. The learning unit might be a course, a module or several learning or assessment activities.

3.3 Ilias [12]

This project supports, creates, imports and exports SQORM/AICC/IMS QTI compatible learning objects. The learning objectives stand an active manipulation. The content manager of this platform is able to define the learning objectives for the whole course. The learning objectives are freely issued free and are not related to certain taxonomies for LOs. The basic delivered services are e-mail, chat, discussion forum. The learning activities can be grouped in modules. The learning unit could be course, module etc. There is the possibility to stack one module in another module. The learning unit includes all the accessible system resources- reading files, audio or vide files, presentations, discussion forums, assignments etc.

3.4 Moodle [14]

Moodle is the acronym for Modular Object-Oriented Dynamic Learning Environment. It was first released several years ago by Martin Dougiamas, who developed the system, and Peter C. Taylor, who built the first web site running this LMS, both from The Curtin University of Technology, Perth, Australia [1]. Nowadays, Moodle is continually being under development by various groups of researchers, worldwide.

Moodle supports learning objects according to IMS QTT [18]. The system does not support creation of SCORM compatible learning objects, but these items might be imported in the learning material. The learning content can be presented in different formats: .pdf, .txt, .html, .doc, graphical files, flash movie, presentations, interactive simulations etc. Unfortunately, the learning objectives could not be associated with the learning unit, learning activity or the learning resource. The learning objectives are passively carried out only for information of the students. Still, there remains the possibility to manually classify these items in categories and subcategories related to the learning objectives, which are be relevant for some cases. The basic delivered services are: chat, discussion forums- text, audio or video (with the aid of additional modules), workshop for collaborative work, assignment, notice board, events calendar etc.. Learning units can be structured according the following hierarchy: course, module (theme), learning activities and resources.

3.5 Method of Evaluation

Methods used for software evaluation are a real concern and efforts have been carried out worldwide to develop standards and appropriate techniques of assessment.

The procedure of software evaluation is founded on several standards issued by a number of international authorities, among which we name the International Standardization Organization, ISO (www.iso.org). Some of the important standards related are ISO /IEC 14598 -5, ISO/IEC 9126 (NBR 13596) and ISO/IEC 12199 [22]. Their scope is to provide requirements and recommendations for the practical implementation of evaluation of software products, developed or in process of development, as a series of activities defined under common agreement between the customer and the evaluator. Basically, the evaluation process is carried out by simulating a normal operational behavior of the software, starting with the provided tutorials and manuals, installing the product as instructed in the documentation and using it in the most extensive way. During this process, the evaluators assign rates to the product according to the questions from a check-list. The rates or grades belong to pre-established, usually from 0 to 10. Meanwhile, the evaluator also has the obligation to record the time spent with the evaluation and emphasize the major features/flaws of the product. Finally, an Evaluation Report is issued, which should address the major positive aspects of the evaluated product as well of suggestion for its improvement.

According to the standards above, recent researches [7] have used as a Check List the QWS List (Qualitative Weight and Sum), a known method for software products assessment, highlighting the strengths and limitations of the open e-learning platforms. This approach relies on the usage of symbols, with 6 qualitative degrees of importance for the weights. After [7], these symbols are: "E = essential, * = extremely valuable, # = very valuable, + = valuable, | = marginally Valuable and 0 = not valuable". The weight of a criterion determines the array of values that can be used to establish a platform's productivity. For a symbol weighted #, for instance, the item can only be assessed #, +, |, or 0, but not *. This means that lower-weighted criteria cannot overpower higher-weighted criteria. To evaluate the results, the different symbols given to each product are counted [6]. Example results can be 2*, 3#, 3| or 1*, 6#, 1+. The product can now be ranked according to these numbers.

As a conclusion of this evaluation, Moodle has achieved, according to [7], high assessment figures. It is closely followed by ILIAS and Dokeos. aTutor, LON-CAPA, Spaghettilearning, and Open-USS are ranked equally at the fourth position, while whereas Sakai and dotLRN are ranked on the last position, due to the fact that they cover only basic features and functions.

Worldwide, there are over 745,000 courses using Moodle [1]. To emphasize its recognition, we would like to briefly account some of the numerous academic organizations that have been employing Moodle LMS in successful e-learning cases in last years: a) in Europe: The University of Glasgow, The Birmingham City University, The Open University UK, University of Kuopio Finland, Dublin

City University Ireland, Universitatea Tehnica din Brasov Romania and others; b) in US: University of Washington, University of Oakland, Missouri State University, University of Victoria, University of Minnesota, Idaho State University, Lewis University, Marywood University, Drew University, Aurora University, Humboldt State University, DePauw University, Lawrence University, Alaska Pacific University, The National Hispanic University from California and others; c) in other countries: University of Regina and University of Alberta Canada, The Chinese University of Hong Kong, The Monash University Australia and others.

4 A Comparison of Proprietary versus Open Source LMS

According to the literature [7], Moodle in on the edge of the open source wave and leads among non-proprietary learning management systems (LMS). Our assessment could not claim to be bias-free without a comparison between the two systems, open-source versus corporate. Blackboard has been selected for the second category, due to its time-proven reliability and wide spread in the academic world.

Recent research [19] shows representative outcomes from the parallel usage of both systems. The evaluation method follows the ISO /IEC 14598 -5, ISO/IEC 9126 (NBR 13596) and ISO/IEC 12199 [22] and uses adapted checklist and mixed teams of teacher/student evaluators. The checklist with student preferences used in a comparison between Moodle and Blackboard [19] shows the followings, from a total of 10 respondents: a) general navigation around the site: 20% prefer Blackboard, 40% prefer Moodle, and 40 % other options; b) forums, discussion boards and email: 20% prefer Blackboard, 50 % prefer Moodle, and 30% other options; c) accessing grades: 50% prefer Blackboard, 10 % prefer Moodle, 40 % other options. The overall assessment of Moodle versus Blackboard [19] shows that, from a total of 10 respondents, eight declared that they prefer Moodle and two declared that they prefer Blackboard. In conclusion, "the students seem to prefer Moodle to Blackboard [19] on most counts". Although they liked the grade book provided by Blackboard, "the final comments show that the students were acutely aware of the open source nature of Moodle, and approved. While Moodle is receiving their endorsement, "it may be what computing students see that makes open source as inherently "good [19]".

5 Moodle Case Study

5.1 Project Team Competences and Responsibilities

Some authors [20] consider that e-learning may be implemented in an organization using one of two major approaches in terms of strategy and time management. The former strategy requests professors and students to use the system by making hierarchical calls and is named Top-Down Strategy. The latter adapts the implementation to the field circumstances and urges professors and students to demand the platform on their own call, which is known as Adaptive Strategy [20].

In its last section, this chapter presents a case study carried at the Computer Networks and Distributed Systems Laboratory from The Faculty of Electronic and Telecommunication at The University "Gheorghe Asachi" of Iasi. This project is carried by a multi-disciplinary team consisting of the following positions:

- The Project Manager determines the technical requirements of the project, using object-oriented methodologies for detailed analysis. The project manager is also responsible for defining the product specifications;
- Several Analysts insure that the final design is in accordance to the specifications determined by the project manager his team;
- The Test Engineer plans and executes system tests and evaluations. The Test Engineer also tests different modules and the code generated during the project and participates in the technical review of the project;
- Several Programmers actually do the programming for the project and are also responsible for the installation and configuration Moodle LMS;
- The Curricula Manager is in charge with the educational payload and deploys the syllabi of the curricula in the Moodle Database after prior approvals from the head of the discipline.

5.2 Concerns That Base Our E-Learning Design

Installing Moodle allows us just to deploy academic resources on a LMS server and manage their content on an organized basis. Furthermore, a key role is to be accomplished by the project team tutors and academic councilors who have to decide the shape and the content of all the pedagogical activities supported by the Moodle environment: organizing and deploying the curricula and the syllabi, perform the student assessment and others.

Consistency in Tools and Layout
When we have designed our Moodle course interface, we attempted to be in the place of the users. Field literature shows that many students already have anxiety about their studies, and even more anxiety when attending web-based courses. In an online course, they often feel like they have to make sense of everything on their own. Thus, we tried to provide the students with an easy-to-navigate, clean, simple design that is consistent throughout all the courses and laboratory works. This consistency in layout and tools is intended to create a familiar approach to different courses, even when they vary in domain, form and content. This is the reason why the authors have selected the layout and tags of the e-learning server interface.

Simplicity
The Moodle environment allows the designer to perform many operations and use a variety of functions, but it is desired to keep the interface functional and simple. We have analyzed and decided what tags, functions, allowances, and resources best fit the content and the LOs for each course on a separate basis, unnecessary buttons being not employed. Furthermore, we have placed all general resources (such as your syllabus, library information etc.) in the top centre block, in sense easy to grasp.

Authority
An authoritative web course has to make it clear who is responsible for presenting the information and outstands the qualifications and credentials of the authors for doing so. The authors have accomplished this goal by stating clearly the.

Syllabi
For online courses, syllabus serves a crucial function. Pedagogical expertise has outlined that distance-learning students have less opportunity to ask questions and engage in conversation with the tutor about assignments, calendar, and assessments. These users view the syllabus as the sole reference guide to the course, needing it to be even clearer in terms of form and content than it might be for a traditional hard-copy course. That is why we have places the syllabus tag in a easy-to-use position.

Actuality of the course
This important requirement states whether we use a static course/laboratory work or we update it on a regular basis. The authors provide the last update on the bottom of the page. We state clearly when the course was conceived, when it was mounted on the web and the last date the page was last modified.

6 Further Discussion and Conclusion

This article suggests that an increasing number of organizations are integrating e-learning and on-line education into their environments. E-learning allows students and tutors to upgrade their skills on a regular basis, remain more competitive, extend classroom training and access higher quality academic resources. While the acknowledged industry leader in course management and e-learning software is Blackboard, a pattern seems to emerge when using open source alternatives and that is Moodle. The chapter paper focuses on the importance of using open source Learning Management Systems (LMS) when producing and deploying E-Learning projects in a large array of configurations. The guiding philosophy of the used case studies is that LMS, the core component for any teaching platform, should be transparent, widely available to assessors and developers, low cost and subject to improvement and customization in order to fit the large number of web based collaborative learning and teaching systems. From the related works, we have found evidence that Moodle is an appropriate candidate for this goal and it is used by a large number of universities and learning establishments all over the world.

It is our belief that the constructivist pedagogy that drives open source platforms in e-learning is a consequence of how students and teachers have become more reflective about the way they learn and perform in the educational process. Given the fact that Moodle and other open systems are being advocated by international organizations like UNESCO [21], we endorse this alternative form of learning as being a good opportunity to dismantle cultural barriers and to harmonize human knowledge.

While the described e-learning approaches have proved to be appropriate in several contexts, there are good reasons to extend the perspective to other important questions. Are the e-learning open platforms adaptive or trainable? Do they learn? Are these systems intelligent? Why? Which? How? How much? What could we expect? How should they develop? Subsequently, to attempt to provide brief answers to these questions.

In our opinion, the goal of an intelligent e-learning system should be highly structured learning objects that are to a large scale under automated authority. Within this structure, the intelligence of the system often emerges in the shape of flexible sequencing or personalization of the educational material, instructions for navigation, or interactive queries. All of these methods rely basically on an indexed stack of pre-processed learning objects.

How should intelligent e-learning open platforms develop? We suggest that some possible intelligent features are proactive links and recommended links. Proactive queries are based on the principle of identifying the navigational patterns of the user and issuing directions automatically to provide extra links to eventually relevant resources. The search function of these applications has a global sight of the available contents and could emphasize structural connections between various sources of information. Thus, this function would provide the user with an enhanced awareness of the available resources. In a similar approach, recommended links are potentially relevant materials that can facilitate a better knowledge of the subject and rely on the same behavioral scrolling patterns of the user.

The final issues of our study are related to commercial aspects of the e-learning open platforms. Are they commercially a success? Why? Why not? Not yet? Not enough? How are they patented? How much money carries their market?

The market for e-learning products is financially important. According to some authors [23], the global e-learning market is expected to surpass $52.6 billion by 2010. Just the U.S. e-learning market has exceeded $17.5 billion in 2007, according to the same source [23]. While Europe is somehow overcome in e-learning adoption compared to the United States [23], U.S. e-learning adoption accounts for 60 percent of the market, while Europe's accounts for 15 percent, overall usage of e-learning in Asia is expected to reach a compound annual growth rate of 25 percent to 30 percent through 2010, according to the report [26]. Worldwide, "that rate should hit between 15 percent and 30 percent", states he same source [23].

According to the literature [24], there is "a growing market demand for Open Source learning management system (OS LMS) products". They claim the fact that Open Source LMS platforms will be competitive with similar corporate products when two market conditions are fulfilled "The market for commercial platforms reaches the commodity stage and OS LMS products exceed the level of innovation of the commercial systems" [24]. After the same authors [24] "Commoditization (for any product) occurs when demand is very high, there are firmly entrenched vendors supplying high-quality products, and competing products lack significant differentiation in the perception of customers. Customers expect high quality but shop for price". The conclusion of the above study is that, at least in US, the LMS have reached the commodity phase. Still, according to the source [24] "... in a commodity market customers will rarely switch brands or substitute products unless there is a clear perception of higher value. This is known as "the threat of substitution" in Porter's Five Factors Model. Essentially, rival vendors arrive on the market with products that can replace the dominant products". The market for open source LMS at the end of 2007 [24] reveals that Moodle gathers a community of approximately 6112 sites and 50,000 users across 126 countries. Claroline seems to be used by some 470 organizations in 65 countries, and ILIAS holds a dominant position [24] the German university market.

Patentability deserves a brief discussion. The European law regarding computer-related patents [25] is affected by the lack clarity in the existing rules. The European Commission still debates these issues. I quote: "The current rules in the European Patent Convention are out of date and leave a very wide decision-making power in the hands of patent examiners. There can be different interpretations as to whether an invention can be patented [26]". Until further regulations, it is our opinion that a customized e-learning platform is due to be copyright-protected.

The theoretical part of this chapter aims to decide whether or not open source e-learning platforms can compete with proprietary ones. Consequently, this evaluation - following international standardized methods - attempts to find a suitable open source LMS to be used in a particular project.

The practical part of this chapter extends the selected platform to a particular case, the author being in process of developing an e-learning laboratory. A project team has customized the e-learning server and adapted the Moodle environment to the unique strengths, learning objectives, knowledge levels, and learning characteristics of the courses and laboratories from the discipline of Computer Network and Distributed Systems from the Faculty of Electronics and Telecommunications, Technical University "Gheorghe Asachi" of Iasi, Romania. Future work consists in extending similar projects in other departments of our faculty, after having received conclusive feedback from users (both student and teachers) regarding the utility and viability of the Moodle e-learning system and in closed cooperation with the academic stuff.

Acknowledgements

The author gratefully acknowledges the support of the first editor and of the referees, moreover to all those involved in developing eCNDS (Computer Network and Distributed Systems E-Laboratory) at The Faculty of Electronics and Telecommunications from The Technical University "Gheorghe Asachi" of Iasi.

References

[1] Dougiamas, M., Taylor, P.C.: Interpretive analysis of an internet-based course constructed using a new courseware tool called Moodle, http://dougiamas.com/writing/herdsa2002/ (visited March 25, 2008)

[2] Ahamer, G.: How the use of web based tools may evolve along three generations of WBT, http://www.doaj.org/doaj?func=searchArticles&q1=elearning&f1=all&b1=and&q2=&f2=all&p=2 (visited March 26, 2008)

[3] SCORM: Best Practice Guide for Content Developers 1st Edition, http://www.dokeos.com/doc/thirdparty/ScormBestPracticesContentDev.pdf (visited March 27, 2008)
[4] http://ieeeltsc.org/ (visited March 28, 2008)
[5] Adascalitei, A.: Instruire Asistata de Calculator. Didactica Informatica, Editura Polirom Iasi, pp. 172–183
[6] http://www.aicc.org/ (visited March 29, 2008)
[7] Graf, S.: Beate List, An Evaluation of Open Source E-Learning Platforms Stressing Adaptation Issues, http://www.informatik.uni-trier.de/~ley/db/indices/a-tree/g/Graf:Sabine.html (visited March 30, 2008)
[8] http://www.atutor.ca/ (visited March 30, 2008)
[9] http://www.claroline.net/ (visited March 29, 2008)
[10] http://www.dokeos.com/ (visited April 1, 2008)
[11] http://dotlrn.org/ (visited April 1, 2008)
[12] http://www.ilias.de/ (visited April 2, 2008)
[13] http://www.lon-capa.org/ (visited April 2, 2008)
[14] http://moodle.org/ (visited February-April 2008)
[15] http://openuss.sourceforge.net/openuss/index.html (visited April 3, 2008)
[16] http://sakaiproject.org/ (visited April 3, 2008)
[17] http://www.docebo.org/doceboCms/ (visited April 4, 2008)
[18] http://www.imsglobal.org/question/ (visited April 4, 2008)
[19] Bremer, D., Bryant, R.: A Comparison of Two Learning Management Systems: Moodle vs Blackboard, http://www.naccq.ac.nz/conference05/proceedings_05/concise/bremer_moodle.pdf (visited April 5, 2008)
[20] Godsk, M., Jørgensen, D.S., Dørup, J.: Implementing E-learning by Nurturing Evolution, http://www.mc.manchester.ac.uk/eunis2005/medialibrary/papers/paper_184.pdf (visited May 1, 2008)
[21] http://www.unesco.org/cgi-bin/webworld/portal_freesoftware/cgi/page.cgi?g=Software%2FCourseware_Tools%2Findex.shtml;d=1 (visited May 2, 2008)
[22] ISO Standards Related to Software Evaluation, http://www.iso.org/iso/search.htm?qt=software+evaluation&searchSubmit=Search&sort=rel&type=simple&published=on (visited May 8, 2008)
[23] A Global Strategic Business Report, http://www.strategyr.com/MCP-4107.asp (visited May 10, 2008)
[24] Adkins, S.S.: Wake-Up Call: Open Source LMS, http://www.learningcircuits.org/2005/oct2005/adkins.htm (visited May 10, 2008)
[25] Patentability of computer-implemented inventions in European Union, http://ec.europa.eu/internal_market/indprop/comp/index_en.htm (visited May 10, 2008)
[26] Statement to the European Parliament on Computer Implemented Inventions, European Parliament Plenary Session Strasbourg, March 8 (2005), http://europa.eu/rapid/pressReleasesAction.do?reference=SPEECH/05/151&format=HTML&aged=0&language=EN&guiLanguage=en (visited May 11, 2008)

16

Software Platform for Archaeological Patrimony Inventory and Management

Dan Gâlea[1], Silviu Bejinariu[1], Ramona Luca[1], Vasile Apopei[1], Adrian Ciobanu[1], Cristina Niţă[1], Ciprian Lefter[2], Andrei Ocheşel[2], and Georgeta Gavriluţ[2]

[1] Institute of Computer Science Iaşi – Romanian Academy, 22A Carol I Blvd, Iaşi, 700505, Romania
{dan.galea,silviub,ramonad,vapopei,cibad, cristina}@iit.tuiasi.ro

[2] Data Invest Iaşi, 26A A.Panu St, World Trade Center Building, Iaşi, 700020, Romania
{ciprian.lefter,andrei.ochesel, georgeta.gavrilut}@datainvest.ro

Abstract. We describe in this paper a complete informational model based on the geographical information system technology for organizing the Romanian territory archaeological information and to put it in an electronically accessible form, while remaining similar in content with a classic archaeological atlas. Geographical and archaeological databases were designed and implemented, along with interfaces for their manipulation and an interface for archaeological atlas consultation. The system has already been implemented using information specific to a narrow archaeological area, in the hydrographical basin of the Bahluiet River in the Iasi County.

Keywords: GIS, archaeology, atlas, map, database.

1 Introduction

This paper is an overview of the research done in the last 4 years by a multidisciplinary team cooperating with archaeologists from the Iaşi Institute of Archaeology of the Romanian Academy. Worldwide archaeologists are interested in studying culture and human behaviour in time and space. GIS offers convenient tools to store archaeological information in a spatial organised manner [12]. GIS integrates dated measurements based on different techniques (seriation, artefact typologies, stratification, C-14, K-Ar, thermoluminiscence), and spatial descriptions like drawings, maps, photographs, aerial photography, penetrating radar and magnetometry results.

Information contained in GIS can be distributed through Internet, which opens the possibility to create online systems based on archaeological information. There are several portals focusing on archaeology, mostly containing photographs and alphanumeric information, some of them using GIS just to show locations on vector maps. However, there is little work done on Internet regarding archaeological atlases, which remain sold mostly as paper books.

H.-N. Teodorescu, J. Watada, and L.C. Jain (Eds.): Intel. Sys. and Tech., SCI 217, pp. 283–297.
springerlink.com

The HGIS project (Institute of European History, Germany) [7] offers digital historical maps of the development of Germany and the European state system. The maps, arranged in thematic strands and combined in series covering important benchmark years, have been placed on an interactive map server. This solution uses the ArcGIS platform. However, map series can only show a limited amount of information for selected dates at fixed scales, and it is generally not possible to attach a large variety of thematic data such as statistical or general historical information to them.

The Great Britain Historical Geographical Information System (GBHGIS, University of Portsmouth, England) [8] is a digital collection of information about Britain's localities as they have changed over time. It uses an object-relational database rather than conventional GIS software to create a system that is simultaneously a toponymic database and a statistical mapping system.

The ARENA Network Information Project [9] is concerned with conservation and presentation of the European archaeological heritage through new information technologies. This project is carried out with the support of the European Community through the Culture2000 programme and has six partner organisations in Poland, Romania, Denmark, Iceland, Norway and the United Kingdom. The lead partner in the project is the Arts and Humanities Data Service Centre for Archaeology and is located at the Department of Archaeology at the University of York in the UK. Archaeological data is regularly collected in digital format and is currently conserved and presented on the World Wide Web on a nation by nation basis by specialised organisations. The ARENA Network partners share and develop expertise in the conservation of archaeological data. The user can specify his information request by choosing sequentially the time period, the theme he is interested in and the area of interest and then the system lists all the archaeological items matching criteria. For the specification of area a simple map with country borders and major towns across Europe can be used and the location is stored as latitude and longitude coordinates.

The Office of Archaeological Research from the University of Alabama [10] uses ESRI's suite of GIS software products for cultural resource surveys and managing the Alabama State Site File. GIS is used to store archaeological sites and several thematic maps are developed.

In Romania, the Institute for Cultural Memory has developed a Map Server for the National Cultural Heritage [11], which covers all Romanian territories and makes possible the access to four main databases: the National Archaeological Record (comprising over 13,000 archaeological sites), the database of the Archaeological Researches in Romania (1983 - 2006) that comprises over 3,000 archaeological reports, the Guide of Museums and Collections in Romania, including over 740 museums in Romania, and Places of Worship in Romania (18,600 entries).

Our system for archaeological atlases modelling has been developed around the GIS (Geographic Information System) concept [5]. It allows the development of a complex archaeological database which is then easily to consult using either a stand alone application or through Internet.

The system manages two different databases [1]. The geographical database (GEO) contains digital maps in vector format and their internal data tables (containing identification and minimal related information). The external archaeological database (ARH) contains extended information about existing archaeological sites and discovered objects (details about their placement in space and time, historical texts and pictures, sources of information, etc.). Both databases are stored on a central server in order to allow multiple users to consult, add or update information concurrently.

The software platform integrates three modules, each with its own functionality:

- the archaeological database management module;
- the geographical database management and map visualization module - the 2D/3D maps for archaeological areas are displayable at different levels of detail; this module also allows the synchronization of information contained in the two databases, GEO and ARH;
- the archaeological atlas consulting module.

The structure of the complete system is shown in Fig. 1.

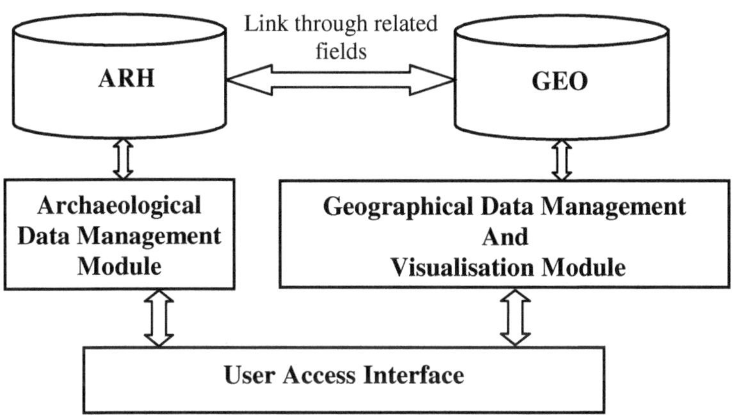

Fig. 1. The structure of the implemented software platform

2 Geographic Information System Design

GIS is a collection of computer hardware, software, and data used for managing, analyzing, and displaying all forms of geographically referenced information.

A GIS application allows to link information (attributes) to location data, such as: people to addresses, buildings to parcels, or streets within a network [5]. The graphical information can be arranged on layers, to give a better insight of how it all works together, the internal databases can be linked with external databases and the access to information is done through interfaces (see Fig. 2).

Placing things on a map, together with features interesting to look for, allows users quickly seeing where to take an action and what action to take:

- *Map Quantities*: features characterized by quantities may be figured on maps, allowing to find places that meet some criteria and take action, or to see the relationship between places. This gives an additional level of information beyond simply mapping the locations of features.
- *Map Densities*: in areas with many features it may be difficult to see which areas have a higher concentration than others. A density map allows measuring the number of features using a uniform area unit, so, a distribution map can be built.
- *Find what's Inside*: GIS can be used to monitor what's happening and to take specific action by mapping what's inside a specific area.
- *Find what's Nearby*: GIS can be used to find out what's occurring within a set distance of a feature by mapping what's nearby.
- *Map Change*: map the change in an area to anticipate future conditions, decide on a course of action, or to evaluate the results of an action or policy.

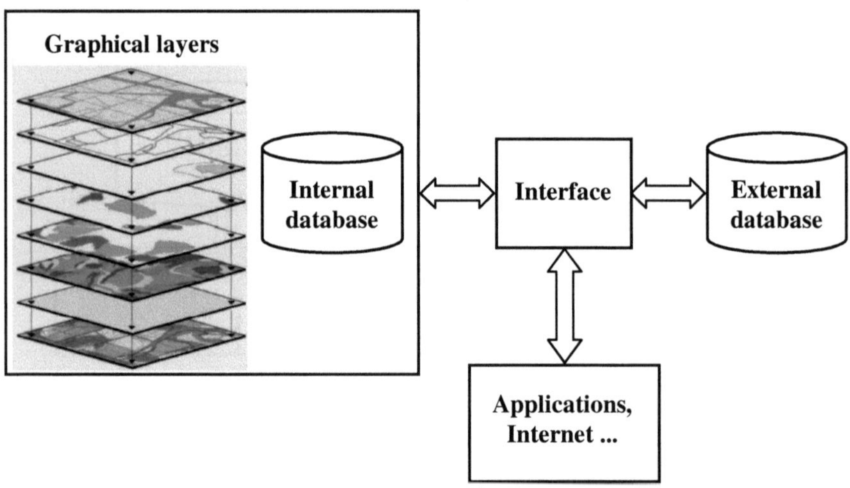

Fig. 2. The general structure of a GIS

When a GIS must be designed, or a GIS must be chosen for use, there are several factors to be considered: which data model to be used, if the features are organized in layers or not, how geographic features are represented, if the 3rd dimension is used or not, and the method used for topologic relationship building.

2.1 Geographical Data Models, Raster and Vector

In a GIS the information is usually linked to a map. The GIS is able to handle both raster image data and vector information. One of the most important problems is to select the proper raster model or vector model for particular features of the GIS [6].

- The raster model uses digital images of the interest area (aerial photos, radar images or scanned maps). The real world is represented through a grid of cells having the same length and width, with values of the cells standing for real-world features at the location of each cell.

- The vector data model represents the world as multiple elementary geometrical figures, like points, lines and polygons. Each figure may consist of one or several connected points. Points are usually connected through straight line segments. Curved lines are represented as multiple short connected line segments. Each point is described by an x,y coordinate pair where the *x* value represents usually the longitude or *x* dimension of the point in the chosen coordinate space and the *y* value represents usually the latitude or *y* dimension of the point. Other features, including the elevation (the z coordinate), may be optionally attached to each point.

One can think of five possibilities for combining the raster and vector data models when implementing a GIS [6]:

- *Vector model with raster underlay.* In this case the vector model is the principal data model, but for completeness and better reference there is a raster image in background. The raster image can be a digital version of a classic aerial photography, more recently a digital orthophotoplan, or an image resulting from processing satellite information. This option is more common nowadays because compression techniques for the background digital images have improved and GIS software allows their rapid manipulation on the screen.
- *Raster model with vector overlay.* This model combination is needed when remotely sensed imagery constitutes the primary source of data for the GIS or the directly processing of raster information is required.
- *Raster model only.* This is theoretically possible, but not really used because GIS software is always capable of drawing vector data.
- *Vector model only.* In this case data is stored only as vectors, with topological relationships of connectivity and adjacency known or being computable. We might not find this combination anymore now, because the capability to show a raster image behind vector data is an useful advantage of modern GIS software.
- *Full vector and raster.* This combination is necessary when the raster model is required and the topological vector information is needed in the same time. In this case, because both possibilities are available, the user will decide which model is more convenient in each analysis procedure.

2.2 Database Organization, Layers and Objects

The *layer based data structures* are known to be used in early GIS implementations. In that period it was common to use layers to produce classical maps. Manual cartographers used to work in layers even if the maps where printed only in black and white. The final layer superposition was photographed for map reproduction [6].

Structuring data in layers has a lot of advantages:

- The layers can be turned on/off to make some layers more visible than other layers,
- The drawing order may be changed to assure that some layers are not overplayed by other layers.

The idea of *object-oriented databases* appeared from the powerful influence of the object-oriented programming languages paradigm. Extending object orientation to databases was a normal development of the programming style based on objects. The features in a database can be seen as objects, properties can be assigned to them, and also objects can communicate with other ones in the database. This functionality elevates the feature from the status of a passive database attribute to the status of an object with properties and functions it can perform on itself or other objects [6]. Object-oriented geographic databases are now common in GIS implementation.

2.3 Representing Geographic Features

There are basically three classes of geographic vector features: points, lines, and polygons. There exists more types of geographic features, but they are not available in all geographic information systems in use. It is possible for some types of geographic features to be extremely useful in some applications and not important at all for other applications [6].

2.4 Topologic Relationships

In the case of data sets with topology the relationships of adjacency and connectivity are explicitly stored in the database.

There is also the possibility for the GIS to be able to compute the topology data, i.e. adjacency or connectivity relationships, at the very time they are needed, in order to be capable to execute complex spatial queries. Such queries are not possible without computing or previously storing topology data.

For instance, in the case of polygons, the following usual geographic functions require knowledge of the topologic relationships between them [6]:

- Clipping out a subset of polygons with another polygon,
- Overlay of polygon layers,
- Dissolving lines between adjacent polygons that are similar on some characteristic.

Topology can be created for lines and this gives advantages in a GIS. Line topology consists of connectivity knowledge (knowing which lines have points in common which other lines). Topologic linear data layers are constructed such that whenever lines touch each other a node is created. The value of this kind of topology is principally in network analysis. Several functions can be performed on data layers with linear topology [6]:

- Find the shortest path between two locations.
- Location-allocation.
- Find an alternate route.
- Create an optimum route to multiple nodes.

Vector GISs missing the capability to store or create topology relationships can be used only for map displaying on computer screens and possess few analytical capabilities; Modern GIS software packages provide vector topology as a standard feature.

2.5 The Third Dimension

Both paper and PC displays are two dimensional, while the world is three dimensional. A solution to this problem was to conceive conventions for representing the third dimension on paper. Cartographers introduced the use of contour lines (hypsography), based on a straight-down projection of the third dimension onto the two-dimensional paper surface [6]. A more complex 3D rendering technique, performed only by crafted and skilled technicians, uses shaded relief and analytical hill shading.

The possible elements of the third dimension needed to be modeled in a GIS are the following:

- *Elevation*. The height of a point in real world above a chosen three-dimensional reference point (e.g., mean sea level).
- *Slope*. The change in height between two points over a given distance.
- *Aspect*. The compass direction a part of the real-world faces.

3 The Archaeological Atlas

3.1 The Archaeological Database and Its Management Module

The archaeological database was built on the MySQL relational database system. We have named it ARH, containing five related tables [1]:

1. *Date_obiectiv (Objective_data)* - contains information about the geographical position of the archaeological sites, the names of the archaeologists that worked on the site, the time period of the research, type of research, etc.; it contains the *cod_ob* field serving as its primary key and as an external key in the rest of related tables; this table contains also 4 special fields: *Link_proj* - the full system path to the GIS project containing the map of the archaeological site; *Link_map* – the corresponding map name; *Link_layer* – the corresponding layer and *Link_obj* – the foreign key in "one to one" relation with the *Link_BDG* field in the layer internal database. These fields

are used for linking the information about each archaeological site with the corresponding location on the maps contained in the geographical database;
2. *Bibl_obiectiv (Objective_library)* - contains information available in scientific papers resulted from the research activity related to archaeological sites or referring to it.
3. *Stratigrafie (Stratigraphy)* - stores information about archaeological periods and cultures to which archaeological sites (registered in the Date_obiectiv table) and discovered archaeological objects (registered in the Patrimoniu table) are assigned.
4. *Patrimoniu (Patrimony)* - contains detailed information about each discovered archaeological object, such as: name of the object, materials and techniques used to make the object, etc.
5. *Bibl_patrimoniu (Patrimony_library)* - contains information available in scientific papers resulted from the research activity related to discovered archaeological objects or referring to them.

The archaeological database was developed and it is maintained and updated through a MySQL server. The access to the information contained in the archaeological database is done through WEB pages consulted with a common Internet browser. These WEB pages are generated by a PHP program [1] which interact with the WEB Server and the MySQL Server.

3.2 The Geographical Database and Map Visualisation Module

The geographical database was created on the NetSET Geographical Information System platform developed by Data Invest Iaşi. The NetSET platform consists of several software modules providing the tools for fully developing a GIS project, starting from processing the input images (like scanned maps, aerial images, etc.) and up to publishing the resulted 2D maps on the Internet through a special WEB server or generating their 3D models, if elevations are known. The NetSET platform works with the shape file format (.shp ESRI standard) for storing graphical information on each layer of a map, while alphanumerical information is stored in database files (.dbf). The link between the two kinds of information is assured through a .shx index file.

The geographical database has been designed to contain several levels of detail, starting with the vector map of the whole Romania, passing through the maps of historical regions, the maps of the counties and ending with local maps for proximal areas to archaeological sites and even more detailed archaeological site plans.

We have already developed a local map for a pilot area in the hydrographical basin of the Bahluiet River in the Iasi County. To do this we have digitized the scanned versions of 15 topographic maps at the 1:25000 scale (see Fig. 3 for their relative placement in geographical coordinates).

The scanned versions of the topographic maps were first pre-processed with the NetSET Raster Processor, to make them more suitable for the vector editing

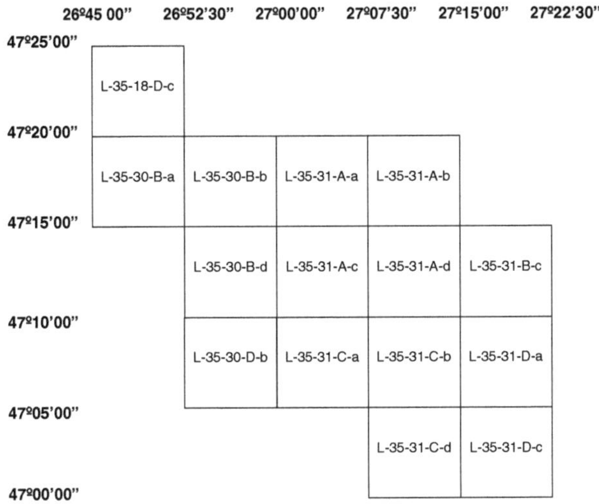

Fig. 3. The placement of the processed topographic maps in geographical coordinates

process. The NetSET Raster Processor provides a series of pre-processing routines like: decomposition/composition of the image in the 3 fundamental colour planes, conversion to black and white with a threshold for each fundamental colour plane, noise cleaning, median filtering, erosion, dilation, global histogram equalisation or applied to each fundamental colour plane, edge detection and contour tracing.

Then the NeSET Image Rectifier tool was used to georeferentiate each scanned topographic map and thus they were automatically composed into a correct bigger map of the whole pilot region. The NetSET Image Rectifier provides conversion of coordinates for all usual systems and projections (we have used the Romanian Stereographic 1970 system of coordinates with the Krassovsky reference ellipsoid). Also there are four algorithms available for georeferentiation, based from 2 to 8 reference points. We have used from 4 points of reference for each topographic map.

A new NetSET Project was developed and the composed map was integrated in it as a reference for further vector editing with the NetSET Map Editor. This complex GIS editing application provides several toolbars with instruments for drawing graphical objects on the map layers. We have extensively used the toolbars for managing map layers (creating layers, loading saved layers, saving layers, converting layers), for map navigation (pan, horizontal and vertical movements, zoom in/out, special zoom), for global editing (cut, copy, paste and find), for drawing, for advanced and precision editing, for symbols management, for internal database management, for external database connection, and some miscellaneous tools. The project consists of one map with the following information organized in several layers:

Table 1. Layer included in the NetSET GIS project

Layer name	Type
Contour lines	3D Polylines
Rivers	2D Polylines
Lakes	2D Polygons
Marshes	2D Polygons
Roads	2D Polylines
Railways	2D Polylines
Localities	2D Polygons
Churches	2D Polygons
Town Halls	2D Polygons
Other important buildings	2D Polygons
Agriculture areas	2D Polygons
Forests	2D Polygons
Archaeological sites	2D Polygons

The "Contour lines" layer was used to generate and navigate through the 3D model of the pilot zone with the 3D NetSET Viewer. Each contour line present on the topographic map was manually drawn in vector format. Then we assigned to each contour line the number representing its elevation, resulting in a very precise 3D model. The 3D NetSET Viewer is capable of reading 2D vector maps developed with the NetSET Map Editor, to find elevation data present in the input layers and to automatically generate a 3D model based on these elevations. All the layer structure of the 2D map is preserved, some layers are taken as reference for their elevations and the rest are drawn on top of the reference layer.

The "Rivers", "Lakes", "Roads" and "Railways" layers were important for an easy orientation when navigating through the map, while the other layers focus on different details assuring a realistic aspect for the resulted vector map.

The layer "Archaeological sites" is used as a link to the external archaeological database ARH. The corresponding shape file in the geographical database contains a graphical element for each archaeological site. In the internal database table of this layer we have created fields for the name of the archaeological site, for the name of the locality to which the site is associated and one special field for synchronization, *Link_BDG*. This field is linked with the filed *Link_obj* (referred in paragraph 2) in the archaeological database ARH and thus the synchronization between the two databases is assured.

The global result of the vector editing process is shown in Fig. 5. The mapped region is part of the Iasi County (in pink) and the most important towns are labelled with blue letters. Parts of the neighbouring Counties are also shown.

In order to see and navigate through detailed maps, a special visualization interface was designed and implemented. It is based on the *GISMapControl* developed by Data Invest, for drawing 2D maps, and on the *Scene_3D* internal ActiveX control, specially created using the OpenGL library, for drawing 3D models. The *GISMAPControl* is a collection of embeddable mapping and GIS

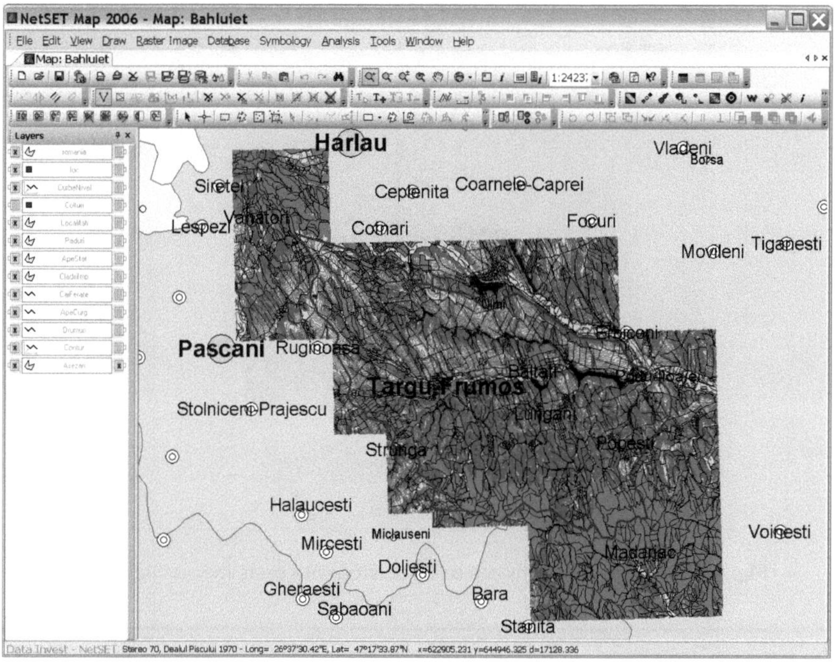

Fig. 4. A global view of the resulted vector map for the pilot region

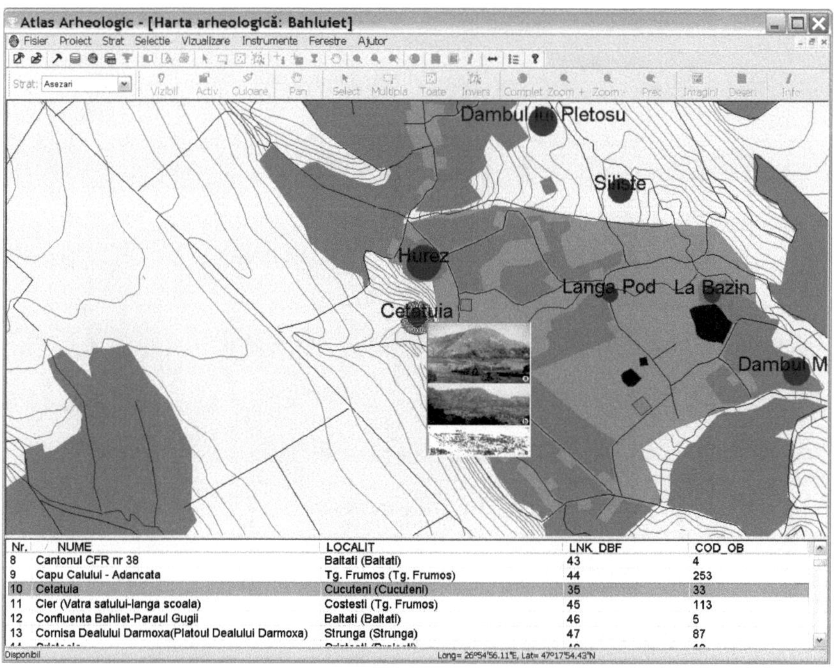

Fig. 5. The interface window for 2D map visualizations

Fig. 6. Various photographs can be associated with each archaeological site

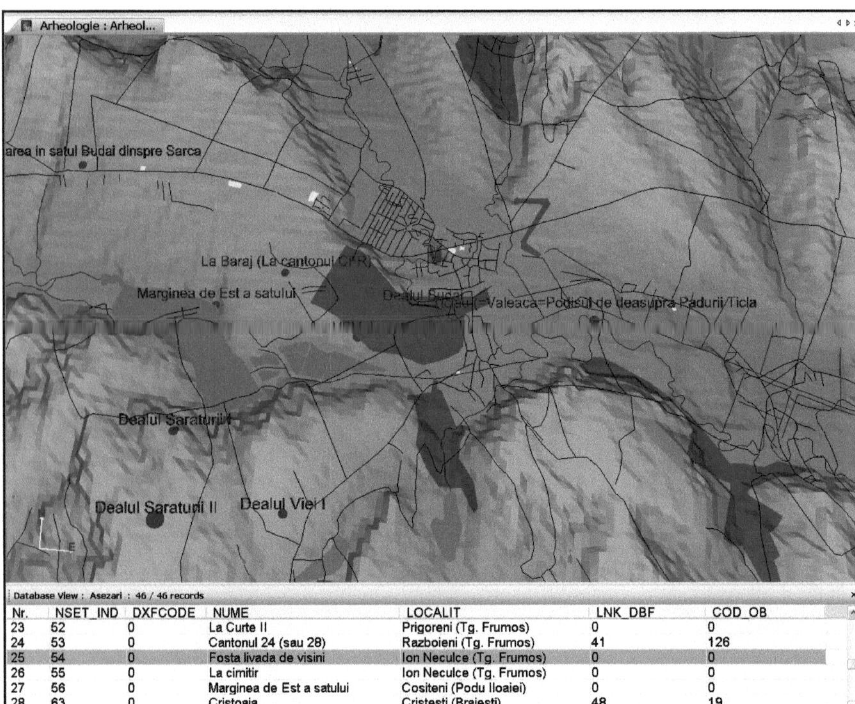

Fig. 7. The interface window for 3D map visualizations

components that can be used to build custom GIS applications. It is especially powerful in drawing functions (points, lines, polygons, rectangles, ellipses, circles) as well as in data management functions, with support for linking external databases of common types.

The figures 5 and 7 present screenshots with 2D and, respectively, 3D map visualization interface windows. In addition, to each archaeological site it is possible to assign various photographs that can be displayed automatically when the cursor of the mouse rests upon it on the map (see Fig. 5 near Cetatuia village). All the photographs associated with an archaeological site can be seen in detail as in Fig. 6.

3.3 The User Access Interface

To assure a synchronized consultation of geographical (GEO) and archaeological (ARH) databases, we have designed a special interface (or desktop window) with three main working areas:

- in the bottom, a scrollable list of archaeological sites identified through: "Objective Code", "Objective Name", "Locality", "Commune" and "County". From this list, one archaeological site can be selected to be displayed in detail in the rest of the window;
- in the right side, a rectangular area for displaying the 2D/3D digital map of the selected archaeological site

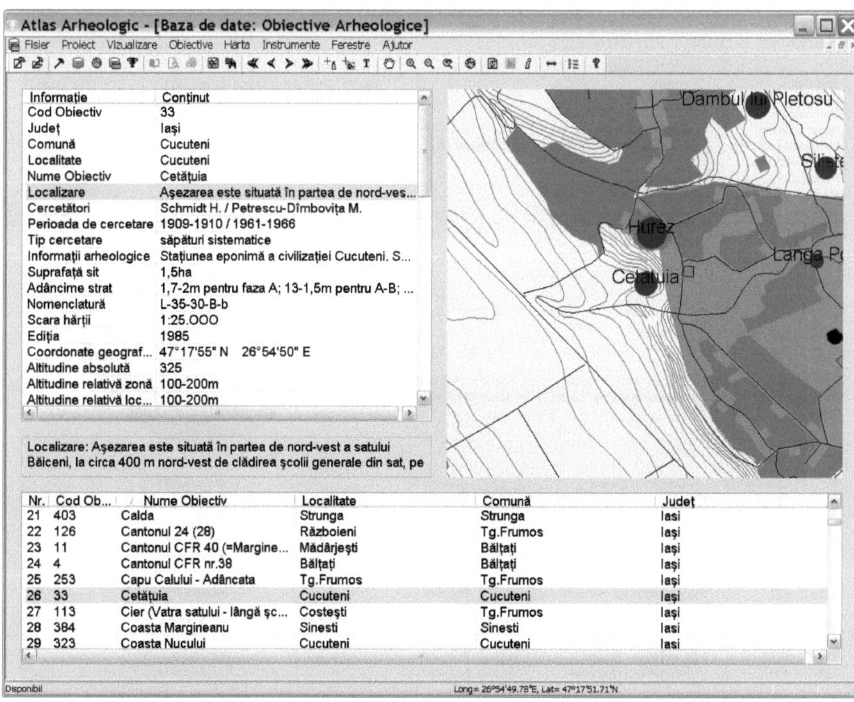

Fig. 8. The interface window for consulting archaeological sites information

– in the left side, a two column table area with detailed information about the selected archaeological site, taken from the *Date_obiectiv* and *Bibl_obiectiv* ARH database tables.

The selection of a new item from the list of archaeological sites immediately produces the focusing of the digital map on the neighbourhood of the selected archaeological site and displays the known information about it (Fig. 8).

A similar interface can be used to view discovered objects of a certain archaeological site through the menu "Objectives/Patrimony". In this case the three working areas of the interface are displaying:

– in the bottom, a scrollable list of archaeological discovered objects identified through "Object Code", "Object Name", "Time Period" and "Cultural Environment";
– in the right side, a rectangular area with the image of the selected archaeological object;
– in the left, a two column table area with detailed information about the selected archaeological object, taken from the *Patrimoniu* and *Bibl_patrimoniu* ARH database tables

The change of the selected archaeological object in the list of the archaeological discoveries produces the change of the information in the other two areas of the window interface (Fig. 9).

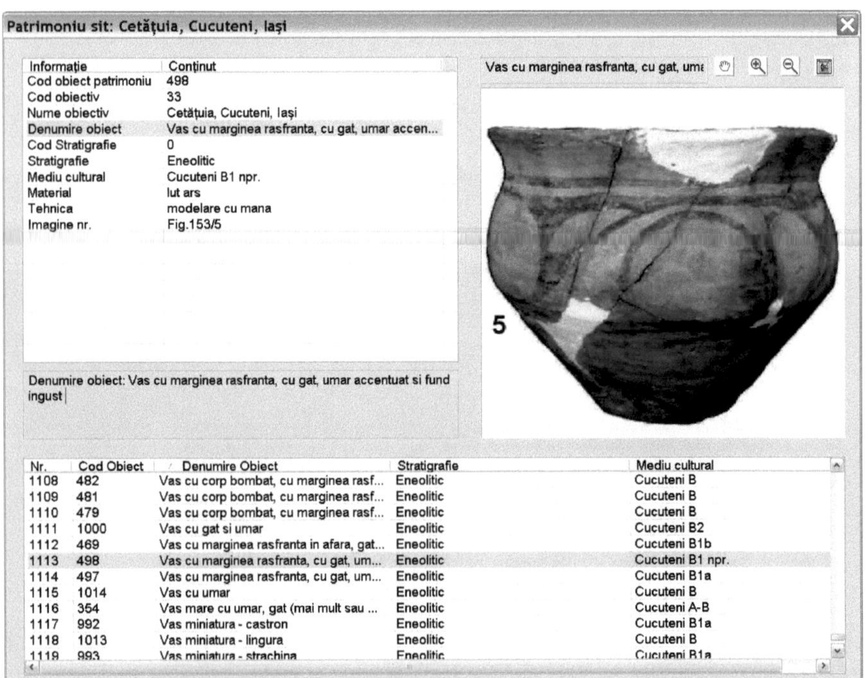

Fig. 9. The interface window for consulting archaeological discovered objects information

4 Conclusions and Further Work

This platform is a research tool available for archaeologists to manage the archaeological patrimony, with support for developing related text based information, photographs and geographical databases focused on archaeological sites and archaeological discovered objects. It provides a user friendly interface to show the known correlated information on a PC screen. The system was tested by covering with maps and archaeological data a Palaeolithic pilot area around the Bahluiet riverbed in the Iasi County [2], [3], [4].

The platform is based on a Geographic Information System that is able to display, with different levels of detail, the 2D and 3D models of the area where the archaeological sites are placed. The maps (developed using the NetSET GIS software) are synchronized with the corresponding archaeological information (managed using the MySQL Server through a WEB interface).

The archaeological information structured with a relational database system integrated with GIS allows complex queries which can determine the position of archaeological objects and objectives in time and space. The results are displayed in form of thematic maps.

We will continue to develop both archaeological and geographic databases and study the system behaviour in the case of large amount of different kinds of data.

Acknowledgments. This system was developed within an interdisciplinary collaborative project with researchers from the Institute for Archaeology, Romanian Academy, Iasi Branch. The maps were developed using the NetSET GIS platform from Data Invest Ltd.

References

1. Luca, R., Niţă, C.D., Muscă, E., Lazăr, C.: Information Management in Archeological Atlases. In: Selected papers of the 4th European Conference on Intelligent Systems and Technologies, Iaşi, Romania (2006)
2. Chirica, V., Tanasachi, M.: Repertoriul Arheologic al judeţului Iaşi (1984-1985)
3. Văleanu, M.C.: Omul şi mediul natural în neo-eneoliticul din Moldova. Helios, Iaşi (2003)
4. Văleanu, M.C.: Aşezări neo-eneolitice din Moldova. Helios, Iaşi (2003)
5. Luca, R., Bejianriu, S., Apopei, V., Niţă, C.D., Lazăr, C., Muscă, E.: Archeological Atlases Modeling using GIS. In: 9th International Symposium on Automatic Control and Computer Science, Iaşi, Romania (2007)
6. Harmon, J.E., Anderson, S.J.: The design and implementation of geographic information systems. John Wiley & Sons, New Jersey (2003)
7. http://www.hgis-germany.de
8. http://www.gbhgis.org
9. http://ads.ahds.ac.uk/arena/project.html
10. http://museums.ua.edu/oar/archgis.shtml
11. http://map.cimec.ro/indexEn.htm
12. http://www.esri.com/industries/archaeology/index.html

Author Index

MIX
Papier aus verantwortungsvollen Quellen
Paper from responsible sources
FSC® C105338

If you have any concerns about our products,
you can contact us on
ProductSafety@springernature.com

In case Publisher is established outside the EU,
the EU authorized representative is:
Springer Nature Customer Service Center GmbH
Europaplatz 3, 69115 Heidelberg, Germany

Printed by Libri Plureos GmbH
in Hamburg, Germany